'[Lynne Olson is] our era's foremost chronicler of World War II politics and diplomacy.'

MADELEINE ALBRIGHT

'[An] excellent book ... [Olson] acknowledges that British bravery and determination were the only things that gave Europe hope during the darkest days of the war. Europeans have always been grateful for this, and continue to be so today. We British have not always shown the same sense of gratitude. Olson's message is simple: when it comes to the Second World War at least, we should give credit where credit is due.'

KEITH LOWE, *The Mail on Sunday*

'[A] pointed volume ... [Olson] tells a great story and has a fine eye for character.'

THE BOSTON GLOBE

'A rip-roaring saga of hair-breadth escape, espionage, and resistance during World War II – Olson's *Last Hope Island* salvages the forgotten stories of a collection of heroic souls from seven countries overrun by Hitler, who find refuge in Churchill's London and then seek payback in ways large and small. In thrilling fashion, Olson shows us that hell hath no fury like a small country scorned.'

ERIK LARSON

'The wealth of evidence [Olson] presents, and the verve with which she tells her stories, many not widely known, is a necessary antidote to the myths with which we all live.'

FINANCIAL TIMES

'For a year, between the fall of France and the German invasion of the Soviet Union, the men and women of the seven countries of occupied Europe were Britain's only non-colonial allies. For them Britain was the only hope. The nature of that unbalanced relationship, so crucial to the postwar future of the Continent, is the subject of this fascinating, uplifting and at times horrifying book ... A complex narrative of need and desperation, tales of almost unimaginable courage, brilliant improvisation, fantastic stupidity and vile treachery ... Exciting and informative.'

DAVID AARONOVITCH, *The Times*

'You wouldn't think that there would still be untold tales about World War II, but Lynne Olson, a master of that period of history, has found some. Not only does she narrate them with her usual verve, but her book reminds us how much we unthinkingly assume that it was the United States and Britain alone who defeated the Nazis in Western Europe. *Last Hope Island* is a valuable, and immensely readable, corrective.'

ADAM HOCHSCHILD, author of *Spain in Our Hearts* and *King Leopold's Ghost*

'A great read, packed with the stories of very colourful characters ... highly recommended.'

MILITARY HISTORY MONTHLY

'In *Last Hope Island* [Olson] argues an arresting new thesis: that the people of occupied Europe and the expatriate leaders did far more for their own liberation than historians and the public alike recognise. Books and films have dramatised individual stories of the resistance, but the scale of the organisation she describes is breathtaking.'

THE NEW YORK TIMES BOOK REVIEW

LAST HOPE ISLAND

Lynne Olson is the former White House
correspondent for the *Baltimore Sun*. In addition
to 2010's *Citizens of London*, she is most recently
the author of *Those Angry Days: Roosevelt, Lindbergh,
and America's fight over WWII 1939–1941*. She is
also the author of *Troublesome Young Men:
the rebels who brought Churchill to power and helped
save England* and *Freedom's Daughters: the unsung
heroines of thecivil rights movement from 1830
to 1970*, and is a co-author of two other books.

BY LYNNE OLSON

Those Angry Days: Roosevelt, Lindbergh, and America's fight over World War II, 1939–1941

Citizens of London: the Americans who stood with Britain in its darkest, finest hour

Troublesome Young Men: the rebels who brought Churchill to power and helped save England

A Question of Honor: the Kosciuszko Squadron: forgotten heroes of World War II (with Stanley Cloud)

Freedom's Daughters: the unsung heroines of the civil rights movement from 1830 to 1970

The Murrow Boys: pioneers on the front lines of broadcast journalism (with Stanley Cloud)

LAST HOPE ISLAND

LYNNE OLSON

SCRIBE
Melbourne • London

Scribe Publications
18–20 Edward St, Brunswick, Victoria 3056, Australia
2 John St, Clerkenwell, London, WC1N 2ES, United Kingdom

Published by Scribe 2017
This edition published 2018

COVER PHOTO CREDITS: All images from iStock; WW2 airplanes
copyright Matt Gibson, White Cliffs of Dover (front cover) copyright
Stevephotos, and (back cover) copyright Lisa Valder. PHOTO CREDITS:
All images are from the Library of Congress with the exception of pages 33,
99, 139, 220, and 363, which are from Corbis; page 87, from the National
Portrait Gallery (UK); page 108, from the Polish Institute and Sikorski
Museum; page 161, from the Bydgoszcz City Council; and page 474, from
Getty Images.

The moral rights of the author have been asserted.

Printed and bound in the UK by CPI Group (UK) Ltd, Croydon, CR0 4YY

Scribe Publications is committed to the sustainable use of natural resources
and the useof paper products made responsibly from those resources.

9781911617181 (UK edition)
9781925322088 (Australian edition)
9781925307986 (e-book)

A CiP entry for this title is available from the National Library of Australia
and theBritish Library.

scribepublications.co.uk
scribepublications.com.au

For Stan and Carly,
as always

Contents

PART TWO: *RULE OF THE TITANS*

Introduction

FOR MUCH OF ITS LONG AND FABLED HISTORY, BRITAIN HAS DONE its best to stay clear of Europe and its entanglements. In the mid-1800s, Prime Minister Benjamin Disraeli declared that his country, with its worldwide empire and mastery of the sea, had "outgrown the continent of Europe." Almost a century after Disraeli's remark, Britons continued to view continental Europeans—and foreigners in general—as nothing but trouble. As the CBS correspondent Edward R. Murrow put it, the British were "sustained by a peculiar quiet arrogance—a feeling that they are superior to other people."

In the 1930s, the British stood quietly by as Hitler rose to power and began his conquest of Europe. For the sake of peace—their own peace—they did little or nothing to prevent country after country from being taken over by Germany. In the case of Czechoslovakia, they actively cooperated in its seizure. Referring to that nation in September 1938, Prime Minister Neville Chamberlain voiced the feelings of many of his countrymen when he complained, "How horrible, fantastic, incredible it is that we should be digging trenches and trying on gas masks here because of a quarrel in a faraway country between people of whom we know nothing."

Then, in the chaos-filled days of May and June 1940, London, to its residents' shock, suddenly found itself the de facto capital of Europe. Every other day, it seemed, King George VI and Winston Churchill,

Chamberlain's successor, were summoned to a London train station to welcome yet another king, queen, president, or prime minister whose country's freedom had been brutally snatched away in the Nazi blitzkrieg of Europe. In less than a month, the British capital had become a haven for the governments and armed forces of six European countries conquered by Hitler—Czechoslovakia, Poland, Norway, Holland, Belgium, and Luxembourg. The self-appointed representative of free France, General Charles de Gaulle, also fled there.

Most of the exiled leaders had initially resisted leaving their nations, feeling the same way about Britain as it did about them. They were horrified by its earlier refusal to confront Hitler and come to their countries' aid. Yet what alternative did they have? Energized in the nick of time by its new prime minister, Britain was the only nation in Europe still holding out against Germany. Only there could the Allied governments join forces and continue the fight.

Ignoring opposition from members of his cabinet and much of the rest of the British government, Churchill warmly welcomed the Europeans. While unquestionably heartfelt, his hospitality contained a strong element of national self-interest. After the fall of France and most of the rest of Europe, Hitler had turned his sights on the British, whose future now verged on the calamitous. They were about to experience the full fury of German power, and they would have to rely on the foreigners they had so disdained—their first allies—to help them survive in the desperate struggle to come.

WHEN MY HUSBAND, Stan Cloud, and I were researching the early years of World War II for our first book, *The Murrow Boys,* we happened to see an old movie about the Battle of Britain, which included a scene about a squadron of Polish pilots. Until then, we had no idea that any but British pilots had flown in that epic fight, and we wanted to find out more. In doing so, we discovered that dozens of Poles not only had participated but actually had played a critical role in winning the battle. We decided that their story, unknown to most Americans, deserved telling. But as we dug deeper, we realized that the importance of the Polish contribution to the Allied victory went far beyond the pilots' exploits. The Poles and their wartime experiences became the subject of *A Question of Honors,* the second book we wrote together.

Over the next ten years, I wrote three more books about World War II, all dealing with various aspects of Britain's struggle for survival in the war's early years. Much of my focus was on Winston Churchill's extraordinary leadership and the courage of ordinary Britons in waging that fight. I also examined Britain's relationship with its two major wartime allies—the United States and Soviet Union.

In exploring these subjects, I made another discovery: Poland was hardly the only occupied European country to have helped the Allied cause. Indeed, most of the captive nations whose governments escaped to London provided aid as well—support that, in the dark years of 1940 and 1941, arguably saved Britain from defeat and, in the latter part of the war, proved of immense benefit to the overall Allied victory.

So why have their contributions been so neglected by historians, who generally portray the victory as an unalloyed American-British-Soviet triumph? Churchill, as it happens, bears much of the responsibility for the omission. Early in the war, he created the image of plucky little England standing alone against the greatest military behemoth in world history. He ceaselessly promoted that idea throughout the conflict and afterward, broadcasting to the British people on V-E Day: "After gallant France had been struck down, we, from this island and from our united Empire, maintained the struggle single-handed until we were joined by the military might of Soviet Russia and later by the overwhelming powers and resources of the United States." Churchill's claim overlooks the fact that the occupied countries, from their base in London, were still at war, too. Without their help, the British might well have lost the Battle of Britain and the Battle of the Atlantic and might never have conquered the Germans' fiendishly complex Enigma code—all essential factors in Britain's survival.

At the center of this rich, intensely human story is its host of larger-than-life characters, from monarchs and scientists to spies and saboteurs. Some, like de Gaulle, are well known. Most, however, are not. The heroic King Haakon VII of Norway and the feisty Dutch queen, Wilhelmina, are two of the book's prominent figures. So is the Earl of Suffolk, a swashbuckling British aristocrat whose rescue of two nuclear physicists from France helped make the Manhattan Project possible. Among others playing noteworthy roles are Marian Rejewski, a Polish cryptographer who cracked the Enigma code long before the involvement of Alan Turing and Bletchley Park, and Andrée de Jongh, a pretty, tough-minded young Belgian whose escape network smuggled hundreds of downed British and American airmen out of enemy territory and back to freedom.

While this book provides a detailed account of these and other European wartime exploits, it also describes how much the occupied countries received from Britain in return. To captive Europe, the mere fact of British resistance to Hitler was a beacon of hope, a talisman against despair. For as long as the war lasted, Europeans engaged in a precious nightly ritual: they retrieved their radio sets, which had been outlawed by the Germans, from a variety of hiding places—beneath the floorboards, behind canned goods in the kitchen cupboard, secreted in the chimney. Then, in whatever the setting, the owners of the sets switched them on and tuned to the BBC in time to hear the chiming of Big Ben and the magical words "This is London calling." During and after the war, Europeans described those furtive moments listening to BBC news programs as their lifeline to freedom. A Frenchman who escaped to London late in the war recalled, "It's impossible to explain how much we depended on the BBC. In the beginning, it was everything."

For a young Dutch law student, hope took the shape of two Spitfires flashing over a beach near The Hague early in the war. He stared up in wonder at the planes, their RAF markings bright in the sun. "Occupation had descended on us with such crushing finality," he later wrote, "that England, like freedom, had become a mere concept. To believe in it as something real, a chunk of land where free people bucked the Nazi tide, required a concrete manifestation like a sign from God: England exists!" Less than a year later, he would escape to Britain and become an RAF pilot himself.

Another escapee, a Belgian journalist who'd managed to flee a Nazi concentration camp, arrived in London "drunk with happiness." "Do you know I have been dreaming of this moment for months?" he exclaimed to a British friend. "Isn't it wonderful to be here! Why, millions of people all over the continent are thinking at this very moment of London!" A young Polish resistance fighter echoed that sentiment, declaring that "getting to London was like getting to heaven." Polish pilots who flew with the RAF during the war referred to Britain as "Last Hope Island."

YET FOR ALL THE SUPPORT that the British and Europeans provided each other, their relationship, more often than not, was a tempestuous one, fraught with conflict and misunderstanding. Thrown together in desperate times under extraordinarily stressful circumstances, they grappled with culture clashes and language differences even as they struggled to survive the German military steamroller bearing down on them. To many exiled Europeans, the British seemed arrogant and insensitive, knowing little of the world outside their island and failing to understand the ruthlessness of the German occupation of the Continent. The British, meanwhile, had little patience with the constant feuding, rivalries, and demands of the foreigners crowding their shores.

Nonetheless, as the war reached its climax, most were able to put aside their differences and work closely together toward their mutual goal: defeating Hitler. At the end of the conflict, an RAF air marshal voiced a common sentiment when he remarked of the European pilots who flew under his command, "Together, we have formed a brotherhood."

A similar sense of fraternity developed among the Europeans themselves. "No matter our varied origins and uncertain futures, we stood shoulder to shoulder," a Dutch intelligence agent noted about the Poles, French, Norwegians, Belgians, and Czechs he met in London. "Beyond the society of Dutchmen with which I earlier had so passionately identified, a wider brotherhood emerged and received me with open arms."

As the war progressed, members of the various European governments in exile also forged tight-knit bonds, both official and personal. The trauma of defeat and occupation had convinced them that their nations must band together after the war if Europe hoped to achieve any kind of future influence, strength, and security. Their cooperation in London planted the seeds of the campaign for European unification that followed the conflict—an extraordinary effort that helped lead to more than half a century of peace and prosperity for western Europe.

The two eastern European allies—Poland and Czechoslovakia—were not so fortunate. When the Soviet Union and United States entered the war in 1941, the solidarity between Britain and occupied Europe gave way to the exigencies of *realpolitik*. Joseph Stalin was determined to gain postwar control of Poland and Czechoslovakia; Franklin D. Roosevelt and the guilt-ridden Winston Churchill eventually acceded to his demands. For those countries, World War II would not really end until communism crumbled in eastern Europe and the Soviet Union more than forty years later.

Britain, meanwhile, reverted to its traditional aloofness from Europe following the war and refused to participate in the movement toward European integration. Although it did finally join the European Economic Community (precursor to the European Union) in 1973, it did so reluctantly. It was similarly skittish about its later membership in the European Union. The issue reached a boiling point in June 2016, when a majority of Britons voted in a national referendum to leave the EU.

The shock and bitterness of that vote—and Britain's impending divorce from Europe—stands in sharp contrast to the resolve and hope of the crucial war years, when Britain joined forces with Europe to help defeat the mightiest military force in history. To French journalist Eve Curie, the daughter of Nobel laureates Marie and Pierre Curie and an exile herself, the grandeur of wartime Britain was embodied by Churchill and the Europeans who joined him in London—"all those insane, unarmed heroes who defied a triumphant Hitler."

PART ONE

FIGHTING ON

CHAPTER 1

"Majesty, We Are at War!"

Hitler Invades Norway

ON A CHILLY APRIL NIGHT IN 1940, LEADING OFFICIALS OF the Norwegian government were invited to the German legation in Oslo for the screening of a new film. The engraved invitations, sent by German minister Curt Bräuer, directed the guests to wear "full dress and orders," which indicated a gala formal occasion. But for the white-tie, bemedaled audience seated in the legation's drawing room, the evening turned out to be anything but festive.

Horrific images filled the screen from the film's beginning: dead horses, machine-gunned civilians, a city consumed in flames. Entitled *Baptism of Fire,* the movie was a documentary depicting the German conquest of Poland in September 1939; it portrayed in especially graphic detail the devastation caused by the bombing of Warsaw. This, Bräuer said after the screening, was what other countries could expect if they dared resist German attempts "to defend them from England." Appalled by the harrowing footage, Bräuer's guests were puzzled as to why the German diplomat thought it necessary to show the movie to *them.* What could any of this have to do with peaceful, neutral Norway?

Four nights later, just after midnight, those same officials were awakened by urgent phone calls informing them that several ships of unknown origin had entered the fjord leading to Oslo. A sea fog blanketing the fjord made it impossible to identify the ghostly armada's markings. Within minutes, however, the mystery of their nationality was solved when reports of surprise German attacks on every major port in Norway and Denmark began flooding Norwegian government offices.

Aboard the German heavy cruiser *Blücher,* General Erwin Engelbrecht, who commanded the attack force heading for Oslo, reviewed his orders with his subordinates. In just a few hours, more than a thousand troops, equipped with minutely detailed maps and photographs of the Norwegian capital, were to disembark from the *Blücher* in Oslo's harbor. Their assignment was to slip into the sleeping city and storm government buildings, the state radio station, and the royal palace. Before noon, King Haakon, Crown Prince Olav, and the rest of the royal family would be under arrest and the Norwegian government under German control. A band, also on board the *Blücher,* would play "Deutschland über Alles" in the city's center to celebrate Germany's triumph, while German military officials took over administration of the country and its two most important material assets—its merchant marine and its gold.

When a Norwegian patrol boat spotted the flotilla and had the temerity to issue a challenge, the boat was machine-gunned and sunk. Farther up the fjord, two small island forts, alerted by the patrol boat, also fired on the ships, but the heavy fog made accurate sighting impossible and the vessels swept on untouched. Shortly before 4 A.M., the convoy approached Oscarsborg Fortress, an island stronghold built in the mid–nineteenth century and Oslo's last major line of defense. The *Blücher*'s captain was as unperturbed by the sight of the fortress as he had been by the pesky patrol boat. On his charts and maps, Oscarsborg was identified as a museum and its two antiquated cannons described as obsolete.

The maps and charts were wrong on both counts. The fortress was operational, and so were the old cannons, fondly called "Moses" and "Aaron" by their crews. The fog lifted a bit, and as the darkened silhouettes of the ships came into view, a searchlight on the mainland suddenly illuminated the *Blücher*. Moses and Aaron erupted at point-blank range, their shells crashing into the 12,000-ton heavy cruiser. One shell smashed into the *Blücher*'s bridge, destroying its gunnery and navigational controls, while another slammed into a storeroom filled with aviation fuel. Shore batteries also began firing. Within seconds, the *Blücher* was ablaze, the flames leaping high into the air, burning off the fog, and lighting up the snow-covered banks of the fjord.

With a great roar, the ship's torpedo magazine exploded, and less than an hour later, the *Blücher,* commissioned only seven months before, rolled over on its side and sank. Nearly one thousand men went down with her, including most of the elite troops assigned to capture the royal family and government officials. General Engelbrecht was one of the several hundred survivors who escaped the burning oil covering the fjord's surface and swam frantically to shore.

Throughout that day—April 9, 1940—Hitler's audacious, meticulously planned invasion of Denmark and Norway had gone almost exactly as planned. By early afternoon, virtually all the Führer's major objectives along the 1,500 miles of Norwegian coastline had been taken—all, that is, except Oslo, the political, economic, and communications center of Norway and the key to the operation's eventual success.

AT 1:30 A.M. ON APRIL 9, the man atop Germany's most wanted list of Norwegians was awakened by his aide-de-camp. "Majesty," the aide said urgently, "we are at war!" The news came as no surprise to King Haakon VII. He had been expecting—and dreading—it for years. In 1932, he had told the British admiral Sir John Kelly, "If Hitler comes to power in Germany and manages to hold on to it, then we shall have a war in Europe before another decade is over."

Hitler *had* come to power, but Norway's political leaders had ignored the king's repeated urging to strengthen the country's shockingly weak defenses. Like other Scandinavian nations, Norway had long since abandoned its bellicose Viking heritage: peace, not war, was deeply rooted in its psyche. Norwegians had little admiration for military heroes, of whom their country, in any case, had few. Much more esteemed were the winners of the Nobel Peace Prize, chosen annually by the Norwegian parliament. "It was very difficult to be a military man in prewar Norway," noted one of the few army officers on active duty in April 1940.

In the late 1930s, this seagoing country's navy had only seventy ships: its two largest were the oldest ironclads in the world, affectionately called "my old bathtubs" by the naval chief of staff. The tiny Norwegian army, armed with vintage rifles and cannons, had no submachine or antiaircraft guns. The cavalry was supposed to be equipped with tanks, but the money appropriated by the government was so infinitesimal that only one tank had been purchased, "so that Norwegian soldiers could at least see one sample in their lifetime." Field maneuvers had not been held for years—they had been abolished as a way of saving money—and many brigade commanders had never even met their men.

Norway's military vulnerability, however, was of little concern to its government leaders. The country had been at peace for well over a century, had successfully maintained its neutrality during World War I, and intended to remain neutral in the future. Money should be spent on social reforms, Norway's leaders believed, not on building up the military. In the view of most Norwegians, "war was the kind of thing that happened in other parts of the world," noted Sigrid Undset, a Norwegian novelist who won the 1928 Nobel Prize in Literature. "How many of us had ever seriously believed it could happen in Norway?"

Having made a close study of Hitler, including reading *Mein Kampf* in the early 1930s, the sixty-seven-year-old king was far less sanguine. If war broke out, his peaceable northern kingdom, though militarily defenseless, would have great strategic importance. Facing Britain to the west, it provided a gateway to the North Atlantic. To the south, it had access to the Baltic Sea and the German coast. Not least, it controlled the northwest sea route through which iron ore from Sweden was shipped to Germany, the ore's main customer. And then there was Norway's far-flung merchant marine fleet, a glittering prize for Hitler or any other belligerent.

But whenever Haakon raised these and other points, government leaders disregarded them—and him. Most Norwegian officials scorned the monarchy as a useless relic of a bygone age and believed it should have no influence in government matters. Many thought there should be no monarchy at all. As much as he loved Norway, Haakon sometimes felt unwelcome there, at least in government circles. Not infrequently, he felt like the foreigner he once had been.

UNTIL HE BECAME KING OF NORWAY, Haakon VII, the second son of the crown prince of Denmark, had barely set foot in the country. He did not learn to speak Norwegian until the age of thirty-three, shortly before his reign began. Known as Prince Carl in Denmark, he had been a modest, unassuming young royal who grew up believing he would never be king of anything, for which he was profoundly grateful. His mother had reportedly pressured him to marry the young Queen Wilhelmina of the Netherlands, but he had resisted, wanting nothing to do with the pomp and formality of official court life. Instead, he wooed and won his first cousin Maud, the sports-mad daughter of King Edward VII of Britain, who was as anxious for a quiet life, out of the limelight, as he was.* At the time of his marriage, Carl, who sported a tattoo of an anchor on his arm, was an officer in the Danish navy and planned to make it his career.

But in 1905, Norway's declaration of independence from Sweden turned the life of the sailor prince upside down. The century-old union between the two countries had never been an equal one: Sweden, whose kings ruled both nations, had been the dominant partner from the beginning, and Norway had been growing increasingly restive. To lessen the chance of forceful Swedish opposition to their peaceful rebellion, Norwegian leaders said they would welcome a junior member of Sweden's royal family as the country's new monarch. Prince Carl, whose maternal grandfather was the king of Sweden and Norway, was the obvious choice.

The prince, however, was appalled at the idea. Not only did he want to remain in the Danish navy, he knew virtually nothing about Norway and its people. He was also acutely aware that many citizens of

---

* Carl's father, King Frederick VIII of Denmark, was the brother of Queen Alexandra, the wife of Edward VII.

Norway, which had abolished its aristocracy in the nineteenth century, were in favor of a republic, not a monarchy. Under heavy pressure from his father-in-law, Edward VII, among others, he finally agreed—but only if Norway held a referendum on the issue. When 88 percent of the electorate voted for a monarchy, Carl was crowned, taking the ancient Norwegian royal name of Haakon. (His wife, English to the core, refused to renounce her given name: she was known as Queen Maud until the day she died in 1938. She continued, as she always had done, to address her husband as Charles, the Anglified version of Carl. "I actually have plans to make him completely English," she confided to her diary early in their courtship.)

With Haakon as monarch, Norway boasted the most egalitarian kingdom in the world. Sir Frederick Ponsonby, an aide to Queen Maud's father, once said that Norway was "so socialistic that a King and Queen seemed out of place." After a visit to Oslo in 1911, Theodore Roosevelt wrote to an acquaintance that the insertion of a royal family into the most democratic society in Europe was like "Vermont offhandedly trying the experiment of having a King."

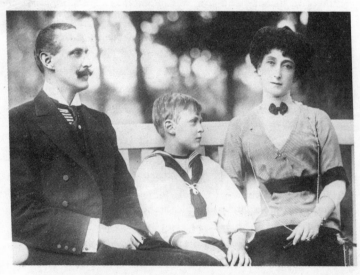

*King Haakon VII with his wife, Queen Maud,
and son, Crown Prince Olav.*

Haakon, who frequently described his position as that of "a very democratic president for life," was known to his people as "Herre Konge" ("Mr. King") rather than "Your Majesty." The royal family lived simply, with Queen Maud often doing her own shopping. In his frequent tours of the country and travels abroad, Haakon impressed those he met with his friendliness and wry sense of humor. Once, at a gathering of the British royal family at Windsor, he noticed a youthful distant cousin of his, Lord Frederick Cambridge, standing awkwardly by himself in a corner. He marched over and vigorously shook the young peer's hand. "You don't know me," he said. "Let me introduce myself. I'm old Norway."

As close as he was to his British relatives and as much as he loved their country, Haakon was horrified by the refusal of Prime Minister Neville Chamberlain's government to confront Hitler over his repeated aggressions in the 1930s. After World War II began in September 1939, Norway, like other neutral European countries, made clear that it wanted no part of a military alliance with a nation that, along with France, had handed over much of Czechoslovakia to the Führer and then, having declared war against Germany for invading Poland, had failed to do anything to aid the Poles. "All the small nations now understand that we in the future have to look after ourselves," Haakon wrote chidingly to his nephew, Britain's King George VI.

Until the spring of 1940, the war was a conflict in name only. Chamberlain and most officials in his government had no interest in and no intention of fighting a real war. They had imposed an economic blockade against Hitler and seemed to think that this would be enough to bring him to his knees.

Winston Churchill, Chamberlain's first lord of the admiralty and the British Cabinet's only bellicose member, strongly disagreed with Chamberlain's "phony war" strategy. From the war's first day, he demanded that Britain take the offensive against Germany—but not on German soil. The confrontation, he said, should come in the waters of Norway. He repeatedly urged the British government to stop the shipment of Swedish iron ore, vital to Germany's armament industry, along Norway's coastline. When both Norway and Sweden protested that idea, Churchill was infuriated by their reluctance to become battlefields for the warring powers. "We are fighting to re-establish the reign of law and to protect the liberties of small countries," he told the War Cabinet (a claim that both Poland and Czechoslovakia might have found hard to stomach). "Small nations must not tie our hands when we are fighting for their rights and freedoms. . . . Humanity, rather than legality, must be our guide."

After hesitating for months, Chamberlain finally gave in to Churchill's pressure. At dawn on April 8, 1940, British ships began sowing mines along the Norwegian coast. Hitler, who weeks earlier had said he would forestall any British move on Norway, had already ordered his high command to implement carefully prepared plans for the following day's surprise attack and occupation of both Norway and Denmark.

In most respects, Germany's land, sea, and air assault on the two Scandinavian countries was a brilliant success. Before it began, Hitler had decreed that the kings of Norway and Denmark must be prevented from escaping "at all costs." In Copenhagen, the Germans had no trouble finding King Christian X of Denmark, Haakon's sixty-nine-year-old brother, who capitulated as ordered. But bad weather and the sinking of the *Blücher* had upset the split-second timing of the assault on Oslo. When German troops finally entered the royal palace, government buildings, and the Bank of Norway that afternoon, they found only frightened low-level government employees and piles of papers burning in furnaces and fireplaces. The bank vaults lay empty, with no trace of the country's gold bullion. The king and government leaders had vanished, too.

WHEN THE RESIDENTS OF OSLO awoke on April 9, they found their world, so neatly ordered the day before, in chaos. Although the Germans had not yet entered the city, Luftwaffe bombers crisscrossed the sky overhead, and the muffled crump of bombs could be heard in the distance. Columns of thick black smoke from the burning government documents spiraled upward. Beautiful Oslo, with its verdant parks, hills, and forests, now lay open to an enemy it never knew it had.

A few hours earlier, as the *Blücher* had steamed toward Oslo in the predawn darkness, German minister Curt Bräuer had confronted the Norwegian foreign minister, Halvdan Koht, with a demand for Norway's surrender, emphasizing "how completely senseless all resistance would be." Though dazed by the sudden attack, Koht had the wit to remind Bräuer of Hitler's remark about Czechoslovakia's capitulation after Munich—that "the nation that bowed meekly to an aggressor without offering resistance was not worthy to live." With that, he rejected the German demand.

Before joining the king and other government officials aboard the special train that spirited them away that morning, Koht told a radio journalist that Norway was at war with Germany, that the king and government had escaped, and, incorrectly, that a general mobilization was now in effect. In response to his announcement, which was broadcast throughout the country, thousands of young men, suitcases in hand, headed for the nearest military center, only to be informed that it was all a mistake. "Reservists and volunteers left recruiting stations in tears when told there were no arms for them," a British diplomat recalled. In the Norwegian capital, dazed crowds gathered in front of newspaper bulletin boards, swapping fears and rumors.

Some Oslo residents, however, already had all the information they needed: the Germans were coming, and they must flee the capital before the invaders arrived. Sigrid Undset hurriedly packed a couple of suitcases and left. The fifty-seven-year-old novelist had been a strong and vocal critic of Hitler since the early 1930s; her books had been banned in Germany shortly after the Führer took control of the government.

A young German émigré named Willy Brandt was also in danger. The twenty-six-year-old Brandt, who had battled Nazis in the streets even before they came to power in Germany and had been deprived of his citizenship as a result, had found asylum in Norway seven years earlier. After studying at the University of Oslo, he had become a journalist and was closely allied to top figures in the country's ruling Labor Party. He was also heavily involved in the work of German émigré groups trying to stimulate opposition to Hitler in their homeland. Early that morning, he had been awakened by an urgent phone call: he must get out of the city as quickly as possible. A few minutes later, he was, as he wrote later, "in flight again"—whisked away to a safe house in the suburbs of Oslo, where he was picked up by a couple of leading Norwegian politicians and driven to safety. Brandt, who would become one of Germany's most eminent postwar chancellors, was eventually spirited out of the country to neutral Sweden. He spent the war there as a journalist and propagandist for "the cause of a free Norway."

Oslo residents watch as German troops march into the city in April 1940.

KING HAAKON AND THE government leaders, meanwhile, escaped to Hamar, a town eighty miles north of Oslo. As their train left the capital early that morning, a long line of trucks idled outside the Bank of Norway, a black granite building near Oslo's harbor, waiting to be loaded with hundreds of boxes and barrels containing the lifeblood of Norway's economy: fifty tons of gold bullion totaling $55 million ($915 million today). Far more foresighted than government officials, the bank's director had made plans months before to evacuate the gold reserves to a secret bombproof vault in the town of Lillehammer, 114 miles north of Oslo, in case of attack.

In Hamar, the Norwegian parliament immediately went into session in the local movie theater, while hundreds more people—civil servants, businessmen, journalists, and foreign diplomats—poured into the little town, occupying all its hotel rooms and crowding its slushy streets. Government ministers bought up Hamar's entire stock of paper and pencils to transact Cabinet business, and government clerks began unpacking the crates of documents they had brought from Oslo. "I think we all, subconsciously, expected to settle down comfortably there," recalled Florence "Daisy" Harriman, a former New York socialite turned social reformer whom President Franklin D. Roosevelt had appointed U.S. minister to Norway. "We were not yet ready to imagine that the King and Government would be hunted like wild animals."

In midevening, the fragile sense of security was shattered when the parliament's president, Carl Hambro, interrupted a debate to announce that German forces were heading toward Hamar and that a train was waiting at the station to take the king and government away. Officials grabbed their hats and coats and ran for the door. Ten minutes later, the train departed, heading for Elverum, a town higher in the mountains and closer to the Swedish border.

For most of the exhausted government leaders, that nightmarish first day of war finally came to an end in Elverum. Only the king, his family, and several key ministers pushed on to the tiny snow-covered hamlet of Nybergsund, which was considered safer from German bombs. From there, Crown Prince Olav sent his wife, Princess Märtha, and their three children to Sweden, the princess's homeland.

The following day, Haakon agreed to meet with Curt Bräuer in Elverum. Employing a mixture of flattery and threats, the German minister promised the king that if he accepted the German demands, he would retain the honors and perquisites of his position and Norway would escape further destruction; if he did not, all resistance would be ruthlessly crushed. The demands, Bräuer said, included not only capitulation but the naming of Vidkun Quisling, the fifty-two-year-old leader of Norway's minuscule Nazi Party, as the new prime minister of Norway.

Haakon was astonished and outraged at the thought of Quisling heading the Norwegian government. Quisling's party, which had never won more than 2 percent of the vote in any election, was considered a joke in Norway, while he and his men, in the words of Sigrid Undset, were thought of as "hysterical half-men." His voice filled with fury, the king told Bräuer that he "could not appoint a government that did not enjoy the confidence of the Norwegian people, and several elections had shown that Quisling did not enjoy such confidence."

Returning to Nybergsund that night, Haakon briefed his son and ministers on Bräuer's demands. On the run for more than twenty-four hours, the unshaven, disheveled men were suffering from both physical and emotional exhaustion. Several of them, badly frightened by the Germans' efforts to capture them and despondent over the apparent hopelessness of Norway's situation, believed they should give in and negotiate for peace without delay. The country was totally unprepared to fight Germany, they argued; if it tried to resist, it would be committing national suicide. They noted that Britain had sent word that its forces would come to Norway's aid as soon as possible. But after what had happened in Czechoslovakia and Poland, how could anyone in his right mind put trust in British promises? The government must capitulate *now*.

The tall, ramrod-straight Haakon was well aware that his ministers had never paid much attention to his advice and counsel in the past. This time, though, the future of his adopted country was at stake, and he was determined to follow his conscience and speak his mind. "The government is free to decide," he said in an unsteady voice, "but I shall make my own position clear: I cannot accept the German demands. This would conflict with everything I have considered to be my duty as king ever since I came to Norway almost thirty-five years ago." If the government chose otherwise, he would abdicate, renouncing the throne of Norway for himself and his family.

"That instant burnt itself into my memory," recalled Trygve Lie, the minister of supply and a future secretary general of the United Nations. "Having said these words, the King gazed intently at Prince Olav. For a long while, he was unable to resume, then he bent over the table and burst into tears. Prince Olav also had tears in his eyes." Raising his head, Haakon struggled to control his emotions. "The government must now make its decision," he said.

His unequivocal stand ended all talk of capitulation. Swayed by his resolve and his willingness to sacrifice his throne for principle, the ministers, including the most defeatist, voted to reject the ultimatum. While Halvdan Koht phoned the news to Bräuer, Haakon and his prime minister signed a proclamation, which was broadcast over Norwegian radio, rejecting the German demands and calling on Norwegians to resist the invaders with all their might. When news of Haakon's intransigence reached Hitler the following morning, the German leader flew into one of his trademark rages. How dare "this ridiculously small country and its petty king" defy him! The time for talking was over, Hitler declared. Haakon VII of Norway must be tracked down and killed.

The following day, April 11, Haakon was conferring with his ministers at an inn in Nybergsund when the pastoral quiet was suddenly shattered by the honking of a car horn—the prearranged signal for imminent danger. The king, his son, and the ministers ran out of the inn toward a nearby grove of trees, flinging themselves onto the ground as six German dive-bombers whined overhead and raked the village and grove with machine-gun fire. Wheeling around, the planes made several more passes, strafing and dropping incendiary bombs as they went. When the attack finally ended, Norway's leaders—wet, cold, and bloody from scratches—rose stiffly to their feet. Nybergsund itself was ablaze, but remarkably the raid had resulted in only two casualties, both of them villagers. When one of the bombers was shot down several days later, the pilot's diary was found with this entry: "The king, the Government, all annihilated . . ."

Once again, Haakon and his party headed north, into the wild, mountainous, and glacier-bound landscape of central Norway. Their convoy of vehicles, painted white for camouflage, crept along rough, narrow hairpin roads bordered by sheer peaks and seemingly bottomless mountain chasms. Cars broke down or were stuck in the snow, and more than once during the next two weeks, Haakon, who was in a car driven by his son, was separated from the government ministers, neither party knowing where the other was or whether anyone else was still alive. Ceaselessly tracking them, the Germans bombed and strafed every place they were thought to be. At the first sight or sound of a

plane overhead, the king and his party headed for the nearest cover—
trees, a rock, anything they could use to hide below or behind.

Throughout these long early-spring days, they stopped whenever
they could—to rest, conduct government business, attempt to find out
what was going on in the rest of Norway, and consult intermittently
with the British, French, and American diplomats who were doing their
best to keep up with them. Inevitably, however, reports of approaching
German troops or Luftwaffe aircraft would force them onward.

NOTWITHSTANDING THEIR VIKING HERITAGE, the Norwegians were
"not good haters," as one of their leaders later noted. But it didn't take
long for most of them to develop a fierce hatred of the Germans—"an
army of marauders," in the words of Sigrid Undset, "who came to live
where they have not built, to reap where they have not sown, to rule
over a people they have never served." The king's defiant rejection of
Germany's demands served as a stimulus for national resistance. Germany had conquered all of Norway's major ports but not the interior
of the country, and once their initial shock and confusion had worn
off, the Norwegians began to fight back. In the days following the invasion, young men by the thousands wandered the countryside, trying to
find army units they could join.

At first, the army was also in turmoil: its demoralized commander
in chief favored surrender or negotiation with the Germans. Backed by
King Haakon, the cabinet replaced him with General Otto Ruge, a tall,
craggy-faced former army chief of staff, who cobbled together a
40,000-man army from the fast-growing stream of citizen volunteers
who poured out of their cities and towns by foot, on skis and bicycles,
in cars, trucks, and buses. Although many were excellent marksmen
who had brought along their own rifles and pistols, they had no artillery, tanks, antitank weapons, or air support in their skirmishes with
the well-equipped, well-trained Germans.

Ruge's strategy was to play for time, to contain the Germans long
enough in the south to allow an orderly retreat and stabilization of the
front in central Norway. "With the weak and improvised forces we had

at our disposal, it was impossible for us to engage in any decisive battle before the Allies came to our aid," he later recalled. "Our small forces fought without respite, without reserves, always in the front line, against heavy artillery, tanks, and bombers. . . . For three weeks, our divisions held out until at last, the Allies began to arrive."

Despite increasingly desperate appeals from Norway, it took Britain almost a week to cobble together an expedition to go to its assistance. Even though, as Ruge noted, the British must have realized that their mining of Norway's waters would provoke a German response, the Chamberlain government was as stunned by the invasion as the Norwegians. "The idea of an operation of this scope against Scandinavia had never entered my head," confessed General Hastings "Pug" Ismay, secretary of the Committee of Imperial Defence, whose members included the War Cabinet and the heads of the military services. "So far as I knew, we had not a vestige of a plan to deal with it."

Making matters worse, the British military commanders knew almost nothing about Norway and its terrain. "There were no maps," said one officer. "We had to tear them out of geography books and send [someone] to the Norwegian travel agency to buy a Baedeker. From the Norwegian embassy and a series of tourist agencies, we gathered an armful of travel advertisement folders." The photos in those brochures, he added, "provided the only clues as to what our prospective theater of operations looked like." The British, a Norwegian historian later observed, hadn't the slightest "elementary knowledge of things Norwegian."

When British land forces finally put ashore in central Norway, Norwegian officers were astonished at the raw troops' lack of equipment and training. Although deep snow and ice still covered the ground in much of the country, almost none of the British soldiers had been provided with snowshoes or skis. They were also without almost everything else they needed—transport, artillery, antiaircraft weapons, communications gear, fighter cover, medical equipment, even food.

Pounded by the Wehrmacht and strafed by the Luftwaffe, the green British troops were overwhelmed. "We've been massacred! Simply massacred!" a young lieutenant exploded after a battle that ended in a rout of British forces. "It's been bloody awful!" Leland Stowe, a *Chicago Daily News* reporter who covered the British campaign, later called it "one of the costliest and most inexplicable bungles in military history." Echoing that sentiment, the disconsolate General Edmund Ironside, who as chief of the Imperial General Staff headed the British Army, wrote in his diary: "Always too late. Changing plans and nobody directing. To bed very upset at the thought of our incompetence."

Shortly after the British landed in Norway, Chamberlain's government canceled a planned attack on the key port of Trondheim. Late in April, without informing the Norwegian government or army, it decided to pull out of central Norway altogether, only nine days after its troops had arrived. When, against orders, the British commander shamefacedly informed Ruge on April 28 of the evacuation, the Norwegian general retorted, "So Norway is to share the fate of Czechoslovakia and Poland! But why? Why? Your troops haven't been defeated!" Overcome with fury, he left the room. After regaining his composure, he returned and calmly said to his British counterpart, "Please tell me what help I can give you to carry out your orders." For the next forty-eight hours, Ruge's forces supported the British in their retreat to the coast.

The day after Ruge learned of the evacuation, the British government sent the cruiser *Glasgow* to pick up King Haakon and his ministers from the pretty coastal town of Molde, their latest refuge, and take them north to Tromsø, a little polar community some two hundred miles north of the Arctic Circle. "You are killing us!" exclaimed Foreign Minister Halvdan Koht when he and other members of the government were informed of the British troop withdrawal. Yet Koht and the others had little choice but to leave. Molde, under heavy German bombardment for more than a day, had become an inferno. High-explosive and incendiary bombs screamed and thundered through the air, reducing houses, shops, churches, and factories to little more than rubble.

Late that evening, cars carrying the royals and government officials sped through town, dodging walls of flame and showers of shattering glass. It was, one official later said, "like driving through Hell." Much of the harbor was also ablaze, and when the king's party pulled up to the quay at which the cruiser was moored, the ship's fire hoses were pouring water on the conflagration.

As Haakon and his companions boarded the *Glasgow,* dozens of British seamen and Norwegian soldiers worked furiously to load hundreds of boxes of gold bullion—Norway's gold reserves—into the cruiser's hold. After being smuggled out of the Bank of Norway and dispatched on a cross-country hegira as perilous as that of the king, the gold had been stored in the cellar of a Molde textile factory. That night, as the factory burned, Norwegian civilians and troops fought smoke, flames, and falling timbers to retrieve the gold and load it onto trucks, which then sped to the harbor.

Only about half the gold had been loaded aboard the *Glasgow* when the jetty to which the ship was moored caught fire, too. Ordering a halt to the transfer, the *Glasgow*'s captain put his vessel full astern. Taking half the jetty with her, the *Glasgow* made her escape, zigzagging wildly down the fjord toward the open sea.*

WITH HAAKON SAFE FROM the Germans, at least for the moment, and with British troops on their way home from central Norway, Ruge's forces surrendered to the Germans on May 3. In Britain, Neville Chamberlain's announcement of the evacuation stunned his countrymen; the realization that the world's greatest sea power had been humiliated by Germany ignited a wave of fury and fear throughout the nation.

Aware that they were facing political disaster, Chamberlain and his ministers now cast about for a scapegoat. At a meeting of the War Cabinet, Winston Churchill, who as first lord of the admiralty had been the main architect of the slapdash Norwegian campaign, asserted that "blame should be attached not to us but to the neutrals, and we should take every opportunity of bringing this point up." Following his own advice, Churchill declared in the House of Commons, "The strict observance of neutrality by Norway has been a contributory cause to the sufferings to which she is now exposed and to the limits of the aid

---

* The rest of the gold was loaded onto small fishing boats, which finally made it to Tromsø as well. From there, all the gold reserves were dispatched to the United States and Canada for safekeeping.

which we can give her." Many members of Parliament, however, refused to accept Churchill's argument. Dissatisfaction with the Chamberlain government's dilatory conduct of the war boiled over in a vitriolic two-day debate in the House on May 7 and 8; at its end, the prime minister barely won a vote of confidence.

Meanwhile, the Norwegians continued to resist. Although the war in southern and central Norway was over, an Allied force in the far north, composed of British, French, Polish, and Norwegian troops, was slowly gaining the upper hand over the Germans in a struggle for the crucial port of Narvik. Then, on May 10, a tsunami of events consigned the war in Norway to oblivion. Early that morning, millions of German troops, accompanied by swarms of tanks and aircraft, swept into the Netherlands, Belgium, and Luxembourg in a lightning assault from the North Sea to the Moselle River. After being tested in Poland and Scandinavia, Hitler's blitzkrieg was now slicing through the very heart of Europe.

That same afternoon, Neville Chamberlain, persuaded that he no longer commanded the confidence of the majority of his party and informed that neither Liberal nor Labour MPs would join a coalition government under his leadership, advised King George VI to send for Winston Churchill as the next prime minister. Years later, Churchill would acknowledge that "considering the prominent part I played [in the Norway disaster] . . . it was a marvel that I survived." But, as the most prominent prewar opponent of Chamberlain's appeasement policy, he was seen, quite rightly, as the only major political figure in Britain with the energy, drive, and determination to lead the country in wartime.

On May 13, he proved his mettle as a warrior in a soaring speech to the House. "You ask what is our aim?" he said. "I can answer in one word: victory. Victory at all costs; victory in spite of all terror; victory, however long and hard the road may be." That single word, as far-fetched as its achievement seemed to be in those dark early days, would remain his touchstone for the war's duration.

CHAPTER 2

"A Bold and Noble Woman"

Holland Falls but
Its Queen Escapes

IN THE PREDAWN HOURS OF MAY 10, 1940, QUEEN WILHELMINA OF the Netherlands gently shook her daughter awake. "They have come," she told Princess Juliana.

This time, the early-morning invaders arrived from the air. They dropped by the thousands over bright green polders and fields ablaze with red and yellow tulips, over steeples and windmills, over the orange-tiled roofs of peaceful Holland. Awakened by the roar of aircraft overhead, the Dutch, many still in nightgowns and pajamas, poured from their homes and peered upward. While milkmen distributed their wares door-to-door and housewives headed to market, German parachutists were landing in country gardens and city streets. To some of the children looking on, it seemed like a fascinating new game.

Queen Wilhelmina knew otherwise. Like King Haakon, she had been warning her government for years of the growing danger of Hitler and Germany, but, as in Norway, government officials had paid no heed to their monarch. The queen "had been expecting Nazis for a long time," recalled Erik Hazelhoff Roelfzema, a law student at Leiden University at the time of the German invasion. "In this she was almost alone in Holland." Even after all the other countries fell, "the entire Dutch nation refused to believe that we would be next. When war engulfed us, we didn't have the slightest idea what to do about it."

Except for a brief war in 1830, when Belgium had risen up and gained independence from the Netherlands, the Dutch had lived in peace for 125 years, ever since they had joined the British in fighting Napoleon at the Battle of Waterloo. Like the Norwegians, they had managed to stay neutral during the Great War, and until May 10, they had clung to the hope and belief that they could do so in this war, too.

But even if the unthinkable happened and Germany did attack, the Dutch were sure they could bottle up the invaders by using the defensive tactics that they had successfully employed against Spain and France centuries before. In the event of an incursion, thousands of acres of land in the northeast and south would be flooded while the Dutch army pulled back to defend Fortress Holland—the country's western provinces, containing its major cities: Amsterdam, Rotterdam, Utrecht, Leiden, and The Hague.

The plan, however, took no account of the fact that airborne troops could now leapfrog flooded areas and land right in the heart of the fortress—precisely the scenario that Germany now followed. On May 10, German paratroopers seized key bridges at Rotterdam and over the Maas River at Moerdijk and Dordrecht, as Wehrmacht troops, spearheaded by tanks and motorized infantry, poured over the border between the two countries. A second airborne force was to land at The Hague, the center of government, and capture the queen, her ministers, and the top military command.

Known for its leafy parks and wide boulevards, The Hague was a quiet, elegant, cultured city a few miles from the Dutch coast and the North Sea. Its swift occupation was a key objective for Hitler, who was particularly determined to seize Wilhelmina within hours of the initial German assault. He had instructed the commander of his airborne force in Holland to treat the queen with the utmost deference and honor. He even ordered that she be presented with a bouquet of flowers after she was taken prisoner. No harm must come to Wilhelmina, "who is so popular with her people and the whole world," he declared.

Having failed so miserably in his attempts to win over the king of Norway to the Nazi cause, the Führer was now set on wooing the Dutch queen, whose mother, husband, and son-in-law had all been German. His doomed efforts showed how little he understood this feisty fifty-nine-year-old monarch, who, despite her German connections, considered the Third Reich "an immoral system" and denigrated Hitler and his followers as "those bandits." In November 1939, *Time* magazine noted that "no one has given Wilhelmina more trouble in recent years than Hitler, and from no ruler has the Führer taken at times such straight talk." Wilhelmina had long made clear, to Hitler and everyone else, that "anyone who threatened the interests of my people and country was my personal enemy." Less than four hours after the German attack began, she declared via Dutch radio, "I raise a fierce protest against this flagrant violation of good faith, this outrage against all that is decent between civilized states."

AS A LITTLE GIRL, Princess Wilhelmina had taken to heart the declared aim of her English governess—to "make a bold and noble woman out of you." Her dream from childhood had been to perform "great deeds," like those of her famed ancestors: William the Silent, who had led Holland's fight for independence against Spain in the sixteenth century, and William of Orange, who had defended both Holland and England against the French a century later.

But to her great dismay, she saw no possibility of realizing that dream. The youngest and only surviving child of her elderly father, King William III, she had grown up in what she despairingly called "the cage"—the oppressively formal, strict atmosphere of the Dutch royal court that precluded, she later said, "any kind of initiative" and provided "no opportunity to show vigor and courage." The shy, serious little royal, who became queen at the age of ten after her father died, was raised with few friends or companions her own age. No one was allowed to address her familiarly, and when she went ice skating in the winter, the designated pond or canal was cleared of people and she was forced to skate alone. She was once overheard scolding one of her dolls, "If you are naughty, I shall make you into a queen, and then you won't have any other little children to play with." Years later, she made sure that Juliana—the only progeny of her unhappy marriage to a playboy German princeling, Heinrich of Mecklenburg-Schwerin, who died in 1934—had as normal an upbringing as she could give her.

From the beginning of her reign (she assumed her queenly duties at

*Queen Wilhelmina
early in her reign.*

eighteen), the strong-willed Wilhelmina was determined to break out of the "cage" and make her mark on the world. At nineteen, she offered one of her palaces in The Hague as a place where countries could come to arbitrate their differences rather than resort to war—an offer that led to the foundation of the International Court of Justice. In 1900, in the middle of the Boer War, the twenty-year-old queen ordered a Dutch warship to defy a British blockade of South Africa and rescue Paul Kruger, the embattled Boer president. Eighteen years later, after the end of World War I, she granted asylum to Kaiser Wilhelm II of Germany and later refused Allied demands to extradite the kaiser as a war criminal.

The Allies were incensed by her effrontery. Wilhelmina and her country, in their view, were acting as if the Netherlands were still one of the most powerful nations in Europe instead of the second-tier country it had become. There was much truth to that opinion. Although Holland's Golden Age—during which it had dominated world trade, produced great artists such as Rembrandt and Vermeer, and controlled an empire as vast as any other on Earth—had ended more than two centuries before, the queen and her people strongly believed that their country still had an important international role to play. While it was true that Holland's overseas possessions had been greatly reduced, it retained control of its great colonial treasure chest: the Dutch East Indies, a chain of islands stretching from Burma to Australia, whose cornucopia of riches included rubber, oil, coffee, tobacco, tin, and gold. And despite its small size in both land and population, Holland remained one of the world's leading banking and trade centers, home to such blue-chip conglomerates as Philips, Royal Dutch Shell, and Unilever.

For her part, Wilhelmina was determined to uphold the greatness of the House of Orange, Holland's ruling dynasty since the sixteenth century. But as her ministers repeatedly made clear, she no longer possessed the power of William of Orange and her other famed predecessors. Since the mid–nineteenth century, Holland had been a constitutional monarchy, much like Britain and Norway, which meant that Wilhelmina, to her great frustration, had only the same types of rights held by George VI and Haakon: to encourage, to warn, and to be consulted and informed. But, as was true in Norway, the leaders of the coalition governments presiding over the country during her rule did not consult her, and when she gave them unsolicited advice, they usually paid little or no heed.

Much more imperious and outspoken than Haakon, Wilhelmina had no qualms about venting her anger over being ignored. "In certain respects, she resembled Queen Victoria, who, in her moments of displeasure, could make such aristocrats as Lord Salisbury quake and Prince Bismarck sweat," noted the British historian John Wheeler-Bennett. (During the later years of World War II, Winston Churchill would remark, "I fear no man in the world but Queen Wilhelmina.")

As the threat of war increased in the mid- to late 1930s, the queen was given a multitude of opportunities to display her formidable temper at the blindness of her ministers and people in respect to Nazi Germany. "By the spring of 1938, when Hitler invaded Austria, [it] was plain to me that German policy would result in a European catastrophe," she later noted. But the Dutch were "quietly asleep on a pillow called neutrality. . . . Shortly before the war it was necessary for me to point out that Hitler had written a book, and that it might be of some use to examine its contents."

Most of the Dutch were strongly anti-Nazi; the country's small fascist party, the National Socialist Movement, had won only four seats in each of the two houses of parliament in the 1937 general election. At the same time, many Dutchmen, complacent in their peace and prosperity, considered Hitler's rise to be a purely German affair, with little potential impact on or danger for Holland. Germany was also Holland's most important trading partner, and Dutch business interests believed it essential not to rile their powerful next-door neighbor, so vitally important to their country's economy. As one young Dutchman wrote, "Our little speck of land . . . conducted its business as usual, as if war were a foreign product unsuited to the Dutch marketplace."

It wasn't until Germany occupied the Sudetenland in October 1938 that the Dutch government began, albeit reluctantly, to prepare for war. It strengthened the country's water defenses and tried to modernize its armed forces, but the attempt came much too late. Dutch factories were incapable of producing all the aircraft, arms, and equipment that the country needed, and Britain and other countries had none to spare. As a result, the Dutch army, which numbered 300,000 men when fully mobilized, was armed with little but nineteenth-century carbines and equally antique artillery. Of the air force's 118 planes, just a handful were of recent vintage. Only the navy, whose main task was to protect the Dutch East Indies, had more than a modicum of up-to-date vessels and equipment when the war began.

Urged on by the queen, the Dutch government mobilized the country's armed forces in August 1939, just days before Britain and France declared war on Germany. At the same time, Wilhelmina, who had been an ardent advocate of world peace since her teens, joined the leaders of five other neutral nations in Europe—Belgium, Norway, Sweden, Finland, and Denmark—in offering their "good offices" to find a nonviolent solution to the crisis over Poland. The leader of the appeal,

King Leopold III of Belgium, declared that he and the others wanted to establish "a great new power—a moral and spiritual power capable of arousing world opinion through the union of the small states. . . . Let the conscience of the world awake!"

This desperate attempt by the neutrals to ward off what they saw as looming disaster served only to annoy Britain and France. When war was declared, the Allies' irritation increased when the neutrals, particularly Holland and Belgium, refused to enter into official military talks. Behind the scenes, however, the British and Dutch military, particularly the two admiralties, quietly exchanged information and formulated contingency plans in the event of a German assault on Holland.

WHEN GERMANY LAUNCHED ITS blitzkrieg, the queen, her daughter and son-in-law, and her two small granddaughters (the younger only nine months old) were staying at her country palace not far from The Hague. Her security forces told her that German paratroopers were landing only a few miles away. Indeed, one of the palace guards had just shot down a German plane, which crashed in a nearby park.

Wilhelmina's bodyguards bundled the royal family into cars and, threading their way through massive traffic jams and enormous crowds jamming the streets, escorted the royals to the relative safety of Noordeinde Palace, the queen's main residence in the center of town.

For the next three days, while war raged throughout the country, the queen and her family took refuge in a little air-raid shelter in the palace's garden. Furious at being sequestered, Wilhelmina insisted that she be allowed to go outside and see for herself what was going on. Her palace guards refused, explaining that they were under orders not to let her leave the palace. She was further incensed when she was unable to contact her ministers, who showed little sign of wanting to contact her. Once again, she found herself trapped in the hated "cage."

Within hours of the invasion, the Dutch government appealed to Britain for arms and air support. The foreign and colonial ministers, "bewildered and dazed as if they simply did not know what had hit them," flew by naval plane to London to make the plea in person. General Pug Ismay, representing the British chiefs of staff, told them that

their request was impossible to fulfill. Britain, whose small expeditionary force was now fighting in Belgium, had no spare planes or men to
send to Holland. "Even if the troops had been available, we had no
means of getting them there in time," Ismay recalled. "I remember observing that, alas, we had no magic carpet." Winston Churchill, who
would become prime minister later that day following Neville Chamberlain's resignation, gave the Dutch officials, whom he described as
"haggard and worn, with horror in their eyes," the same bad news.

By May 13, the Luftwaffe had destroyed most of the Dutch air force,
and German troops, despite a determined defense by the ill-equipped
Dutch army, had gained control of much of the country. Early in the
invasion, Dutch soldiers had succeeded in routing German airborne
troops from three airfields outside The Hague, thus saving the queen
and government from imminent capture. But The Hague's defenders
couldn't halt the swelling German tide for long: the city was now par-
tially surrounded, and there was hand-to-hand fighting in many of its
streets.

At five o'clock that morning, Wilhelmina made a desperate personal
plea for aid to her fellow sovereign George VI. A police sergeant on
duty at Buckingham Palace woke the British king with the news that
the queen of the Netherlands was on the phone. The startled monarch,
who had never met Wilhelmina, thought it was a joke, but he took the
call and found, as he wrote in his diary, that it indeed "was her. She
begged me to send aircraft for the defence of Holland. I passed this
message on to everyone concerned & went back to bed." While George
slept, Wilhelmina prepared to flee. Shortly after her appeal to the king,
she was told by General H. G. Winkelman, commander in chief of the
Dutch forces, that German units were now on their way to the palace
to capture her. She must leave at once.

The night before, the queen had dispatched the thirty-one-year-old
Juliana and her family to England aboard a British destroyer—the fruit
of prewar talks between the Dutch and British admiralties that had resulted in the drawing up of evacuation plans for the royal family and
government in case of invasion. Reluctant to leave the country herself,
Wilhelmina decided to travel to Zeeland province in southwest Holland, where the Dutch were continuing to hold out against the enemy

with the help of French troops. After hurriedly filling a briefcase with official papers, she was driven in an armored car to Hook of Holland, a port near Rotterdam, where thousands of people, many of them Jews, were milling about on the docks, dodging German bombs while desperately seeking a passage out.

Once aboard the British destroyer HMS *Hereford,* the queen, "calm and unruffled" in a life jacket and steel helmet, asked the captain to set his course for Zeeland. He informed her that, according to his instructions from the Admiralty, he was not allowed to make contact with the shore and must go directly to England. Wilhelmina was devastated. She had hoped to join her troops on the battlefield, as her illustrious ancestors had done so many years before and as King Albert I of Belgium, Leopold's father, had done in the Great War. If the worst happened, she was prepared, in the words of William of Orange, to "be the last man to fall in the last ditch." She found it impossible to accept that these "great deeds" were to be denied her. She would follow the Admiralty's instructions and go to England, she decided, but as soon as she landed, she would demand an immediate return to Holland, as well as more aid for her beleaguered country.

When Wilhelmina phoned George VI a few hours later from the British port of Harwich, his answer to her requests was a polite no. The king explained that the military situation in Holland had worsened considerably since she had left that morning and that it was unthinkable for her to go back. He added that a train was waiting to bring her to London. For a second time that day, the tough, imperious queen was close to tears. How could she abandon her country and people during the worst moment in their history? She knew that the Dutch would not understand the circumstances of her departure, that it would make a "shattering impression" on everyone back home. Still, what alternative did she have but to follow the king's instructions and board the train for London? She was well aware that she was following in the footsteps of William of Orange, who had taken up residence in the English capital more than two centuries before. But he had come in triumph as the new king of England, while she was going into exile.

When Wilhelmina's heavily guarded train arrived at London's smoke-blackened Liverpool Street station late that afternoon, George

VI and an honor guard of khaki-clad British soldiers were on hand to meet her. The somber queen stepped from the carriage, a gas mask over her shoulder and clutching the steel helmet she'd been given aboard the destroyer. After kissing her on both cheeks, King George escorted her to Buckingham Palace. "She was naturally very upset & had brought no clothes with her," he later wrote in his diary.

Wilhelmina still held out hope that she might return, but the fire-bombing of Rotterdam the following day put an end to her plans. On the afternoon of May 14, a wave of Heinkel bombers, flying wingtip to wingtip, dropped hundreds of bombs on Rotterdam's downtown, incinerating much of the area and killing nearly a thousand residents. Told by German officials that other Dutch cities would suffer the same fate unless Holland surrendered, General Winkelman signed a letter of capitulation that afternoon.

Once again, however, not everything had gone according to the Germans' carefully worked-out plans. Most of Holland's gold bullion had vanished, already on its way to Britain and the United States. Also missing were more than $1 billion worth of international securities, as well as diamonds worth tens of millions of dollars. But the most significant disappearance was that of the rebellious Dutch queen and her government—another German misstep, like Haakon VII's escape, that Hitler would soon come to regret.

CHAPTER 3

"A Complete and Utter Shambles"

The Collapse of Belgium and France

UNLIKE THE SOVEREIGNS OF NORWAY AND THE NETHERLANDS, King Leopold III of Belgium was not whisked away to safety during Germany's invasion of his country. Instead, as commander in chief of the Belgian army, the thirty-eight-year-old king took charge of Belgium's defense as soon as the Wehrmacht crossed the border.

Under the Belgian constitution, Leopold possessed more power and governmental responsibility than any other western European monarch: in addition to his role as commander in chief, he also acted as president of the Cabinet. When Hitler dispatched a note to him and his ministers warning that resistance to German occupation might well mean the destruction of Belgium, Leopold responded, "When it is a question of sacrifice or dishonor, the Belgian in 1940 will hesitate no more than his father did in 1914."

Like Norway and Holland, Belgium had been neutral before the 1940 German attack, as it had been before World War I. Indeed, when the Belgians won their independence from Holland in 1831, they had been promised permanent neutrality and "inviolability of Belgian territory" by Europe's great powers. That promise had not been kept; virtually from the day the treaty was signed, several of the powers had sought at one time or another to breach Belgian neutrality in order to promote and protect their own interests.

While Norway and Holland had been able to keep their neutral status during the Great War, Belgium was the first country invaded by Germany. In their August 1914 incursion, German troops were simply continuing the centuries-old tradition of warring powers using Belgian territory as a convenient passageway to reach whichever country was their main target at the time—in this case, France. When the Belgians put up a stubborn and surprisingly strong resistance, ending Germany's hope of a quick victory over the French, the invaders were ferocious in their retaliation, launching a campaign of wholesale executions and destruction. The medieval city of Louvain, like a number of other Belgian towns, was pillaged; many of its buildings, including its world-famous university library, were burned to the ground. At the end of the war, much of the country was left in a state of devastation, with most of its roads, industry, and railway system destroyed.

Only twelve years old when the Great War began, Leopold was profoundly affected not only by its impact on Belgium but also by the widely lauded wartime actions of his father, King Albert I. One of the most admired military and political figures of the conflict, Albert refused numerous invitations from the Allies to flee to Britain or France. Instead, he rallied his vastly outnumbered troops in October 1914 to win the crucial Battle of the Yser, which halted the German advance in Belgium and gave the king and his forces control of a small sliver of Belgian territory on the North Sea coast for the rest of the war. That victory in turn allowed the Allies to maintain control of several key French ports nearby.

Thanks to his boldness and resolution, Albert did more than benefit the Allies; he also helped restore the reputation and popularity of the Belgian monarchy, which had been blackened by his predecessor and uncle, Leopold II. In the late nineteenth century, Leopold had acquired and exploited for his private gain what later became known as the Belgian Congo, arousing international outrage for his henchmen's horrific treatment of Congolese workers, millions of whom had died as a result. When Albert assumed the throne in 1904, he instituted reforms in the Congo to try to ameliorate the human and physical damage wreaked by his rapacious forebear. Leopold III followed his father's reformist example.

Modest and soft-spoken, the younger Leopold hero-worshiped Albert and, from early childhood, modeled his life on that of his father. When the Great War began, the crown prince was sent with his two siblings to England, where he was enrolled at Eton. But he persuaded Albert, who remained in Belgium throughout the war, to allow him to train and then serve as a soldier during his school breaks. For the war's duration, Leopold led a curious life—a student at Britain's most exclusive boys' school for most of the year, coupled with stints as a private in the Belgian army during Eton's vacation periods.

King Leopold III with his bride, Princess Astrid of Sweden.

In 1934, Albert, still vigorous at the age of fifty-eight, was killed in a mountain-climbing accident, and the thirty-two-year-old Leopold assumed the throne. Blond and boyishly handsome, the new king and his pretty Swedish wife, Astrid, along with their three young children, brought youth and a touch of glamour to the monarchy. Just over a year later, however, the twenty-nine-year-old queen died in an automobile accident in Switzerland when her husband, who was driving, lost control of their car on a winding mountain road. Bereft by the back-to-back deaths of the two people to whom he was closest, the sensitive, reserved Leopold was also haunted by his own culpability in the accident that had killed his wife. Attempting to assuage his grief and sense of guilt, he threw himself into his kingly duties with a new intensity.

In leading the country, he closely emulated his father, particularly in foreign affairs. After the Great War, Albert had pursued two major foreign policy objectives: retaining Belgium's neutrality while strengthening its defenses to allow it to put up a better fight against Germany or any other likely aggressor. During his son's reign, Belgium spent almost a quarter of its budget on national defense, far more than any other European country except Germany. At the outbreak of World War II, more than half of all Belgian men between the ages of twenty and forty—a total of 650,000—were under arms. When the Belgian army was mobilized in May 1940, its numbers had swelled to 900,000, compared to 237,000 soldiers in the British Expeditionary Force in France.

At the same time, however, Leopold, like his father, insisted that the country remain neutral, much to the dismay of France and Britain, which had pressed Belgium hard in 1939 and early 1940 to enter into a military alliance that would permit their troops to enter Belgian territory before any fighting broke out. Leopold and his government suspected that the Western Allies' eagerness for a military partnership was prompted by a desire to keep the war as far away as possible from their own soil. Indeed, France's commander in chief, General Maurice Gamelin, acknowledged as much when he wrote in a confidential memo that the strategy behind sending Allied troops into Belgium at the start of the fight was to "carry the conflict out of our northern industrial provinces . . . and hold off the enemy threat from Paris."

EVEN THOUGH IT WAS better prepared militarily than most of the other countries invaded by Germany, Belgium still found itself overwhelmed by the sheer ferocity of the enemy's initial assault. By nightfall on May 10, seemingly endless waves of Luftwaffe bombers had wiped out much of the Belgian air force, and glider-borne troops had captured the linchpin of the country's defense system, the strongly fortified Eben-Emael fortress on the Belgian-Dutch border.

Still, after the first shock, Belgium's twenty-two divisions regrouped to mount what the American historian Telford Taylor called "a determined and well-directed defense," retreating to the west while making the Germans pay for their gains. The Belgians "fought like lions, from house to house," the CBS correspondent William L. Shirer wrote at the time. Telford Taylor later observed that "if the quality of the Belgian performance had been duplicated in other lands, the German march of conquest might have been shorter."

Meanwhile, French and British forces, this time with the Belgians' agreement, had crossed the Belgian-French frontier to take up positions behind a fortified defense line in the center of the country. During its strategic retreat, the Belgian army headed for the same defense line; there it would take its place alongside British troops, to help fend off what France and Britain believed would be the chief German offensive.

Instead, on May 13, the main enemy force, consisting of more than 1.5 million men and 1,800 tanks, thundered through the heavily wooded Ardennes Forest, farther south in Belgium. Outflanking the vaunted Maginot Line, France's supposedly impenetrable chain of fortifications, the Germans smashed into the least protected sector of the French frontier, routing the ill-equipped reservists assigned to guard it and crossing the Meuse River into France.

In just three days, the German offensive had split the Allied forces in two, sealing off French, British, and Belgian troops in central Belgium from the bulk of the French army. With France's defenses breached beyond repair and German armored columns now racing through the French countryside, a wave of panic enveloped the government and military. At that point, French premier Paul Reynaud picked up the phone to call Winston Churchill.

For the new British prime minister, May 1940 was filled with one nightmarish phone call after another, each bearing news of the latest military disaster. But none was as shocking as the one he received from Reynaud in the early morning of May 15. "We have been defeated!" Reynaud exclaimed as soon as his British counterpart answered the phone. When Churchill, still groggy from sleep, failed to respond, Reynaud, speaking in English, rephrased his dire message: "We are beaten! We have lost the battle!" Finally finding his voice, Churchill said he found that impossible to believe: though the Germans had certainly had the element of surprise, they would soon have to stop for supplies and regroup, giving the French forces the chance to counterattack. Reynaud went on as if Churchill hadn't said a word. "We are defeated," he said again, his voice breaking. "We have lost the battle."

Churchill was dumbfounded. This was not one of the little neutral countries he had criticized so often: it was understandable that they would fall like a house of cards before the German blitzkrieg. This was France—Britain's chief ally, supposedly the mightiest military power on the Continent! Yet when France's turn had come for invasion, its army of 2 million men had proved as unprepared and overwhelmed by Germany's new, stunningly fast style of warfare as had the smaller nations. How could *anyone* cope with tanks slicing through defense lines as if they weren't even there or with clouds of aircraft bombing bridges, roads, and train stations, strafing troops and civilians alike?

AS ITS CLOSEST EUROPEAN NEIGHBOR, France had much in common with Britain. They shared similar liberal values and were the most democratic of the leading countries of Europe. In World War I, they had joined hands as allies against Germany and Austro-Hungary. That partnership, however, had masked deep rifts and rivalries. The difficulties Churchill and Reynaud had in communicating with each other during their traumatic May 15 conversation were emblematic of the misunderstandings, suspicion, and antagonism that had existed for centuries between these two once mighty imperial powers and hereditary enemies.

Janet Teissier du Cros, a Scottish writer living in France during World War II, observed that both nations were noted for their suspicion of foreigners and their feeling of superiority to other countries. "It is so much a second nature . . . to believe there is no one like them that the notion scarcely even reaches the level of conscious thought," she wrote. "It is probably one of the reasons they find each other so oddly irritating."

Far from bringing them closer together, Britain and France's shared victory in World War I drove them further apart. Each had made huge contributions to that victory, but neither would give the other credit for its efforts, sacrifices, and achievements. "The real truth, which history will show, is that the British Army has won the war," Field Marshal Sir Douglas Haig, commander of the British Expeditionary Force in France, wrote in his diary after the November 1918 armistice. "I have no intention of taking part in any triumphal ride with [French commander Marshal Ferdinand] Foch, or with any pack of foreigners, through the streets of London."

Although only twenty miles of the English Channel separated the two countries, the psychological and cultural distance between them was as vast as an ocean. Each was also indelibly marked by its own geography: Britain was an island nation that had escaped invasion for more than eight centuries, while France, with its vulnerable borders, had endured repeated invasions, defeats, and occupations. At the 1919 Paris peace conference, French premier Georges Clemenceau explained to British prime minister David Lloyd George and U.S. president Woodrow Wilson the rationale behind his country's demands for draconian restrictions on Germany: "America is far away and protected by the ocean. England could not be reached by Napoleon himself. You are sheltered, both of you. We are not."

As the historian Margaret MacMillan noted about the post–World War I period, "France wanted revenge and compensation, but above all it wanted security." Between 1814 and 1940, Germany had invaded and occupied all or part of France five times. During World War I, in addition to turning much of France into a battlefield and charnel house, the Germans had pillaged the territory they occupied; in northern France, for example, most of the machinery and equipment of the country's textile industry had been taken to Germany.

In the years following the 1918 armistice, British policy makers, showing little sympathy with or understanding for French security concerns, increasingly insisted that the Versailles Treaty had been overly punitive to Germany and that the new German republic should be conciliated and even strengthened. The French were fiercely opposed to any such tolerance, arguing that the resurgence of German militarism was still very much a possibility. They pressed the British hard for a new Anglo-French military alliance, to no avail. To the British, France was being paranoid and vindictive.

When Hitler seized power and began rearming Germany, the governments of prime ministers Stanley Baldwin and Neville Chamberlain persuaded themselves that the Führer was simply trying to correct the injustices of Versailles and must be placated. By the mid-1930s, France, having given up hope that Britain would stand firm, had accepted the British policy of appeasement. For both former allies, the idea of another war was unthinkable. It had been only twenty years since their young men had begun marching off to battle, the cheers of their countrymen ringing in their ears. Four years after that, more than 700,000 Britons lay dead. France's battlefield losses were double that number—1.4 million men, the highest proportion of deaths per capita of any of the great power combatants. Still suffering from the war's devastating psychological and economic toll, the French doubted they could survive another conflict.

When Hitler occupied the demilitarized Rhineland in March 1936, first Britain, then France, turned a blind eye. They did the same after German troops goose-stepped into Austria in March 1937. But when Hitler threatened Czechoslovakia in the summer of 1938, France showed signs of standing up to him. The French, who had a military pact with the Czechs, began mobilizing their forces; they told the British they would resist dismemberment of their eastern European ally. Yet when Neville Chamberlain declined to join them in confronting Germany, the French fell into line. Following Chamberlain's lead at the Munich Conference, they ordered Czechoslovakia to give in to German demands for the Sudetenland, a vital area containing most of Czechoslovakia's fortifications and major centers of industry.

At Munich, Hitler had promised the British and French that he had no further designs on Czechoslovakia; six months later, he seized the rest of the country. It was then that Chamberlain, his appeasement plan clearly in ruins, announced to Parliament one of the most dramatic reversals of foreign policy in modern British history: Britain, he declared, would go to the aid of Poland, next on Germany's hit list, if it were invaded. It would also join France in a new military alliance, after years of scorning French fears of a renewed German threat.

The tables were now turned: the French, who had deferred to British diplomatic leadership throughout the 1930s, were unquestionably the senior partner in military matters. With an army less than one-fifth the size of France's, Britain was now suffering the consequences of its leaders' reluctance to rearm in the interwar years. During that period, its relatively small increases in arms spending had been used for the production of fighter planes to protect the country from possible German aerial attack. In early 1939, the British Army, which alone among the major European powers had no conscription policy, numbered only 180,000 men, with another 130,000 in reserve. Even those minuscule numbers were starved for adequate equipment, arms, and training.

When war broke out, neither France nor Britain sent troops to the aid of Poland. Instead, the British Expeditionary Force dispatched four divisions to France, a far smaller number than the French had expected; by the time of the blitzkrieg, there were ten. General Alan Brooke, who commanded two of the British divisions, morosely wrote in his diary that his troops were unfit for combat and that the Chamberlain government had probably sent them to France only as a public relations gesture, to show that some action, however minimal, was being taken. Concurring, the French ambassador in London snapped that "the English have such confidence in the French army that they are tempted to consider their military support as a gesture of solidarity rather than a vital necessity."

As it turned out, both Brooke and the ambassador were right: the British did count on the French army's eighty divisions, as well as its superior artillery and tank force, to counter German might. What the British failed to understand was that, despite its strength in numbers,

France was as unprepared for the coming conflict as they were. In their planning for a new European war, French military leaders had envisioned a relatively bloodless version of World War I—a long slog beginning with an enemy offensive through the flatlands of Belgium. With that in mind, the French high command had sent its best troops and armored units, along with the ten British divisions, to central Belgium. If German forces managed to make it past the Allied troops, they would then wear themselves out attacking the Maginot Line, the French believed. No one had anticipated the German breakthrough at the Meuse.

In truth, Britain's misplaced faith in France's military strength was just another example of the two countries' mutual obtuseness. On paper, their alliance had been renewed; in fact, no real partnership existed. "A genuine alliance is something that has to be worked at all the time," observed the French historian Marc Bloch, a Sorbonne professor who fought in the 1940 battle. "It is not enough to have it set down in writing. It must draw the breath of life from a multiplicity of daily contacts which, taken together, knit the two parties into a single whole."

In their dealings, British and French military leaders, few of whom spoke the other's language, were often at cross-purposes, reflecting a mutual distrust, suspicion, and even personal dislike. The French generals treated their British counterparts with condescension, regarding them as "learners in the military arts." The British commanders, for their part, were unhappy about the decision to send their troops into Belgium in the event of a German attack but, conscious of their country's minimal military presence, made no complaint to the French.

By the time the Germans finally invaded the Low Countries and France, the Allies' already dysfunctional military relationship had become toxic—as Winston Churchill discovered for himself when he traveled to Paris on May 16, the day after Paul Reynaud's shattering phone call.

UNLIKE MOST BRITONS, Churchill had been head over heels in love with France since boyhood. An admirer of Joan of Arc, Napoleon, and other French historical figures, he had visited the country more than a hundred times and spoke its language, albeit in a highly idiosyncratic way. During World War I, he had spent several months in the frontline trenches of France as commander of a battalion of Royal Scots Fusiliers and had been greatly impressed by the courage and resolution of the French troops fighting alongside his men. "Ever since 1907, I have in good times and bad times been a true friend of France," he would write in 1944.

Churchill's support, however, was not quite as unequivocal as he made it seem. In the 1920s and early 1930s, like most British politicians, he had advocated conciliation of Germany and opposed new commitments to France, suggesting that the country be left to "stew in her own juice." In 1933, he expressed the hope that "the French will look after their own safety" and that Britain would be free to stand aside from any new European conflicts. Churchill did support a vigorous rearmament policy for his own country but only to make Britain strong enough to defend its neutrality. Not until Hitler stepped up his march to war in the mid- to late 1930s did he push for a firm line against Germany and a close military alliance with France.

Much of the alliance's appeal for Churchill lay in his belief in the superiority of the French army—"the finest in Europe," he called it. His faith would remain unshakable until he arrived at the Ministry of Foreign Affairs on the Quai d'Orsay on the afternoon of May 16 and saw "utter dejection written on every face" of the officials with whom he met. In the gardens outside, clouds of smoke billowed up from bonfires stoked by official documents that government workers were heaping on the flames.

The French military leaders summarized for Churchill the disastrous news of the previous four days: the German breakthrough at the Meuse and the onrush of tanks and troops "at unheard-of speed" toward the northern French towns of Amiens and Arras. When Churchill asked about plans for a counterattack by reserve forces, General Gamelin shrugged and shook his head. "There are none," he said. Churchill was speechless: no reserves and no counterattack? How could that be? Gamelin's terse response, Churchill wrote later, was "one of the greatest surprises I have had in my life."

The British prime minister's shock and confusion, his failure to grasp the speed and immensity of the German onslaught, were no different from the dazed reactions of French and British officers and troops in the field. Years later, General Alan Brooke would write dismissively, "Although there were plenty of Frenchmen ready to die for their country, their leaders had completely failed to prepare and organize them to resist the blitzkrieg." Brooke didn't mention that he and his fellow British commanders were as guilty as their French counterparts in that regard—a point repeatedly made by General Bernard Law Montgomery, a subordinate of Brooke's in France. In his diary of the campaign, Montgomery, who commanded a British division in the battle, was scathingly critical of General John Gort, the British Expeditionary Force commander. Later Montgomery would write, "We had only ourselves to blame for the disasters which early overtook us in the field when fighting began in 1940."

Trained for static defensive warfare, the Allied military simply did not know how to react when the blitzkrieg—"this inhuman monster which had already flattened half of Europe," in the words of an American observer—burst upon them. Coordination and communication between the French and British armies broke down almost immediately; within a few days, most phone and supply lines had been cut, and the Allied command system had virtually ceased to function. The only way army commanders could communicate was through personal visits.

While French and British units functioned without information or orders, their tanks and aircraft were running out of fuel and ammunition. An RAF pilot called the situation "a complete and utter shambles"; a British Army officer wrote in his diary, "This is like some ridiculous nightmare." Back in London, Churchill told one of his secretaries, "In all the history of war, I have never known such mismanagement."

With Allied losses escalating and French and British troops in retreat, Paul Reynaud and the French high command begged Churchill to send ten more RAF fighter squadrons to France, in addition to the ten already there, to counter the Luftwaffe dive-bombers that were decimating their forces. Churchill eventually agreed to the request, arousing the impassioned opposition of the RAF's Fighter Command, which insisted that sending any more fighters abroad would pose a grave danger to Britain's own security.

Just six days into his tenure as prime minister, Churchill was faced with an agonizing choice: whether to give France as much material assistance as possible to bolster its morale and resistance or to withhold such support so that it could be used in Britain's own defense. As the French saw it, the British had nothing to lose by pouring all their resources into France, because if France went down, Britain would soon follow. The pugnacious Churchill did not share that view. Once the ten squadrons were dispatched, France would get no more, despite repeated appeals from Reynaud. And, unbeknownst to the French, on the day he returned from his May 16 trip, Churchill ordered plans drawn up for a possible evacuation of the British Expeditionary Force.

Increasingly doubtful of France's will or ability to fight back and fearing the encirclement and annihilation of his troops, General Gort was also contemplating evacuation. By the last week of May, the British forces had begun their retreat toward the beaches of Dunkirk, pursued by German troops and strafed by dive-bombers as they fled down dusty roads and lanes leading to the port. Churchill renewed his appeals to the French to stand and fight, never telling them until after the evacuation began that his own troops were leaving the field of battle.

ALSO LEFT IN THE DARK was the Belgian army, which had borne the brunt of Germany's aerial and tank juggernaut, shielding British and French troops in Belgium from much of its fury. Churchill's failure to inform the Belgians of the British retreat was not an oversight; he was counting on them to help keep the German forces at bay while British troops boarded the armada of small boats and large ships now being dispatched to Dunkirk.

In fact, the Belgian army—pummeled relentlessly by German dive-bombers, tanks, and artillery for more than two weeks and running out of food and ammunition—was already in the throes of disintegration. When the British began their westward retreat toward Dunkirk, the Belgians agreed to guard their flank but repeatedly warned both the

British and French commanders that their reserves were nearly depleted and that unless the Allies came to their assistance, they would soon have to surrender. In London, Churchill was given the same message by Admiral of the Fleet Sir Roger Keyes, a flamboyant British war hero and close friend of Churchill's, who was serving as the prime minister's personal liaison with King Leopold. But the Belgians' pleas for help carried no weight with Churchill, who told the War Cabinet that "the Belgian Army might be lost altogether, but we should do them no service by sacrificing our own Army."

When Colonel George Davy, the BEF's liaison officer with the Belgian army, asked General Gort and his deputy, General Henry Pownall, if Belgian forces would be allowed to participate in the Dunkirk evacuation, Pownall scoffed at the idea. "We don't care a bugger what happens to the Belgians," he said. Seemingly oblivious to the stalwart defense being waged by the Belgians, Pownall wrote in his diary on May 15, "Belgian morale, already thoroughly bad from top to bottom. They are simply not fighting." He later referred to them as "rotten to the core" and "lesser breeds."

On May 26, the Belgian commander in chief sent his last request for aid to Britain and France. Like his earlier pleas, it went unanswered. Instead, Churchill instructed Roger Keyes to emphasize to Leopold the importance of his troops remaining in the field. Obviously, the Belgians would have to capitulate soon, Churchill told a subordinate, but only "after assisting the BEF to reach the coast." He bluntly added, "We are asking them to sacrifice themselves for us."

The exhausted Belgians, however, believed they had done enough sacrificing. Abandoned and isolated by their allies, lacking everything they needed to keep fighting, they felt they had held off the Germans for as long as humanly possible. On May 27, the Belgian government, in an official communiqué, informed France and Britain of its imminent surrender to Germany: "The Belgian Army has totally exhausted its capacity for resistance. Its units are incapable of renewing the struggle tomorrow." Leopold sent an envoy to the Germans, and early on the morning of May 28, a cease-fire was announced.

The Belgians' surrender was a purely military act, a laying down of arms, but it was complicated by Leopold's decision to remain in Belgium. His fateful choice followed more than a week of soul-searching discussions with his government ministers about whether to go or stay. Prime Minister Hubert Pierlot and his colleagues informed the king of their plans to escape to France and urged him to accompany them. As head of state, they argued, it was his duty to continue Belgium's resistance in exile. Under no circumstances should he be taken prisoner by the Germans.

Leopold, however, saw his duty very differently. In that, he was guided by the example of his father. During the Great War, Albert, in his role as commander in chief, had repeatedly declared that he would never leave Belgium, even if the Germans conquered all of it. "Never would King Albert have consented to take refuge abroad," leaving his troops to their fate, Leopold told his ministers. Like his father, he believed that his responsibilities as commander in chief trumped those of head of state.

Pierlot and the others contended that according to the Belgian constitution, it was Leopold's duty to follow the wishes of the government. They added that if he stayed behind, the Germans would make political use of him whether he cooperated with them or not. The king rejected all their arguments. He would not, he said, become "an idle refugee monarch, cut off from the Belgian people as they bow under the invader's yoke." To abandon the army, he added, "would be to become a deserter. Whatever happens, I must share the fate of my troops."

At the time of the surrender, Leopold pledged not to have any dealings with the enemy while his country was in German hands. "For the duration of the occupation," he declared, "Belgium must not do anything in the military, political, or economic sphere which could harm the Allied cause." He asked to be put in a prisoner-of-war camp, along with his captured troops, but Hitler confined him instead to his palace in Laeken, on the outskirts of Brussels.

Leopold had been scrupulously correct in his handling of the surrender, but the French and British erupted in fury, joining forces to whip up a campaign of violent verbal abuse against the Belgians and their king. "Defeat arouses the worst in men," Irène Némirovsky noted in Suite Française, her posthumously published novel about the fall of France. As one historian put it, "When one is fighting a war and things are going badly, one cannot afford the luxury of being generous or even fair to an ally who has ceased to be of any use. If the only usefulness he retains is that of a scapegoat, then a scapegoat he must be."

Seeing a way to evade responsibility for France's looming defeat, French and British leaders put the onus on Belgium for all their troubles. To General Maxime Weygand, who had replaced Gamelin as French commander in chief on May 17, the capitulation of Belgium was actually a "good thing," because "we now shall be able to lay the blame for defeat on the Belgians."

In covering up their own ineptitude, the Allied commanders resorted to outright lies. Both Weygand and Gort made the patently false claim that they had been given no warning of Belgium's impending surrender. Accusing the Belgian army of cowardice, Gort also charged that its withdrawal from the fight had endangered the lives of his troops in their flight to Dunkirk. In reality, as the British military historian Brian Bond wrote, "the Belgian Army, virtually without air cover, bore the brunt of the German . . . attack while the BEF had a comparatively easy withdrawal to the French frontier. Indeed, but for the prolonged resistance of the gallant Belgian Army, the evacuation of the BEF from Dunkirk would have been impossible."

French premier Paul Reynaud went even further in his diatribes against Leopold and the Belgians. One of the few French politicians to oppose the appeasement of Hitler in the late 1930s, Reynaud, who had headed the government for just two months, was nearing the end of his emotional tether. In the early days of the German invasion he had aligned himself with Churchill, arguing that France should continue to hold out. But as the military situation worsened, he began yielding to the defeatist mood of many of his ministers, prominently including Marshal Philippe Pétain, the eighty-four-year-old architect of the failed Maginot Line strategy who was now deputy premier. Since Reynaud had vowed he would never agree to a surrender, he knew he would soon have to hand over power to Pétain—an act that would infuriate the British. In Belgium's capitulation, he saw a golden opportunity to shift the blame from himself and his government to the hapless Leopold.

"There has never been such a betrayal in history!" Reynaud exclaimed to his ministers when he heard of the Belgian surrender. "It is monstrous, absolutely monstrous!" In a May 28 broadcast to the French people, he accused Belgium of capitulating "suddenly and uncondi- tionally in the midst of battle, on the orders of its King, without warn- ing its French and English fellow combatants, thus opening the road to Dunkirk to German divisions."

Before he made the broadcast, the premier bullied the Belgian gov- ernment officials who had just arrived in France to support him in his attack upon their king. If they didn't, Reynaud said, he couldn't answer for the safety of the more than 2 million Belgians who had fled to France after the German invasion.

The Belgian ministers, who apparently feared that Leopold was thinking of establishing a new government in cooperation with the Germans, gave in to Reynaud's blackmail. In doing so, they made far graver and equally false accusations against Leopold, charging him with "treating with the enemy"—in effect, accusing him of treason. Instead of preventing acts of violence against their countrymen, their denun- ciation only added to the French fury against Belgian refugees, who were jeered at, spat upon, beaten up, and ejected from restaurants and hotels. A number of Belgian pilots who had escaped to France were handcuffed and thrown into jail, while several thousand young Bel- gians undergoing military training in France were imprisoned in their barracks.

Kept in the dark about the ineptitude of the British and French mil- itary response to the German blitzkrieg, public opinion in Britain read- ily accepted as truth the accusations against Leopold and Belgium. In London, the *Daily Mirror* ran a front-page cartoon depicting the Bel- gian king as a snake wearing a swastika-topped crown; the *Evening Standard* called him "King Quisling." One British newspaper columnist wrote that no child would be christened Leopold in Britain or any- where else for the next two hundred years. Mollie Panter-Downes, the *New Yorker*'s London correspondent, told her American readers that "for the space of a day, Hitler had to give up his title of most-hated man to Leopold III of the Belgians," who apparently "would rather be a live Nazi than a dead Belgian."

In the midst of all the vituperation, only a few lonely voices spoke up for Leopold. "The king's capitulation was the only thing he could do," the U.S. military attaché in Belgium reported to his superiors in Washington. "Those who say otherwise didn't see the fighting, and they didn't see the German Air Force. I saw both."

Admiral Keyes and Colonel Davy, the two British liaison officers assigned to the king and the Belgian military, also strongly defended the actions of Leopold and his army. Both were appalled when they returned to Britain on May 28 to find that Gort and his staff were heaping blame on the Belgians for their own incompetence. Particularly galling to Keyes and Davy was the fact that Gort himself was guilty of what he falsely accused the Belgian king of doing—withdrawing from the fight without warning his allies that he was going to do so.

Both officers, however, were forbidden by the British high command to make any public statements about their mission in Belgium. Furious at being muzzled, Davy wrote an account of what actually occurred there and gave copies to Keyes and the War Office for use in preparing the British official history of the war after it had ended. In a cover letter, he declared that the "savage and lying attacks" made on Leopold by "prominent military persons who found in him a profitable and unresponsive scapegoat" (i.e., Gort and Pownall) had prompted him to act. He added that "the truth should not be suppressed forever."

Keyes, for his part, mounted a passionate defense of Leopold in a letter to Churchill, urging him to put a stop to British officials' "vilification of a brave king." At first, the prime minister seemed to heed his friend's admonition, telling Parliament at the end of May that the Belgian army had "fought very bravely" and that the British should not pass "hasty judgment" on Leopold's surrender.

His forbearance was short-lived. Annoyed that Leopold had chosen to remain in Belgium, Churchill was still riding his hobbyhorse of anger at the European neutral countries for not joining Britain and France in preinvasion military alliances. Refusing to acknowledge that the neutrals might have had valid reasons for shying away from such ties, he repeatedly made statements blaming their alleged cowardice for Germany's military successes. He told Keyes privately that Leopold's surrender had "completed the full circle of misfortune into which our Allies had landed us while we had loyally carried out our obligations and undertakings to them"—a comment that could not have been less true.

Churchill's already strong prejudice against Leopold was exacerbated by growing pressure on him from Paul Reynaud to join France's scapegoating of the king. Reynaud accused the British of being too subdued in their expressions of outrage against Leopold and the Belgians, and Churchill, desperate to keep France in the war, finally gave in to the French premier's arm-twisting. On June 4, in a speech announcing the success of the Dunkirk evacuation, Churchill employed all his formidable rhetorical skills in a fierce denunciation of Leopold. "Suddenly, without prior consultation . . . he surrendered his army and exposed our whole flank and means of retreat," the prime minister thundered, as the MPs around him cried "Shame!" and "Treachery!" "Had not this ruler and his government severed themselves from the Allies, had they not sought refuge in what has proved to be a fatal neutrality, the French and British armies might well at the very outset have saved not only Belgium but perhaps even Poland."

The sheer absurdity of Churchill's statement—that Belgium's neutrality, not Germany's military prowess, had been responsible for the defeat of Poland and other European countries—registered with Roger Keyes but with few others in Churchill's parliamentary audience. An MP himself, Keyes listened to the prime minister's diatribe with mounting anger and disbelief. Instead of praising the Belgians for having protected the BEF from the worst of the German onslaught, Churchill was echoing Reynaud in accusing them of having endangered the British evacuation, as well as causing the encirclement and surrender of thousands of French troops.

Yet, in retrospect, Churchill's harangue, though unjustified, is understandable. Prime minister for only four weeks, he considered his political position at that point to be extremely tenuous. Many Conservative MPs, whose party dominated Parliament, had not yet reconciled themselves to his succeeding Neville Chamberlain; indeed, a fair number were openly hostile to him. "Seldom can a prime minister have taken office with the establishment so dubious of the choice and so prepared to have its doubts justified," noted John Colville, one of Churchill's private secretaries.

With his country now facing the greatest challenge in its history, Churchill was eager not only to fortify his own position but also to draw a veil of secrecy over the incompetence of his top generals as well as the other grave shortcomings of the British military's performance thus far in the war. What better way to do so than to pin the blame on a smaller ally whose king and commander in chief was unable to defend himself?

Roger Keyes, however, refused to fall into line. In early June, he filed a libel suit against the *Daily Mirror* for a story accusing him of abetting what the *Mirror* called Leopold's treachery. Determined to exonerate himself as well as the Belgian king and his military, Keyes pressed for a public trial. Before the case was finally heard in March 1941, the *Mirror* acknowledged that it had erred in its statements about Leopold and Keyes and agreed to apologize to both. Declaring that "the public interest would not be served" by publicizing the matter, Churchill and his government pressured Keyes to accept an out-of-court settlement rather than go to trial. Keyes agreed, but, in settling the case, his lawyer outlined in open court what had really happened in Belgium the previous May; in the same hearing, the newspaper's attorney conceded that the *Mirror* had done the king "a very grave injustice."

The story of Leopold's vindication made front-page headlines in Britain. K.C. CLEARS KING LEOPOLD'S NAME: LONDON TOLD OF SURRENDER PLAN, one blared. Another noted, KING LEOPOLD WARNED BRITAIN OF SURRENDER. But the BBC, under pressure from the War Office, suppressed the news of the king's exoneration; it remains relatively unknown to this day. In the seventy-plus years since 1940, many if not most historians who have written about the battles in France and Belgium have accepted as true the charges made by the British and French against Leopold and his country.

Yet even during the chaos of May 1940, there was one celebrated Briton who knew better and who refused to participate in the mudslinging. King George VI was said to be furious at the campaign aimed at the Belgian sovereign, who was a distant cousin of his and whom he had known and liked since the teenage Leopold had attended Eton during the Great War. When British officials proposed that Leopold be dropped from the Roll of the Knights of the Garter, Britain's highest order of chivalry and one of its most prestigious honors, George, who keenly understood the excruciating dilemma faced by his fellow monarch, rejected the idea.

As George's biographer, the historian John Wheeler-Bennett, has pointed out, the choice confronting the heads of state of German-occupied countries was "one of hideous complexity, [with] little time for calm consideration. To leave their homeland and follow their Governments into exile leaves them open to the charge of desertion by those who remained behind; yet to remain [in their countries] involves the risk of their being held hostage for the submissive conduct of their peoples."

The day before Belgium surrendered, Leopold wrote a fond letter to George, whom he addressed as *mon cher Bertie*—a diminutive of his given name, Albert, which was used only by members of the British king's family and a few others close to him. In the letter, Leopold explained his rationale for staying in Belgium, declaring that his overriding duty was to share the ordeal of German occupation with his troops and the rest of the Belgian people and to protect them as much as possible. "To act otherwise," he told George, "would amount to desertion."

As it happened, King George did not agree with Leopold's choice. When Harry Hopkins, Franklin Roosevelt's closest aide, visited London in early 1941, George told him he thought that Leopold had gotten his two jobs—king and commander in chief—"mixed up." In a memo to FDR, Hopkins observed that George had "expressed a good deal of sympathy for the King of the Belgians and had little or no criticism of him as C-in-C of the Army, but as King . . . he should have left the country and established his government elsewhere." Yet while questioning the wisdom of Leopold's decision, George never doubted that his cousin was following his conscience and keen sense of duty in staying behind.

Ironically, George himself had taken the same vow made by Leopold: under no account, he said, would he leave his country if it were invaded by Germany. Fortunately for him and for Britain, he was never called upon to make that choice.

CHAPTER 4

"We Shall Conquer Together— or We Shall Die Together"

The European Exodus to Britain

JUST DAYS AFTER LEOPOLD DECIDED TO REMAIN IN BEL-
gium, King Haakon VII of Norway found himself wrestling with
the same dilemma. For most of May 1940, war had continued to rage in
the far north of Norway, even as the world's attention turned to the
collapse of France and the British evacuation from Dunkirk. Having
played a major role in bringing Winston Churchill to power just one
month before, the Norwegian conflict "already seems incredibly remote," Mollie Panter-Downes observed in the *New Yorker*.

Although Germany had swiftly conquered southern and central
Norway in April, an Allied force fighting in the north finally captured
the key port of Narvik on May 27 and drove the Germans there back to
the Swedish border. As the Allied troops celebrated their victory, they
had no idea that the British government, facing what it believed was an
imminent German invasion of its own soil, had ordered yet another
evacuation: all British troops in Norway were to be on their way home
by early June. Once again, the British were doing what they had accused Leopold of doing: deciding to withdraw without consulting
their allies. "One feels a most despicable creature in pretending we are
going on fighting when we are going to quit at once," General Claude
Auchinleck, the commander of British forces in Norway, wrote to a
colleague.

The news hit King Haakon like a thunderclap. From their refuge in

Tromsø, he and Norway's government ministers had vowed to fight on. When told of the withdrawal, the king reluctantly agreed to go to London as the British government requested. But he immediately had second thoughts. On the night of June 5, he sent a letter to Britain's envoy to Norway, Sir Cecil Dormer, declaring that he could not abandon his soldiers or people and that "on further reflection, he felt it his duty to remain behind in Norway."

At five o'clock the next morning, Dormer drove to the farmhouse outside Tromsø where Haakon and Crown Prince Olav were staying. It was an unseemly hour to be calling on royalty, the British diplomat knew, but his mission was urgent. Wasting no time on diplomatic niceties, he told the king that Norway was doomed and that if he stayed, he would become a German pawn, "unable to help his people in any way or even be in touch with them. The Germans would issue orders purporting to be made with his approval: in short he would be playing into their hands." But if Haakon escaped to Great Britain, he and his government could join the other Allies and renew the fight.

There was yet another reason for Britain's determination to spirit Haakon to safety, one that Dormer left unmentioned. Twenty-three years before, Haakon had offered to send a warship to rescue his first cousin, Tsar Nicholas II, and the tsar's wife and children after Nicholas was dethroned in 1917. King George V, Haakon's brother-in-law and another first cousin of Nicholas's, had told the Norwegian king not to bother, that he would send a ship for the Russian royal family himself. The British vessel, however, had not been dispatched, and Haakon had reportedly never forgiven his brother-in-law for abandoning Nicholas and his family, who were murdered by the Bolsheviks in 1918. George V's son, the current British king, was determined not to repeat his father's mistake; he urged his government to do everything in its power to save his uncle from the Germans.

Dormer told Haakon that the last British ship to leave Norway, HMS *Devonshire,* would depart from Tromsø harbor promptly at 8 P.M. the following day; he hoped that Haakon and his government would be on board. The king replied that he would have to consult with his Cabinet before making a final decision.

That day and the next, Haakon agonized about what to do. A few hours before the *Devonshire* was to sail, he met with key government officials at the creamery that had become their makeshift headquarters in Tromsø. After a heated debate, they decided to leave. As the session ended, Haakon, his voice trembling, struggled to speak. He was still not sure they were doing the right thing, he said. What could they possibly accomplish in England? Were they cutting themselves off forever from Norway? His voice trailed off, and, looking at his son, he was barely able to utter the words "God save Norway!" The others in the room, most of them in tears, stood up and repeated, "God save Norway!" As the ministers drifted out of the room, the king said in a broken whisper, "I am so afraid of the Norwegian people's judgment."

That evening, as government officials and foreign diplomats boarded the *Devonshire,* Dormer stood on the dock and anxiously scanned the road leading to the harbor. Finally, just a few minutes before eight, a car pulled up, and, to Dormer's intense relief, Haakon and Olav stepped out, accompanied by their aides-de-camp. The king paused for a moment, looking back at the snow-clad mountains towering over the town and then out to sea, as if imprinting the breathtaking views on his memory. After another glance around, he slowly climbed the ship's accommodation ladder, his head bowed. He seemed "extremely depressed," recalled a Norwegian official who watched Haakon from the ship's railing. "It was certainly one of the most painful hours of his life."

At precisely 8 P.M., the *Devonshire* weighed anchor. Within minutes, it had disappeared into a heavy mist, setting its course for Britain.

WHEN HAAKON REACHED LONDON on June 9, still dressed in his winter uniform and high boots, he found a jubilant city, whose residents were acting as if their country had just scored a major military triumph. Over the previous two weeks, more than 220,000 British troops had been snatched from the beaches of Dunkirk, far more than anyone had thought possible.

For many if not most Britons, the unexpected success of the effort to save what was thought to be a doomed army completely overshadowed the ongoing military catastrophe in France. The triumph of that massive rescue, aided by hundreds of pleasure boats and other small craft that had crossed the English Channel, was viewed, then as now, as one of the most heroic epics in British history. "So long as the English tongue survives, the word Dunkirk will be spoken with reverence," the *New York Times* declared at the time. "For in that harbor . . . the rages and blemishes that have hidden the soul of democracy fell away. There, beaten but unconquered, in shining splendor [Britain] faced the enemy."

With the curtain now drawn over the fiasco across the Channel, the British and their new prime minister now focused on their own struggle for survival. "We shall fight on the beaches," a defiant Winston Churchill declared to Parliament on June 4. "We shall fight on the landing grounds, we shall fight in the fields and in the streets, we shall fight in the hills. . . . We shall *never* surrender."

It was one of Churchill's most magnificent wartime speeches, but it meant nothing to the French. For them, Dunkirk, far from being a triumph, was a tragedy and an act of betrayal. The British had not informed the French that they were evacuating until tens of thousands of BEF troops had already left. When finally informed on May 31 of the mass departure, a French admiral angrily told the British general who gave him the news, "So you are admitting that the French army alone will cover the embarkation of the English army, while the English army will give no help to the French army in covering its own withdrawal. . . . Your decision . . . dishonors England." At that point, the British agreed to take off French troops, too. More than 100,000 French soldiers were eventually evacuated, but for the British, they were clearly an afterthought.

As the admiral's remark indicated, the BEF's escape was greatly aided by a stalwart defense of Dunkirk's perimeters by French forces, along with several British regiments. In and around the city of Lille, about fifty miles southeast of Dunkirk, one French division, fighting house to house, kept seven German divisions at bay for four vital days during the evacuation. The Germans were so impressed by the French troops' valor that they allowed them to keep their arms during the surrender ceremonies. Thousands of French soldiers were killed and wounded in the fighting around Dunkirk, while some 30,000 to 40,000 were taken into captivity. In the overall fighting in France, more than 90,000 Frenchmen lost their lives (compared to 11,000 British), and more than 200,000 were wounded.

After Dunkirk, the myth grew in Britain and elsewhere that "the gutless collapse of the French" and the "sheer spinelessness" of the French army had been largely responsible for the BEF evacuation. It's true that a great many French soldiers did lay down their arms, but many others kept fighting, despite the growing defeatism of their civilian leaders and their geriatric military command. As the Canadian military historian John Cairns noted, "the French war effort was substantial," despite "political, economic, social, and finally critical military mistakes." Echoing that view, the British historian Julian Jackson has written that French troops fought "as bravely as those of 1914 when they were properly led and equipped. The failure of 1940 was above all a failure of military planning."

Privately, Winston Churchill acknowledged that the French had indeed borne the brunt of the struggle. On June 14, he told the British War Cabinet that "very few British divisions have fought in France. At the end, very few indeed. French losses have been out of all proportion to ours, in every sphere." To General Pug Ismay, the prime minister declared that "our contribution to the battle in France had been niggardly. So far the French have had nine-tenths of the casualties . . . and endured 99/100s of the suffering." Publicly, however, he said little about the magnitude of the French losses or the valor of the French army—an oversight that brought bitter complaints from the French ambassador in London.

Not surprisingly then, the last two meetings between Churchill and French leaders, held during and after Dunkirk, were utter failures. In the course of his desperate attempt to dissuade Paul Reynaud and his demoralized government from capitulating to the Germans, Churchill declared that Britain would never give up: whatever happened, it would fight "on and on and on, *toujours,* all the time, everywhere, *partout, pas de grâce,* no mercy. *Puis la victoire!*" He begged the French to remain in the struggle, too—if not in Paris, "then in the provinces, down to the sea, and then, if need be," from their colonial possessions in North Africa.

The French were not impressed by either his eloquence or his arguments. Although Reynaud was still marginally in Churchill's camp, General Weygand and Marshal Pétain angrily accused the British of caring only for themselves. They noted that while Churchill had refused to send more RAF fighter squadrons to France before Dunkirk, he had dispatched every British fighter available to cover the BEF's withdrawal.

Besides, they argued, how could Britain possibly win against Germany when it had so few troops and had shown so little resoluteness up to this point? In Weygand and Pétain's view, the war in Europe would be decided in France, not Britain. Any continued resistance by the British would be futile, they declared; according to Weygand, Britain would have "her neck wrung like a chicken" within a month.

In the midst of all this political wrangling, some French forces fought on, inflicting a large number of German casualties in their defense of a last-ditch line along the Somme and Aisne rivers in northern France. "This was the great battle of 1940, largely forgotten in France and never heard of in England," observed the historians Robert and Isabelle Tombs, who noted that one German panzer division lost two-thirds of its tanks in two days. Resolute as the French were, however, they were no match for the superior German forces, which soon smashed through the overextended line.

By the first week of June, France, deeply split by social and political divisions, was in profound shock, and the government was in shambles. On June 9, Reynaud and his ministers fled Paris, having made no arrangements for its defense or evacuation, and took refuge first in Tours, then in Bordeaux. Millions of Parisians also stampeded out of the city, on foot and in every imaginable kind of vehicle. Overall, more than 6 million French citizens, "like an anthill that had been knocked over," flooded south—the biggest single movement of people in Europe since the Dark Ages. The scene was pure chaos, filled with "all the ugliness of panic, defeat, and demoralization" of a disintegrating society, remarked the American diplomat George Kennan, who witnessed the mass exodus.

On June 14, German troops goose-stepped into Paris, just as they had done in Vienna, Prague, Warsaw, Oslo, Copenhagen, The Hague, and Brussels. Two days later, Reynaud resigned, and Pétain took over as premier of France. On June 17, the aged hero of the Great War's Battle of Verdun ordered French troops to lay down their arms and petitioned the Germans for an armistice—actions that French officials had denounced King Leopold of Belgium for taking just three weeks earlier.

In a broadcast to his countrymen, Pétain attributed France's defeat to "too few arms, too few allies" and to France's own moral failures, which included a lack of discipline and an unfortunate "spirit of pleasure." Ignoring Pétain's reference to a scarcity of allies, the British, along with much of the rest of the world, readily accepted his condemnation of France's grave internal shortcomings—social, psychological, economic, and political—as the primary reason for the country's astonishing collapse.

In doing so, however, British leaders conveniently overlooked their own country's role in the defeat. As Robert and Isabelle Tombs wrote in their magisterial history of the Anglo-French relationship, Britain's contribution to the two countries' World War II alliance was "shamefully feeble." And before the creation of that uneasy partnership, the British government's two-decade-long policy of "deliberate estrangement" from France had had a profoundly negative effect on French diplomacy, military strategy, and sense of confidence and security.

In the end, each ally blamed the other for the French fiasco— a pointing of fingers that exists to this day. "The fact is that they let each other down," Robert and Isabelle Tombs noted, "and neither has entirely forgotten it."

AS THE CBS CORRESPONDENT Eric Sevareid was departing France by ship for England in late June 1940, he sensed a distinct change of mood among the British journalists also on board. "They seemed almost happy," he recalled. "They were British, and their course was clear. They were sticking together now."

Although their country's future appeared impossibly bleak, many other Britons shared that same sense of relief and exhilaration. As an island people, they had never been comfortable with alliances, European or otherwise. Now they were on their own again, and proud to be so.

Even those with strong ties to France felt that way, among them General Edward Spears, Churchill's personal liaison with the French government, and General Alan Brooke, one of Britain's top commanders in France. Both Spears and Brooke had been born and raised in France, spoke its language fluently, and always considered it a second home. Yet Brooke told a fellow British officer in mid-June that he was determined "not to remain in this country an hour longer than necessary." Spears, for his part, noted that "a lifetime steeped in French feeling, sentiment, and affection was falling from me. England alone counted now."

Similarly, when King George VI asked War Secretary Anthony Eden why he was in such good spirits, considering the direness of Britain's situation, Eden replied, "Now we are all alone, sir. We haven't an ally left." The king himself observed to his mother, Queen Mary, "Personally I feel happier now that we have no allies to be polite to and pamper."

Yet regardless of whether they wanted allies or not, the British desperately needed them. The soaring combative rhetoric of Churchill had inspired his countrymen to fight on, but inspiration by itself would hardly ward off a German invasion. "Certainly everything is as gloomy as can be," Sir Alexander Cadogan, the permanent undersecretary of the Foreign Office, wrote in his diary. "As far as I can see, we are, after years of leisurely preparation, completely unprepared."

Despite the miraculous Dunkirk rescue, Britain's situation verged on the calamitous. Many of the RAF's most experienced pilots—not to mention hundreds of planes and more than 20,000 ground troops— had been lost during the defense of Belgium and France. The country now had only enough men to field twenty army divisions, less than a tenth of the forces mustered by Germany. And that small number had almost nothing to fight with, having left behind virtually all their tanks, armored cars, weapons, and other equipment in France. There were only a few hundred thousand rifles and five hundred cannons in all of Britain—and most of the cannons were antiques, appropriated from museums.

On June 26, Churchill inspected the defense line hastily being cobbled together along the coast of southeast England, where the Germans were expected to land if they invaded. The general in charge of the defenses near Dover told the prime minister that he had only three antitank guns to cover five miles of coastline, with just six rounds of ammunition per gun. "Never," Churchill would later write in his memoirs, "has a great nation been so naked before its foes."

Throughout the spring and summer of 1940, the new British leader made several desperate appeals for aid to the neutral United States and its president, Franklin D. Roosevelt. Though FDR was sympathetic to Britain's plight and wanted to do all he could to help the country survive, he was wary of his isolationist opponents in Congress and skeptical about Britain's chances. Indeed, many in Washington had already written the country off. How could this small island resist an invader that had toppled every country in its path? "One had to be little short of a visionary and mystic to believe that the war might still be won," the Dutch historian James H. Huizinga observed. The *New Yorker*'s Mollie Panter-Downes put it another way: "It would be difficult for an impartial observer to decide today whether the British are the bravest or merely the most stupid people in the world."

EVEN AS SOME OF his countrymen celebrated their lack of allies in the summer of 1940, Churchill did the opposite. Having heaped blame on the occupied countries for their defeats just a few weeks before, he now threw open Britain's doors to their governments and armed forces. With his customary energy and passion, he welcomed anyone prepared to carry on the fight. In doing so, he ignored strong opposition from members of his Cabinet and Whitehall officials, many of whom harbored a strong prejudice against foreigners. The Foreign Office, for one, complained that Churchill was enlisting "every crank in the world." Nonetheless, as France approached collapse in mid-June, the prime minister ordered his government to rescue as many foreign troops and airmen fighting there as possible, regardless of their politics or nationality.

The largest contingent to fetch up in England was, by far, the Poles.

After Poland fell to the Germans in September 1939, its surviving soldiers, airmen, and sailors—tens of thousands of them—had been ordered by their government to escape, however they could, and fight on. In charge of that effort was General Władysław Sikorski, a highly respected war hero and the new Polish prime minister and commander in chief.

Sikorski was named to head the Polish government on September 29, 1939, after the collapse of the authoritarian military junta that had ruled Poland during the 1930s; many of the junta's members had been interned in Romania following their escape from their homeland. As soon as he assumed office, Sikorski, from his base in France, established an elaborate underground network, using forged passports and visas, to spirit Polish forces out of Romania and the other countries to which they had fled and back into combat. "All we knew was that we had to get to the only remaining front [in France] at any price," said one Polish pilot.

During the fighting in France, Polish forces comprised some 85,000 men, of whom about 10 percent were in the air force. The 75,000 Polish ground troops were a motley group that included university professors, coal miners, poets, priests, and college students. On June 13, 1940, a Polish armored brigade repulsed a German attack near the French industrial town of Montbard and mounted a counteroffensive, inflicting heavy casualties. Farther east, near Belfort, the 2nd Polish Rifle Division held the Germans at bay for six days, facing an artillery barrage three times stronger than its own. General Weygand later remarked that if he had had a few more Polish divisions, he might have been able to stem the German tide.

When the French armistice with Germany was announced, Polish airmen and soldiers were almost as distraught as when their own country had fallen. But there was no time to mourn France, a nation that had given refuge to Polish political exiles for almost two centuries. On June 18, Sikorski flew to London for an urgent meeting with Churchill. He asked if Britain would help rescue Polish forces so they could fight again. Churchill's response was swift and unequivocal: "Tell your army in France that we are their comrades in life and in death. We shall conquer together—or we shall die together." Later that day, he ordered the British admiralty "to make every effort to evacuate the Polish forces and personnel." Sikorski, in turn, instructed all Polish forces in France to head immediately to ports in the south. British and Polish ships, he said, were already on their way to pick them up.

In all, some 20,000 Polish soldiers and 8,000 airmen made it to Britain. So did hundreds of sailors, along with three destroyers, two submarines, and other smaller vessels that had sailed for Britain at the outbreak of the war. In fact, the Polish navy had already been in combat against the Germans, fighting with the Royal Navy in the battle for Norway.

Joining the Poles were some 5,000 members of the Czech armed forces, about 1,000 of them airmen. In the weeks following the German occupation of Czechoslovakia in March 1939, thousands of soldiers and pilots had managed to leave the country, most of them heading for Poland. While some had remained to fight alongside the Poles, others had moved on to France. Until war was formally declared in September 1939, the French had assigned them to the Foreign Legion in North Africa. Afterward, they were integrated into the French armed forces and took part in the battle for France, distinguishing themselves in the general fighting as well as in the defense of the Dunkirk perimeter during the British withdrawal.

After France's capitulation, the Czechs, like the Poles, were ordered south for evacuation by British ships. As one Czech army officer prepared to leave the garrison where he was stationed, its French commander barred his way, declaring, "The war is over." "That's true for you, Colonel," the Czech replied. "For us, the war is only beginning." The Frenchman, visibly moved, signed a pass and let him go.

THE ARMED FORCES OF the other occupied nations numbered just two or three thousand men each. Nonetheless, those countries possessed additional resources that would soon become crucial in the British struggle for survival. King Haakon, for example, brought with him only 1,400 soldiers, 1,000 sailors, and 3 pilots (a number that would grow rapidly in the coming months). But he and Norway had another asset that was greatly coveted by both Germany and Britain: the fourth largest and most modern merchant marine fleet in the world. Fast and efficient, Norway's 1,300 oceangoing merchant ships, most of which had been built in the previous decade, totaled more than 4.4 million gross tons and were manned by some 30,000 seamen.

When Germany invaded Norway in April 1940, most Norwegian ships were at sea, and the Germans and Norwegians engaged in a frenzied contest to gain control of them. German-run broadcasting stations in Norway ordered the ships to proceed to Norwegian or other German-occupied ports, while the Norwegian government, over the BBC, directed them to go to Britain, France, or one of the Allies' dominions or colonies. Virtually every Norwegian ship captain obeyed the instructions from London. In late April, the Norwegian government requisitioned the ships from the companies that owned them—more than 1,200 vessels in total—and leased them to Britain, whose own merchant shipping was being decimated by German submarines. The very survival of Britain depended on those ships and the oil, food, and other goods they transported to the island nation.

With the Norwegian ships and crews at its disposal, Britain was able to keep open the crucial Atlantic lifeline—and eventually win the Battle of the Atlantic. In early 1941, a British official declared that the Norwegian merchant fleet was worth more to England "than an army of a million men." The Norwegian navy, together with the Polish and Dutch navies, also helped man the destroyers protecting Britain's merchant marine convoys.

The Dutch government contributed its own sizable merchant fleet of some six hundred ships, as well as some of the rich material resources of the Dutch East Indies. The Belgians, meanwhile, had considerable gold reserves, along with the immensely valuable raw materials of the Congo, including rubber, iron, and uranium.* At the beginning of their wartime relationship, however, the Belgians presented the British with two major problems: their king had remained behind and the two top leaders of their government, who had escaped to France when Belgium fell, now refused to come to Britain.

---

* Over the course of the war, the Belgian government in exile shipped 1,375 tons of uranium to the United States—a stockpile that fueled the Manhattan Project.

When France collapsed, Prime Minister Hubert Pierlot and Foreign Minister Paul-Henri Spaak of Belgium collapsed with it. Having accused King Leopold of defeatism less than a month before, Pierlot announced after the French armistice, "France has thrown in the sponge. We abandon the struggle with her." British officials made repeated offers to spirit Pierlot and Spaak out of France, but the demoralized Belgians just as repeatedly turned them down; they had allied themselves so totally with France that they completely lost heart when it fell. For those two men, Britain was a truly foreign land. Neither of them had ever visited it or spoke its language.

The British were horrified. So were several of Pierlot and Spaak's Cabinet colleagues, who had gone to London immediately after France's capitulation. The Belgian ambassador to Britain passed on to Pierlot and Spaak a message from Lord Halifax, the British foreign secretary, warning that "if you give up now, you lose everything, whereas if you continue the struggle, you safeguard the future."

The two Belgian leaders ignored Halifax's admonition and soon became even more of an embarrassment. They ordered Belgian soldiers and airmen in France not to decamp to Britain; when many defied the order and went anyway, they were tried and sentenced in absentia for desertion. Having falsely accused their king of treating with the enemy, Pierlot and Spaak sent a message to Leopold asking him to do just that—to bring the two of them back to Belgium so they could form a government that would open peace negotiations with Germany. They dispatched a similar message to Berlin.

The Nazis failed to respond to their appeal, while the king, who had refused all contact with Nazi officials, rejected it outright. Most Belgians agreed with their monarch. "You must have no illusions," a Belgian Red Cross official informed Pierlot and Spaak. "Belgium in its entirety is behind the king. You are despised. . . . They consider that you have acted in an atrocious manner." Paul Struye, a Belgian politician who became a leader of the country's resistance later in the war, recalled that "virtually the entire country rallied around the King. This was a manifestation of spontaneity, unanimity and fervour which is quite exceptional in our history."

Scorned by the Nazis, Pétain's France, and their own king and country, Pierlot and Spaak finally realized they had nowhere to go but Britain. Hats in hand, the disgraced Belgian officials left France on October 24, 1940, for London, where they quickly won official recognition as leaders of Belgium's government in exile.

Contemptuous as they were of Pierlot and Spaak's behavior, the pragmatic British were prepared to overlook it in exchange for immediate access to Belgium's gold hoard and the Congo's resources. By the fall of 1940, heavy armaments purchases from the United States had drained Britain of most of its dollar and gold reserves, which made Belgian gold that much more crucial to its defense. To ensure the continuance of arms shipments, the Belgians loaned much of their gold cache, held by the Federal Reserve Bank of New York, to the British Treasury.

Pierlot and Spaak were also compelled to proclaim their wholehearted support of the British war effort and their allegiance to King Leopold. In fact, both were by then acutely aware that their earlier accusations against Leopold were untrue. Yet neither retracted or apologized for them; from then on, they simply behaved as if they had never accused their monarch of anything untoward. As Spaak's biographer noted, both men "were in too deep by now to make a public recantation of their error."

OF THE SEVEN OCCUPIED countries that found refuge in London in the spring and summer of 1940, six presented their British hosts with invaluable dowries of men, money, ships, natural resources, and intelligence information. The lone representative of the seventh nation, France, brought only himself.

Forty-nine-year-old Charles de Gaulle was not a king, a president, or even a high government official. The most junior brigadier general in the French army, he had been appointed undersecretary of war just eight days before his dramatic flight from France on June 17. Shy, unsmiling, and aloof, this minor functionary "lacked all social vices and graces," an observer remarked. Many people found him impossible to get along with. His glacial attitude, his family joked, was the result of his having fallen into an icebox as a child.

Lord Moran, Winston Churchill's physician, would later describe the towering, ungainly de Gaulle as "an improbable creature, like a human giraffe, sniffing down his nostrils at mortals beneath his gaze." Paul Reynaud—a friend and ally of de Gaulle's who, in one of his last acts as premier, appointed him to the War Department post—said that he had "the character of a stubborn pig."

Yet for all his personal flaws, he was the only French official willing to abandon his homeland and cross the English Channel to continue the fight against Hitler. Churchill had first met de Gaulle during the prime minister's final stormy meeting with French leaders, held June 11 at a château near Orléans. He had been enormously impressed with this cool, unemotional brigadier, who, in contrast to his panicked superiors, insisted that capitulation to the Germans was not an option.

Like Churchill, de Gaulle had been a lifelong rebel. Also like the British prime minister, he had viewed himself as a man of destiny from childhood. Writing in his journal at the age of fifteen, he pictured himself as the head of an army embarked on a crusade to save France— a Napoleonic sense of mission he retained all his life. A report on him early in his military career noted that he had "brilliance and talent; [was] very highly gifted, and [had] plenty of character. Unfortunately, he spoils his undoubted qualities by his excessive self-confidence, his severity towards the opinions of others, and his attitude of a king in exile."

In the 1930s, this regal young officer conducted a one-man campaign against the French high command's military strategy, warning against its reliance on prepared fortifications like the Maginot Line and arguing for creation of a fast-moving mechanized army working closely with and supported by aircraft. His was a prescient plan with striking parallels to the blueprint implemented by the Germans in their blitzkriegs of Poland, Norway, France, and the Low Countries.

When army leaders ignored his recommendations that they jettison their emphasis on defense and reform France's obsolescent military machine, de Gaulle made his contrarian views public in a book published in 1934. That, too, had little effect except to heighten the enmity of the military brass toward him.

The German invasion of France resulted in yet another diatribe from de Gaulle, this one openly critical of the General Staff's conduct of the war and addressed to dozens of high-level military and civilian leaders throughout the country. To describe de Gaulle's letter as "an 'undisciplined act' is too weak; 'mutiny' is nearer the mark," Jean Lacouture, a de Gaulle biographer, wrote. De Gaulle's superiors were ready to cashier him, but he had the support of Paul Reynaud, who agreed with his views. On June 6, when Reynaud appointed de Gaulle undersecretary of war for national defense, he told him that "your presence at my side is a sign of our resolve" to carry on the war.

Reynaud's resolve, however, proved as evanescent as a soap bubble. Ten days later, he gave way to Pétain, who had helped orchestrate France's failed defense strategy and was a fierce critic of de Gaulle. On the night of Reynaud's resignation, de Gaulle waylaid General Edward Spears, Churchill's representative to the French government, at the government's temporary headquarters in Bordeaux. De Gaulle told Spears that Pétain and Weygand planned to arrest him the following day. He appealed to the Englishman to help him escape to London so he could "rally French opinion in favor of resistance."

After getting Churchill's blessing, Spears put the escape plan into motion. Early on the morning of June 17, de Gaulle and his aide-de-camp accompanied Spears to a small airport outside Bordeaux, as if to see him off. The two Frenchmen watched as Spears boarded a four-seater RAF plane. Then, as the pilot revved up the engines and began taxiing slowly down the runway, Spears abruptly pulled de Gaulle and his aide aboard. Slamming the door, the British general noted the "gaping faces" of the French pilots and ground crews standing nearby.

A few hours later, Spears and de Gaulle arrived at 10 Downing Street. It was a lovely late-spring day, and Churchill was sitting in the garden, enjoying the sunshine. He rose with a smile to greet his guest, who, at six feet, seven inches, towered over him by a foot.

As warm as the prime minister's welcome was, however, it could not disguise the fact that de Gaulle was, as General Pug Ismay noted, "in a hideously difficult position." Unlike the governments of the other occupied European countries, he and his infant movement were not recognized by Britain—or the rest of the world, for that matter—as the official governing body of his nation. Because Reynaud had legally handed over power, Pétain's regime, now based in the spa town of Vichy, was without question the legitimate government of France. Technically, that made de Gaulle an insurrectionist; indeed, a few weeks after his escape, a Vichy military court sentenced him to death for treason.

Not surprisingly, then, many leading British officials wanted nothing to do with the newcomer. As much as they might disapprove of the actions of the Vichy government, not only was it legal but it also had significant resources—a large naval fleet and far-flung French colonies—that must be kept out of German hands. In the view of some in the British government, appeasement of Vichy might even lure Pétain and his men back to the Allied side.

Whitehall's reaction to de Gaulle himself was, for the most part, condescending and hostile. Alexander Cadogan, for one, referred to the French general as "a loser" and "that ass de Gaulle." After his first meeting with de Gaulle, Cadogan told his colleagues that "I can't tell you anything about [him], except that he's got a head like a pineapple and hips like a woman."

Churchill, however, viewed de Gaulle in a far different light. He had tremendous admiration for the Frenchman's refusal to accept defeat and for his iron-willed determination to fight on in the face of what appeared to be impossible odds—qualities shared by the prime minister himself. But he also saw in de Gaulle what he wanted to see in France. When Anthony Eden told Churchill that France's disgrace was so great it could never recover, Churchill vehemently disagreed. France, he said, would unquestionably rise again. But for now, he and Britain would have to settle for de Gaulle as the lone emblem of an undefeated France.

For a few days after the armistice, British authorities had hoped that at least one or two weightier French political figures—Reynaud, former interior minister Georges Mandel, Édouard Herriot, the president of the Chamber of Deputies, or any of the few other opponents of capitulation—would join de Gaulle in London. But none did, and Churchill finally called a halt to the official foot-dragging over de Gaulle's status. On the evening of June 27, he summoned the general to Downing Street and told him, "You are alone! Well, I shall recognize you alone!" The next day, the British government gave official recognition to "Gen. de Gaulle as the leader of all free Frenchmen, wherever they may be, who rally to him in support of the Allied cause." That announcement, limited as it was, provided the legal basis for de Gaulle's relationship with the British government from then on. The brainchild of Winston Churchill, it was, noted the French historian François Kersaudy, "an act of faith in a solitary man and an abstraction called Free France."

Although de Gaulle's wartime relationship with Churchill would later become highly contentious, he never forgot the great debt he owed the British prime minister for his support in 1940. "I was nothing to begin with," he noted after the war. "Washed up from a disastrous shipwreck upon the shores of England, what could I have done without his help? He gave it to me at once."

With Churchill's recognition in hand, de Gaulle set out on his "magnificently absurd" undertaking to reclaim France. In the beginning, his London headquarters consisted of several small rooms in St. Stephen's House, a dilapidated office building on the bank of the Thames. From this makeshift office, furnished with a table, four rickety chairs, a phone, and a large map of France, de Gaulle and a handful of aides began the long, painful effort of building an army.

To the few Frenchmen who initially answered his call, he didn't mince words about how arduous this quixotic campaign would be: "I have neither funds nor troops. I don't know where my family is. We are starting from scratch."

CHAPTER 5

"Something Called
Heavy Water"

The Rescue Mission That
Changed the Course of the War

IN THE EARLY MORNING OF JUNE 22, 1940, A LARGE GROUP OF rumpled, bleary-eyed passengers clustered together on a platform at London's Paddington Station, surrounded by luggage, wooden crates, and twenty-six metal canisters. The crowds hurrying by paid little attention to the bedraggled travelers, who had just arrived from France. Londoners were focused on more important concerns that day, including the imminent capitulation of France to Germany.

Nothing about the group's members indicated their prominence; they included some of France's most distinguished scientists and engineers, experts in everything from ballistics and chemical warfare research to explosives manufacturing. Also on the platform were two nuclear physicists from the renowned Collège de France in Paris, a leading center of nuclear fission experiments. Unobserved as their arrival was, the physicists—and the precious substance in the canisters beside them—would end up playing a vital role in one of the most momentous developments of the war.

A tall, unshaven Englishman, wearing flannel trousers and a travel-stained trench coat, was tending to the group. Known as Jack to his family and friends, he was Charles Henry George Howard, the 20th Earl of Suffolk and a scion of one of Britain's most ancient and powerful families. Lord Suffolk had scooped up the scientists a few days earlier and spirited them out of France aboard a scruffy Scottish coal freighter. An awestruck Harold Macmillan, then a junior minister in Churchill's government, was introduced to the swashbuckling Suffolk a few hours after the group's arrival in London; he would later describe the thirty-four-year-old peer as a "mixture between Sir Frances Drake and the Scarlet Pimpernel."

Yet while Lord Suffolk had shown considerable daring and ingenuity during his adventure in France, its success was not due solely to him. His partner in arranging the rescue was Raoul Dautry, the French minister of armaments. Unlike most of his colleagues in the French government, the fifty-nine-year-old Dautry, a former head of the French railway system and a man of audacity and vision, was determined to do all he could to help Britain and defy the Nazis.

DAUTRY'S INVOLVEMENT IN THIS cloak-and-dagger episode had begun a few months earlier, when Frédéric Joliot-Curie, the son-in-law of Nobel laureates Marie and Pierre Curie, paid him a series of visits shortly after France and Britain declared war on Germany. In 1935, Joliot-Curie and his wife, Irène, had won the Nobel Prize in Chemistry for their seminal work on artificial radioactivity. Under Joliot-Curie's direction, nuclear physicists at the Collège de France had demonstrated that uranium had the potential to produce an explosive chain reaction and had even designed a workable reactor on paper.

Joliot-Curie told Dautry that his team's research might well lead to the development of a vastly powerful new bomb. But to harness nuclear energy, he said, it would be necessary to find a material that would slow down the rapid chain reaction caused by the splitting of uranium nuclei, which in turn would allow the reaction to become self-sustaining. One such moderator was an extremely rare substance called heavy water, a liquid that looked like ordinary water but contained deuterium—an isotope, or variation, of hydrogen. Only one company in the world—a Norwegian firm called Norsk Hydro—produced heavy water in quantities greater than a few drops, Joliot-Curie said. It was manufactured at a Norsk Hydro electrochemical plant nestled in a narrow, mountain-rimmed valley about seventy miles west of Oslo.

In late December 1939, soon after his meetings with Joliot-Curie,

Dautry received alarming news from officials in France's military intelligence bureau. Norsk Hydro, which produced heavy water as a sideline and sold small amounts of it to laboratories all over the world for various kinds of scientific experiments, had just informed them that I.G. Farben, the huge German chemical-industry conglomerate, had placed an order for its entire heavy water stock. When the Norwegian company had asked the reason for such a large purchase, Farben had declined to answer.

Dautry's fear—that Germany was also pursuing the possibility of producing a nuclear bomb—was in fact correct. Indeed, the German government had created a military department devoted exclusively to the wartime development of nuclear energy. "The country which first makes use of [nuclear fission] has an unsurpassable advantage over the others," observed the German physicist Paul Harteck. Under government sponsorship, Harteck and other physicists had formed what they called "the Uranium Club," conducting nuclear chain reaction experiments in half a dozen laboratories throughout the Reich. Like Joliot-Curie and his team, the Germans had decided on heavy water as the best instrument for controlling and sustaining a nuclear reaction.

News of the Germans' interest in heavy water galvanized Dautry into action. Determined to keep nuclear weapons out of Nazi hands, he organized a secret mission to Norway to bring back all the heavy water its operatives could find. To head the operation, Dautry chose Jacques Allier, a dapper, bespectacled young Frenchman who in peacetime was an officer at the Banque de Paris et des Pays-Bas, one of France's leading banks and a chief shareholder in Norsk Hydro.

On February 28, 1940, Allier, now a reserve officer with the Ministry of Armaments, left Paris for Oslo—and the adventure of his life. Traveling under an assumed name and with a false passport, he carried with him a letter of credit for 1.5 million Norwegian kroner (more than $5 million today).

In the snowbound Norwegian capital, Allier told Axel Aubert, Norsk Hydro's director, of the race between French and German scientists to build a nuclear bomb and of the vital importance of heavy water in that endeavor. When the Frenchman offered the letter of credit in exchange for all of Norsk Hydro's stock, Aubert shook his head. His company, he said, did "not wish to receive one centime" for the heavy water; instead Norsk Hydro would lend it to France, as well as any the company manufactured in the future. "I know that if the [German] experiments succeed and if later France has the misfortune of losing the war, I will be shot for what I am doing today," Aubert said. "But it's a matter of pride for me to run that risk."

With the heavy water in hand, Allier and his team of agents faced a new challenge: how to get it out of the country without German interference. Despite the secrecy of the French mission, the Nazis knew all about it: the Abwehr, Germany's military intelligence agency, had cabled its operatives in Oslo to follow and waylay a suspicious Frenchman named Allier, who was traveling under an alias.

Late one evening in early March 1940, workers at the Norsk Hydro plant poured the heavy water into twenty-six metal canisters, then drove over ice-covered roads to deliver them to Oslo. The following morning, Allier and one of his French colleagues arrived at Oslo's Fornebu Airport. Two passenger planes were revving up on the airfield side by side, one bound for Scotland, the other for Amsterdam. Both were scheduled to depart at about the same time.

In the departure hall, the Frenchmen, under the watchful eye of Abwehr operatives, confirmed their reservations for the Amsterdam plane. As passengers for both flights began boarding, a taxi drove up to the departure gate, was waved through to the tarmac, and stopped between the two aircraft, out of sight of the terminal. The cans of heavy water were hurriedly transferred from the taxi to the Scotland-bound plane as Allier and his colleague headed for the other. Hidden in a swarm of fellow passengers, they abruptly shifted direction, scrambling aboard the airliner flying to Scotland just as its door was closing. Both planes took off and headed out over the North Sea.

A few minutes into the flight, two Luftwaffe fighters intercepted the craft heading for Amsterdam and forced it to land at Hamburg, a port city in northern Germany. As soon as it touched down, Abwehr agents forced its cargo compartment open and unloaded several large wooden crates. Inside, they found loads of Norwegian crushed granite instead of the heavy water they had been ordered to retrieve. By then the canisters were safely in Edinburgh. By March 16, they were in Paris, stored in the cellars of the Collège de France. Three weeks later, Germany invaded Norway.

Raoul Dautry, however, had little time to savor his success. On May 16, he received an urgent call from General Weygand, informing him of the German breakthrough at the Meuse River. Unlike his superiors, Dautry was not focused on the twin specters of defeat and capitulation. He shared Churchill and de Gaulle's belief that his country must stand its ground and fight; if vanquished on French soil, he argued, its army should retreat to France's colonies in North Africa to continue the battle alongside Britain.

His prime concern at the moment, though, was to ensure the safety of the Collège de France physicists and the heavy water they were sheltering. At his urging, two leading members of Joliot-Curie's team—the tall, burly, Russian-born Lew Kowarski and Hans von Halban, a cultivated Austrian who had grown up in Germany—accompanied the precious substance to Clermont-Ferrand, an industrial town some 250 miles south of Paris, where they set up a temporary lab. In early June, as the enemy drew closer to Paris, Joliot-Curie and his wife joined them. By then Dautry had realized the hopelessness of the situation; it was essential to get the scientists and the heavy water out of France before the Germans tracked them down. Just as he was working out plans to do so, Lord Suffolk providentially appeared in his office.

In early 1940, Suffolk, a scientist in his own right, had been sent to Paris as the liaison between Dautry's Ministry of Armaments and the British Ministry of Supply's department for scientific and industrial research. After taking a suite at the Ritz, he set about acquainting himself with the latest French developments in science and engineering, including Joliot-Curie's nuclear fission experiments.

In the first days of June, with France about to fall, Suffolk decided on his own initiative to rescue from the country whatever he could of scientific and industrial value. Dautry wholeheartedly approved and presented Suffolk a letter authorizing him to do so. With that in hand, Suffolk raced around Paris collecting scientists, engineers, state-of-the-art machine tools, and millions of dollars worth of industrial diamonds, dispatched from Belgium and Holland in advance of the German blitzkrieg. When some of the bankers in whose vaults the diamonds were stored refused to hand them over, the English lord pulled out two ivory-handled pistols along with Dautry's letter. The bankers quickly complied.

The Earl of Suffolk

LORD SUFFOLK'S CONFRONTATION WITH the bankers in Paris was not atypical. All his life, Jack Howard, as he preferred to be called, had refused to follow the rules of polite society. His rebellion against conformity and thirst for adventure seemed to have been firmly embedded in his family's DNA. "For twenty generations," an observer remarked, "the earls of Suffolk have done just what they pleased—and what they pleased to do was invariably dangerous." Thomas Howard, the 1st Earl of Suffolk, was given his title by Queen Elizabeth I for the key part he played in defeating the Spanish Armada in 1588. The queen referred to him as "Good Thomas," to distinguish him from his father, Thomas Howard, the 4th Duke of Norfolk, who was executed in 1572 for plotting to dethrone Elizabeth and replace her with Mary, Queen of Scots. Some three hundred years later, Queen Victoria tartly referred to the family as "those mad Howards."

Jack's mother, Daisy, the Dowager Countess of Suffolk, was as disapproving of her son's escapades as Queen Victoria had been of the exploits of his ancestors. Yet Daisy was hardly a model of conformity herself. The youngest daughter of Levi Leiter, a multimillionaire businessman from Chicago, she was one of a flock of American heiresses, labeled "the buccaneers" by the novelist Edith Wharton, who had crossed the Atlantic in the Victorian era to marry members of the English nobility. Daisy's eldest sister, Mary, had scored one of the biggest catches of all—Lord Curzon, the brilliant, temperamental peer who became British viceroy to India and later foreign secretary.

During a visit to her sister in India, Daisy met and fell in love with Henry Paget Howard, the 19th Earl of Suffolk, a dashing sportsman and aide-de-camp to Curzon. The couple was married in 1904. Like her husband, Daisy was addicted to excitement and adventure—fast cars, speeding planes, and hunting to hounds or on safari. The marriage produced three children and lasted almost thirteen years; in 1917, Henry Howard was killed fighting the Turks near Baghdad in World War I.

Eleven years old when he inherited his father's title, Jack Howard had no interest in pursuing what he viewed as the self-indulgent, pleasure-seeking lifestyle of English aristocrats. He hated hunting and shooting and was bored by the thought of devoting his life to the upkeep of the Suffolks' ancestral home—a forty-room Elizabethan mansion and 10,000-acre Wiltshire estate known as Charlton Park. "Jack was a rebel against everything in his own past and against everything the society he was born into stood for," one friend remarked.

When his mother sent Jack to the Royal Naval College at Osborne, the sixteen-year-old lasted barely a year. He loved the idea of going to sea—but on his terms, not the Royal Navy's. A short time later, after attending Radley College, he signed up as a common deckhand on a clipper ship bound for Australia. "I don't see how you can fit yourself to any great position in life unless you have spent time roughing it and learning what the other fellow goes through," he later declared. He spent much of the next six years in Australia in a variety of jobs, from cowboy to sawmill worker. Toward the end of his time there, he became part owner of a large sheep station in Queensland. When he returned to England to take over the management of his estate, he sported a beard, a pet parrot, and tattoos of a snake and skull-and-crossbones on his arms. His mother broke down in tears when she saw him.

Once again, though, his restlessness got the better of him, and he enrolled at the University of Edinburgh to study chemistry. At the age of twenty-eight, the Earl of Suffolk had finally found a field that truly absorbed him. Graduating from the university with a first-class degree, he went to work at the Nuffield Laboratory at Oxford as a research chemist.

When war broke out in 1939, he tried to enlist in the army but was turned down because of a childhood bout with rheumatic fever. In early 1940, his fluent French and scientific expertise won him the job of liaison between Dautry and the British Ministry of Supply. Herbert Gough, who was Suffolk's boss at the ministry, later recalled that he had been "won over completely" by the earl's "tremendous enthusiasm, infectious personality, and buccaneering spirit"—traits that, in the chaos of France's defeat, would yield huge benefits for the British and Allied cause.

IN MID-JUNE 1940, with his cargo of diamonds, machine tools, and scientists secure, Suffolk headed south to Bordeaux, hoping to commandeer a ship that would transport them all to Britain. When he arrived, he discovered a city in chaos. Hordes of fleeing French soldiers and civilian refugees had swollen the population from 300,000 to 900,000 virtually overnight. Food, water, and lodging were extremely scarce, and millionaires from Paris camped in the city's public square alongside shopkeepers and factory workers. "The single thought in everyone's mind was escape," an American journalist observed.

Forced to sleep in his car, Suffolk haunted the Bordeaux docks for three days trying to find a captain willing to make the journey across the Channel. He had no luck until the morning of the fourth day, when, thanks to a tip from the British commercial attaché, he located an old Scottish freighter called the *Broompark*. Its skipper agreed to take the job, provided they left as soon as possible. German planes, which had been strafing refugees on the roads leading to Bordeaux, had begun bombing ships in the harbor; later that afternoon, in fact, a freighter tied up alongside the *Broompark* would be hit by a bomb and severely damaged.

Meanwhile, Joliot-Curie and his colleagues, along with the heavy water canisters, had arrived in Bordeaux, under orders from Dautry to leave with Suffolk. Having fought their way through the massive crowds clogging the docks, the Collège de France team stared in wonderment at the fantastic bearded figure, looking like "an unkempt pirate," who greeted them at the *Broompark*'s gangplank. Stripped to the waist, his arms covered with tattoos, Suffolk ushered them aboard the ship, swinging a riding crop and shouting to the crew to begin loading the heavy water.

Before dawn on June 19, the *Broompark* weighed anchor and headed for England. Lew Kowarski and Hans von Halban were among the couple dozen scientists aboard; Frédéric Joliot-Curie was not. Despite intense pressure from Suffolk, he decided at the last minute that he could not leave his homeland. Above all, he could not bear the thought of abandoning his wife, who was ill with tuberculosis, and his two young children, who were living with relatives. Before the ship sailed, however, he instructed Kowarski and Halban to work closely with the British in continuing their nuclear fission experiments.

Throughout the uneventful voyage, Lord Suffolk plied the scientists with champagne to settle their nerves. On June 21, the *Broompark* tied up at the Cornish port of Falmouth, and Suffolk, with his usual panache, somehow procured a special train and armed guard to transport the scientists, diamonds, machine tools, and heavy water to London.

Early on the morning of June 22, Harold Macmillan, the future prime minister who then was parliamentary secretary to the minister of supply, was awakened at his flat by a phone call and told to report to his office immediately. There he found "a young man of somewhat battered appearance, unshaven . . . yet distinguished by a certain air of grace and dignity." It was his first meeting with Lord Suffolk, who briefed him on the valuable cargo he had rescued in France, which included, in Macmillan's words, "something called heavy water." As Macmillan recalled years later, "I did not know at the time what heavy water was, and I was too confused to inquire." What he remembered most about that encounter was Suffolk himself, whom he described as "a truly Elizabethan character." In his memoirs, Macmillan would write, "I have had the good fortune in my life to meet many gallant officers and brave men, but I have never known such a remarkable combination in a single man of courage, expert knowledge and charm."

The following day, Suffolk escorted Halban and Kowarski to a meeting with leading British scientists at London's Great Western Hotel. In the mid- to late 1930s, a number of physicists in Britain, including several refugees from Nazi Germany and Austria, had also been conducting experiments to determine the possibility of nuclear fission. But once the war broke out, more pressing matters occupied their attention, including creating a radar system to detect the approach of enemy aircraft. Nonetheless, in early 1940, half a dozen of the country's top scientists persuaded the British government that development of a nuclear bomb was a distinct, if distant, possibility. Thus was born the MAUD (Military Applications of Uranium Detonation) Committee, which, like its German government counterpart, began overseeing uranium-related research in laboratories throughout the country. In their discussions with Halban and Kowarski, British scientists realized how much further ahead Joliot-Curie's team was in such work than they were. They immediately invited the Collège de France nuclear physicists to join their efforts.

Meanwhile, Suffolk and his colleagues at the Ministry of Supply were debating where to store the heavy water, which, according to one ministry official, "may prove to be the most important scientific contribution to our war effort." After depositing the canisters briefly in cells at Wormwood Scrubs prison in suburban London, ministry officials finally found the perfect spot for them: with the permission of King

George VI, they were stashed with the British crown jewels in a heavily guarded hiding place somewhere deep inside Windsor Castle.

Kowarski and Halban were given space at Cambridge University's renowned Cavendish Laboratory, which had been largely evacuated because of fears that Cambridge would be in the direct path of an expected German invasion. During the remainder of 1940, as the Battle of Britain raged overhead, the two physicists used some of their cache of heavy water to continue their nuclear fission experiments. By early 1941, their research had convinced British leaders that, with enough uranium and heavy water, a nuclear reactor—and a bomb—could be built in time to affect the course of the war. "I remember the spring of 1941 to this day," recalled James Chadwick, Britain's foremost physicist, who won the 1935 Nobel Prize in Physics for his discovery of the neutron. "I realized then that a nuclear bomb was not only possible. It was inevitable."

But the war-battered British did not have the enormous economic and industrial resources it would require to undertake such a massive project. For that, they had to turn to the neutral United States, where nuclear fission research was also being conducted in a number of laboratories. Only the physicists at the Columbia University lab, Enrico Fermi and Harold Urey among them, could be considered serious rivals to the Collège de France team. Yet although Fermi, Urey, and their colleagues recognized that reactors could be used to produce bombs, they were not yet as advanced in their research as Kowarski and Halban.

In mid-1941, the MAUD Committee sent a report to the U.S. government urging the development of a nuclear bomb. A few months later, Harold Urey visited Britain for discussions with Kowarski, Halban, and British physicists and engineers. That report and those consultations in turn led to the Manhattan Project—and the dropping of the atomic bomb on Hiroshima and Nagasaki in 1945.

"Had the British not taken up fission in earnest in 1940 and 1941 . . . and had they not then pushed the Americans to act, it is likely that no nuclear weapon would have been ready before the end of the war," the science historian Spencer Weart has noted. He added, "If von Halban and Kowarski had not come to Britain in June 1940, there almost certainly would have been no British reactor program at all."

SINCE THE RESCUE OF the French scientists and the heavy water was considered top secret at the time, no public recognition was given to Lord Suffolk for his crucial role in bringing it about. In a closed session of Parliament on June 27, 1940, Herbert Morrison, the minister of supply, informed MPs that a mission had been mounted in France to save valuable materials, "some of them of almost incalculable scientific importance." He added only that a Ministry of Supply official, whom he did not name, had been responsible for the rescue.

Within a few months, Suffolk would take on another vital and even more perilous mission for the government. This time, the British public would be made fully aware of what he had done.

CHAPTER 6

"They Are Better Than Any of Us"

Polish Pilots Triumph in the Battle of Britain

HAVING CONQUERED MOST OF WESTERN EUROPE BY THE END of June 1940, Germany was now ready to direct its "whole fury and might," as Winston Churchill put it, against his small island nation. "What General Weygand called the battle of France is over," Churchill told Parliament on June 18. "I expect that the battle of Britain is about to begin."

By the first days of July, Germany had transferred more than 2,500 fighters and bombers to captured bases in northwestern France as well as to Norway, Denmark, Belgium, and Holland. Luftwaffe head Hermann Göring assured Hitler that his fearsome air fleet would wipe out the RAF by the beginning of autumn. And once that job was finished, he said, Germany would have no trouble bombing Britain into submission or launching the cross-Channel invasion, code-named "Sea Lion," that the Führer was now considering.

As it prepared to square off against this utterly confident, seemingly invincible enemy, RAF's Fighter Command struggled to rebuild its forces, shattered in the debacle in France. Lacking combat experience and steeped in fly-by-the-book procedures, the British pilots sent to fight there had had no idea what they were getting into. For that matter, neither had their superiors. In just three weeks, more than three hundred British fighter pilots had been killed or reported missing—

close to a third of the command's overall strength. More than a hundred had been taken prisoner. During the Dunkirk operation alone, the RAF had lost some eighty pilots and one hundred planes. Altogether almost a thousand aircraft, about half the RAF's frontline strength, had been destroyed.

In mid-July, the Luftwaffe began attacking British ship convoys in the Channel as well as targets on England's southern coast. The RAF warded off these limited assaults fairly well, but against an all-out German air attack, it would be able to send up a combined total of only about seven hundred fighters—Hurricanes and the faster Spitfires. Worse, fewer than two pilots per plane were available. It would take a good many more of both for the British to maintain control of their skies. New Hurricanes and Spitfires were being turned out as fast as possible, and Hugh Dowding, the head of Fighter Command, was doing everything possible to make up the shortfall in men.

Among other things, he was now pirating pilots from the RAF Bomber and Coastal commands as well as ordering mere trainees to prepare for combat—youngsters "with blond hair and pink cheeks," wrote the American journalist Virginia Cowles, "who looked as though they ought to be in school." Many of them had fewer than ten hours of flying time in either Hurricanes or Spitfires. Barely 10 percent had undergone rigorous gunnery practice. Few knew how to sight their guns: when attacking, they tended to open fire at ranges of 500 yards or more, then break away just as they were getting close enough to actually hit something. They learned their lessons quickly in combat, but a good many died before they could put the lessons to use.

Even with all his foraging, Dowding still came up short and had to turn to fliers from other countries to fill his depleted ranks. As a result, fully 20 percent of the RAF pilots who fought in the Battle of Britain were not British. About half that number—250 in all—came from British Commonwealth countries, including Canada, Australia, and New Zealand. Many more were needed, however, and the RAF, much against its will, was forced to use pilots who had escaped to Britain from occupied Europe.

Dowding himself was highly doubtful about the wisdom of what he called "the infiltration of foreign pilots into British squadrons." His definition of "foreign," as it turned out, was highly elastic. Although he clearly would have preferred to use only homegrown or British Commonwealth fliers, he did assign several dozen western European pilots, including thirty Belgians and eleven Frenchmen, to undermanned RAF squadrons early in July.

But that was as far as he was initially prepared to go. Like other top officials in the RAF and Air Ministry, Dowding wanted nothing to do with the Poles and Czechs, who made up the lion's share of the European pilots in Britain. Indeed, he insisted that he would dissolve British squadrons before allowing the eastern Europeans to join them. Dire as the situation was, Britain, in his view, did not need the help of a couple of benighted countries that, to most Britons, were little more than "names on a map."

Neville Chamberlain had spoken the truth in September 1938 when he had noted that to most Britons, including himself, Czechoslovakia was "a faraway country" populated by "people of whom we know nothing." Similarly, Poland was, for most Englishmen, "the other Europe"—exotic, unknown, a bit savage. According to Geoffrey Marsh, an RAF officer who taught English to Polish fliers, the average Englishman imagined that Poland "was some one hundred years behind" Britain and that "its inhabitants lived in a state of superlative ignorance." RAF commanders, for their part, regarded the Poles and Czechs as being on "a rung or two lower on the ladder of civilization."

When Germany crushed Poland in September 1939, its victory merely confirmed British prejudices about the alleged fecklessness of the Polish war effort. As in the United States and most of the rest of Europe, Britons accepted as truth Germany's claims that the Poles had demonstrated both military ineptness and lack of will in their fight against the Reich. Neither charge was true: the Poles in fact had managed to inflict relatively heavy losses, killing more than 16,000 German troops and wounding some 30,000.

Senior RAF officers, meanwhile, were highly doubtful about the flying skills of Polish pilots, who they believed had lost their nerve in confronting the Luftwaffe. "All I knew about the Polish Air Force was that it had only lasted about three days against the Luftwaffe, and I had no reason to suppose that [it] would shine any more brightly operating from England," noted Flight Lieutenant John Kent, one of the RAF's hottest test pilots, who, much to his chagrin, would be named deputy commander of an RAF Polish squadron.

While the Poles were distrusted because of their defeat, the Czechs were scorned because they hadn't defended their country at all when the Germans had occupied it. British officials seemed oblivious to the fact that the Czechs' failure to fight had in no small part been due to the British and French betrayal of their country at the 1938 Munich conference. Before Munich, Czech president Edvard Beneš had declared that his nation would resist if German troops marched into the Sudetenland. The highly trained, well-equipped Czech army—more than a million men in all—had been mobilized, as had the air force. The country's already formidable fortifications had been strengthened; all main roads and bridges had been blocked and mined. "The country was poised for action, calm and determined in full preparedness for the expected bloody struggle," a Czech government official recalled.

But Beneš's will collapsed when he was informed that France and Great Britain would not stand behind him. To the horror of Czech military leaders, the demoralized president ordered the country's forces to stand down. "We are only adding our name to the cowardice of our allies!" a distraught general exclaimed. "It is true that others have betrayed us, but now we alone are betraying ourselves." Beneš paid no heed. The army and air force were disbanded, and the news was announced to weeping crowds in Prague's Wenceslas Square.

After the Sudetenland was surrendered, thousands of Czech soldiers and pilots streamed out of their country; more followed after Germany's occupation of the rest of Czechoslovakia in March 1939. The fact that a number of them joined Poland's air force and army and took part in the Polish fight against Germany meant little or nothing to the British.

In the end, the RAF did change its mind about using eastern European fliers—but only because it had no choice. On August 13, the Luftwaffe launched its all-out air assault against Britain, blitzing airfields, radar installations, and aircraft factories in the southern part of the country. "We have reached the decisive point in our air war against England," Göring declared. "Our first aim must be the destruction of the enemy's fighters." Day after day, hundreds of German bombers, closely guarded by swarms of fighters, swept over the Channel with the intent of blasting British defenses into rubble.

To help counter this onslaught, dozens of Polish and Czech pilots were integrated into already existing RAF squadrons. In addition, the RAF agreed to form two squadrons made up solely of Poles, as well as two all-Czech units. The new squadrons, however, were firmly under British command and subject to British regulations; their pilots wore dark blue RAF uniforms with "Poland" or "Czechoslovakia" flashes on their jacket sleeves. If Polish and Czech "air units in this country are ever to become efficient," said one RAF report, "their command must never be taken out of British hands." The eastern Europeans, especially the Poles, the report added, needed to suppress their "inherent individualism and egotism," show some discipline, and "learn by example" from their British counterparts.

That kind of condescension only compounded the grievances already held by the Poles and Czechs against the country that, along with France, had failed to come to their nations' aid. The fliers also were less than impressed by the quality of the British military effort thus far. "My mind was still reeling from the desperately heroic Polish shambles and the insouciant French shambles," recalled a Polish airman. "Therefore it was with some apprehension that I awaited the first symptoms of some third variety of shambles—a British shambles."

Most of the Polish fliers in Britain were already battle-hardened pilots. The problems they had faced in defending their country had had little or nothing to do with their flying skill. Many had been flying at Polish *aeroklubs* since they were boys, and most were graduates of Djblin, one of the most difficult and demanding air force training academies in the world. Virtually all had flying experience against the Luftwaffe, which was more than most British pilots, including those who commanded them, could claim.

That was particularly true of one of the all-Polish units, known to the RAF as 303 Squadron but called the Kościuszko Squadron by those in its ranks. Named for Tadeusz Kościuszko, the young Polish patriot who became an American hero in the Revolutionary War, the unit contained some of Poland's most skilled fliers. It was assigned to

Polish pilots from the RAF's 303 Squadron.

Northolt, a key RAF station only fourteen miles from central London, which was under the umbrella of the all-important 11 Fighter Group. The twenty-one squadrons belonging to 11 Group would be in the thick of the Battle of Britain, responsible for protecting London as well as the rest of southeast England. As such, the group's squadrons would be crucial to Britain's defense, and many in the RAF had serious doubts that the Poles of 303 were up to the challenge.

Even as German air activity increased markedly over the Channel and the English coast, the RAF insisted that the squadron could not become operational until its personnel learned British tactics and basic English. "I'm not having people crashing round the sky until they understand what they're told to do," declared Group Captain Stanley Vincent, the station commander at Northolt. When they arrived in Britain, none of the Poles or Czechs knew the basic English vocabulary for flying. At language school, they learned RAF code words— "angels" for altitude; "pancake" for landing; "bandits" for enemy planes; "tally ho" for launching an attack. They learned how to count to twelve in English so they could understand the clock-face system for giving bearings—for example, "bandits at twelve o'clock."

During their early training in Hurricanes, the Poles of 303 also struggled with the many unfamiliar intricacies of modern aircraft controls. Having trained in primitive planes and being unaccustomed to having radios in the cockpit, they often violated proper radio-telephone procedures or failed to respond properly. In Polish planes, to accelerate you pulled the throttle back, whereas in British planes you pushed it forward. "We had to reverse all our reflexes," said one of 303's pilots. There were several instances of overshooting the runway, and because most of the fliers had never flown aircraft with retractable landing gear, a number of landings occurred with the wheels still up.

At one point, the Poles were ordered to ride a fleet of oversized tricycles—each fitted with a radio, compass, and speed indicator—in flying formation around a football field. As they rode, they were directed to "interceptions" from an "operations room" at the top of the bleachers. The indignity of it all infuriated the Poles—skilled veteran pilots being forced to ride around a field on trikes.

For almost a month at the height of the Battle of Britain, the British and the Poles of 303 engaged in a tense, stormy conflict of wills. It had been nearly a year since Germany had swept into Poland. In that time the Polish fliers had experienced nothing but frustration, bordering on despair. Their inaction fed their already considerable guilt about not being able to save Poland. They also agonized about escaping and leaving their country, especially their families, to suffer under the twin occupations of Germany and the Soviet Union, which, under a secret codicil to the Ribbentrop-Molotov nonaggression pact of 1939, had invaded and annexed eastern Poland in mid-September of that year. Hungry for combat, the Poles of 303 discovered that, in the words of one of them, "we were not to be let off the leash." Adding to the torture was the knowledge that at least forty other Polish pilots were already operational—in otherwise all-British squadrons.

The pilots of 303 particularly objected to being forced to take orders from British officers, whom they considered arrogant and condescending. They were in no mood to be lectured to about their language abilities or their tactics, least of all by the likes of their squadron leader, Ronald Kellett, a wealthy London stockbroker in civilian life who had never flown in combat for as much as a minute.

The emotional, high-spirited Poles showed their rebelliousness in various colorful ways. They were constantly being reproved for not conforming to regulations in dress (unbuttoned uniform jackets, missing belts, nonregulation shirts and shoes) and for sneaking into Women's Auxiliary Air Force (WAAF) housing at night or smuggling WAAFs into their quarters. "They were a complete law unto themselves," recalled a British air mechanic.

As the training continued, however, the Poles' British commanders gradually began to show more understanding and respect. Despite the belly landings and other early problems, it soon became clear to Kellett and Kent that the pilots under their command were very good indeed. Kent was impressed with both their flying ability and their unusually quick reaction times, and Kellett even developed into something of an advocate for them, quick to challenge any disparaging remarks by other RAF officers—remarks such as he himself had made in the beginning. When Northolt officials offered the squadron the use of a battered old truck, instead of the usual car, to take them from the officers' mess to the airfield, Kellett declared that such a shabby form of transportation simply would not do. He brought in his own Rolls-Royce, a roomy 1924 open touring car that could ferry as many as twelve pilots at a time to their planes.

Meanwhile, the intensity of the battle over southern England was steadily mounting. Every day, the three-hundred-plus Hurricanes and Spitfires of 11 Group hurled themselves against massive enemy bomber and fighter formations. The days were clear and hot, and the RAF pilots scrambled from dawn to dusk, which in Britain at that time of year amounted to about fifteen hours a day. Their lives had suddenly become a madness of activity—fierce combat, followed by frenzied refueling and rearming on fields that German bombing was turning into lunar landscapes. The stress of several sorties a day took an enormous physical and psychological toll: some fliers were so exhausted after a sortie that as soon as they landed they immediately fell asleep in their cockpits. Nerves were frayed, morale began to slip; more and more mistakes, including faulty landings, were made, and accidents occurred.

During this period, both the RAF and Luftwaffe suffered severe losses of both men and planes. The planes were replaceable—and, in the RAF's case, were being replaced by British industry—but the men were another matter. From August 8 to 18, 154 RAF fliers were killed, seriously wounded, or missing in action—more than twice as many as could be replaced. By the third week of August, Fighter Command was short more than two hundred pilots.

"Most people who went into II Group didn't last," recalled one flier fortunate enough to have survived the Battle of Britain. "They couldn't last. They had no chance at all." Many years after the battle, another former RAF pilot was paging through a book listing the names of the pilots who had been in his squadron. "Some I couldn't remember," he said. "They passed through, and had been shot down before I could get to know them."

During this period of "intense struggle and ceaseless anxiety," as Winston Churchill called it, the Germans were relentless. As August drew to a close, they launched raids in such numbers against RAF airfields and radar stations that the controllers in II Group were forced to choose which attacks should get priority attention from their depleted squadrons. "On virtually every occasion that the Germans operated in force, they grossly outnumbered the defending squadrons," noted the official postwar RAF account of the Battle of Britain. For the most part, RAF squadrons went into combat one unit at a time, sometimes engaging enemy formations of fighters and bombers that outnumbered them more than tenfold. Yet despite the crisis, the Poles in 303 Squadron were kept far from any action.

On August 30, the Germans mounted their most concentrated assault yet, cutting the electricity to seven radar stations and knocking them temporarily off the air. Most key air bases in the south were hit as never before. That evening, Ronald Kellett put in a call to Fighter Command headquarters, urging that it make 303 operational. Having lost nearly a hundred pilots in the previous week, the RAF finally gave in and ordered the squadron into battle the next day.

The fighting on August 31, as it turned out, was even more intense than the day before. On that single, white-hot day, the Luftwaffe flew more than 1,400 sorties against the beleaguered airfields and radar stations encircling London. Shortly before 6 P.M., 303 Squadron finally got its orders to fly. It had been one year, almost to the day, since the Luftwaffe had devastated Poland and humiliated the Polish air force.

Now, after twelve months of anguish, anger, and frustration, the time had come to begin settling the score.

Shortly after taking off, the Poles hurtled down on the surprised enemy like avenging furies. In less than fifteen minutes, six of them had each shot down a Messerschmitt in the skies over south London. While the squadron would go on to compile a brilliant overall record in the Battle of Britain, it is doubtful that its contribution was ever more urgently needed than on its first day of combat. For it was on August 31 that Fighter Command suffered its heaviest losses of the entire battle— thirty-nine fighters shot down and fourteen pilots killed. The Germans, however, lost an identical number, with 303's pilots credited with 15 percent of those kills—and no losses of their own.

The RAF senior command, which just the day before had been so loath to let 303 fight, now deluged the squadron with congratulations. "Magnificent fighting, 303 Squadron," cabled Sir Cyril Newall, chief of the air staff. "I am delighted. The enemy is shown that Polish pilots [are] definitely on top." And that was just the beginning. On September 5, 303 was credited with destroying eight enemy aircraft—20 percent of that day's RAF kills.

Two days later, the now-familiar waves of German bombers and fighters failed to head for their previous targets—the coastal defenses and RAF bases of southern England. Following the curve of the Thames, they were aimed straight at London. Hitler had ordered attacks on the British capital in retaliation for a few scattered RAF bombing raids on Berlin. He and Göring had also persuaded themselves that the Luftwaffe had neutralized the RAF and therefore was free to concentrate on London and other British cities. It was a spectacular miscalculation—and no less so for being almost true. In the previous two weeks, the RAF had lost 227 fighters, had seen major damage inflicted on airfields and sector-control stations, and was close to being finished. What Fighter Command needed above all was time to regroup, and Hitler provided just that. Instead of persisting in heavy attacks against RAF installations and communications, the German air force began eight weeks of massive bombing of London—the most intense chapter in the eight-month reign of terror called the Blitz.

On that frenzied first day of the Blitz, the Poles of 303 shot down fourteen German planes in less than fifteen minutes. They also managed to disperse a German bomber formation before it could hit London. With nearly a quarter of the formation destroyed, the surviving bombers turned and headed back to France.

In just over a week of combat, the all-Polish squadron had destroyed nearly forty enemy aircraft—by far the best record in the entire RAF—and in doing so had become unofficial heroes of the realm. Government officials, senior RAF commanders, private citizens, Churchill, and the king himself joined at various times in paying honor to 303's fliers. "You use the air for your gallant exploits, and we for telling the world of them," the BBC's director general wrote the unit. "Long live Poland!"

At Buckingham Palace, King George VI's secretary, Alexander Hardinge, admiringly referred to the Polish pilots as "absolute tigers." In a letter to Lord Hamilton, Hardinge wrote, "One cannot help feeling that if all our Allies had been Poles, the course of the war, up till now, would have been very different." An RAF squadron leader, speaking of the Polish airmen, was quoted as saying, "They are fantastic—better than any of us. In every way they've got us beat."

Again and again, the question was asked: What made the Poles so good? The answer wasn't simple. Generally older than their British counterparts, most Polish pilots had hundreds of hours of flying time in a variety of aircraft, as well as combat experience in both Poland and France. Unlike British fliers, they had learned to fly in primitive, outdated planes and thus had not been trained to rely on a sophisticated radio and radar network. As a result, said one British flight instructor, "their understanding and handling of aircraft was exceptional." Although they appreciated the value of tools such as radio and radar, the Poles never stopped using their eyes to locate the Luftwaffe. "Whereas British pilots are trained . . . to go exactly where they are told, Polish pilots are always turning and twisting their heads to spot a distant enemy," an RAF flier noted.

The Poles' intensity of concentration was equaled only by their daring. British pilots were taught to fly and fight with caution. The Poles, by contrast, had been trained to be aggressive, to crowd and intimidate the enemy, to make him flinch and then bring him down. After firing a brief opening burst at a range of 150 to 200 yards, the Poles would close almost to point-blank range. "When they go tearing into enemy bombers and fighters they get so close you would think they were going to collide," observed one RAF flier. On several occasions, crew members of Luftwaffe bombers, seeing that 303's Hurricanes were about to attack, baled out before their planes were hit.

On September 15, the Poles of 303 were given their biggest opportunity to date to show off their exceptional combat skills. More than a month had passed since the Battle of Britain began. The RAF was still flying, and London, after a weeklong battering, was still defiant. Although Hitler was having doubts about Operation Sea Lion, he decided to give Göring another chance to make *Götterdämmerung* possible. And so Göring gave the order: every available Luftwaffe aircraft was to be unleashed in an all-out push to end RAF resistance. All of 11 Group's squadrons were engaged in combating the furious daylong assault, plus much of 12 Fighter Group, which included the other Polish squadron and a Czech unit. In all, some one hundred Polish and Czech pilots participated in the dogfights of September 15, making up about 20 percent of the total RAF force.

Churchill would later call the September 15 melee "one of the decisive battles of the war." The Luftwaffe had thrown nearly everything it had at the British but had failed to achieve its main objective: elimination of the RAF as a defensive force. There would be more German raids in the future, more destruction and death visited on London and other English cities. Nevertheless, the myth of the Luftwaffe's invincibility was forever shattered. Two days later, Hitler decided he'd had enough and canceled Operation Sea Lion until further notice.

Years after the war, a Battle of Britain historian would write, "Even though it was equipped with the Hurricane, the least effective of the main fighters, 303 Squadron was by most measures the most formidable fighter unit [RAF or Luftwaffe] of the Battle."

In only six weeks of combat, during the battle's most crucial period, the squadron was credited with shooting down 126 enemy aircraft, more than twice as many as any other RAF squadron during that time. Nine of its thirty-four pilots qualified as aces—fliers who had shot down five or more enemy planes. One of them, Jozef František, who flew with a fury that no other flier could match, was the RAF's top gun in the battle, with nineteen kills to his name.

František, as it happened, was a Czech—one of the pilots who had escaped to Poland after Munich—but from the day he arrived there, he had allied himself with Poland. In the days after the German invasion of Poland, he had flown reconnaissance missions for the Polish air force and on at least two occasions lobbed grenades from his unarmed, open-cockpit observer plane at German infantry columns. In England, when asked his nationality, František invariably answered, "I am a Pole." Proud to be in 303, he refused many invitations to join a Czech unit.

In the opinion of a number of RAF pilots and commanders, the contribution of the Polish pilots, particularly those in 303 Squadron, made the difference between victory and defeat in the battle. Perhaps the most telling comment came from no less than Hugh Dowding, initially so reluctant to send the Poles aloft. Shortly after the battle, the head of Fighter Command declared, "Had it not been for the magnificent [work of] the Polish squadrons and their unsurpassed gallantry, I hesitate to say that the outcome of Battle would have been the same." Many years later, Queen Elizabeth II would make the identical point: "If Poland had not stood with us in those days . . . the candle of freedom might have been snuffed out."

BY SEPTEMBER 1940, THE RAF's early doubts and suspicions of the eastern Europeans were nothing but a memory; the same was true of the Poles' and Czechs' qualms about the British. Regardless of nationality, all airmen in the RAF were fused together now, fighting for the common cause of freedom.

Alexander Hess, the senior pilot in one of the all-Czech units, recognized the power of that union when he and his squadron were ordered into the air, along with dozens of other squadrons, on September 15. As the Czechs flew at top speed toward London, the forty-two-year-old Hess felt great strength "in the knowledge that every one of us in the RAF was there for *all* of us—and that *all* of us were there for every single RAF man." As he gazed down through the haze at London's "stripes of streets, tiny rectangles of gardens, and millions of invisible people," he felt, too, that "all of this was mine—*my* house, *my* street, *my* child, *my* future. As if thousands of voices were calling up to me" to protect them.

A few minutes later, Hess's Hurricane was hit by a Messerschmitt; trailing clouds of black smoke, it began plunging toward the ground. Realizing he was coming down over a heavily populated neighborhood in the eastern suburbs of London, Hess frantically maneuvered the plane away from the area. He was barely three hundred feet above roof level when he spied a small open field. Just as he reached it, his plane burst into flames, and he parachuted out. When he opened his eyes after a hard landing, he saw "the stern face and ice cold eyes" of a member of the Home Guard, Britain's volunteer defense organization, who was pressing the barrel of a hunting rifle into his chest.

With a shock, Hess realized that the man thought he was a Luftwaffe pilot. "I am British!" he shouted again and again, but he could see that his heavy foreign accent only intensified the guardsman's suspicions. Just then, a military car roared up, and a couple of RAF officers stepped out.

After a few moments of discussion between the RAF men and the Home Guard volunteer, Hess was gently helped to his feet, put into the car, and taken to the volunteer's home. There he was placed in an easy chair, covered with a blanket, and given a whiskey as well as thanks for guiding his crippled aircraft away from nearby homes. "A feeling of deepest gratitude overwhelmed me from head to foot," Hess remembered. "I was grateful for all the kind and loving care, for the blanket and warm drink, for the people around me."

Alexander Hess would later be awarded a Distinguished Flying Cross for his heroic actions that day.

OF HESS AND THE other pilots who defended Britain, Winston Churchill declared: "Never in the field of human conflict was so much owed by so many to so few." Yet there were countless other heroes who emerged during that chaotic summer and fall of 1940. Chief among them were the members of UXB (unexploded bomb) squads, who, throughout the Blitz, were dispatched to neighborhoods in London where bombs with delayed-action fuses, their timers still ticking, had burrowed underground. In this frantic race against time, the volunteer squads would dig down to the bombs, knowing they could go off at any moment.

Just two months after his rescue mission in France, the Earl of Suffolk organized a unit to perform experiments on various kinds of German bomb fuses and to come up with safer and more effective ways of dismantling them. Working with the Ministry of Supply's scientific research department, he, together with his squad of sappers, transformed a van into a mobile laboratory, retrieving bombs from sites and bringing them back to Richmond Park just outside London, the main proving ground for experimental bomb research.

Possessing a "rare combination of steel nerves and a scientific mind," Suffolk constantly prodded the scientists with whom he worked to greater efforts. "He had us all slapping bombs around as if they were ostrich eggs," one researcher noted, "but we made progress that we'd never have made without him."

It was the perfect job for a man with an unquenchable thirst for adventure and danger. Happily shedding his aristocratic lifestyle, Suffolk turned over his mansion and estate to the government for use as an army hospital and took up residence in a small room at the Royal Automobile Club in London. He spent most of his time with his team of sappers, many of whom were from the city's East End; after a tense day of extricating and defusing bombs, he would often treat them to dinner at his favorite restaurant, a smart place called Kempinski's, in west London. In return, Suffolk's men presented him with a silver cigarette case inscribed with their names.

For eight months, Suffolk and his unit led what others considered a charmed existence, putting themselves daily in harm's way and emerging without a scratch. Then, on May 12, 1941, they drove out to a marsh in southeast London to dismantle a bomb, called "Old Faithful," that had lain there for months without detonating. Just as the thirty-five-year-old Suffolk was withdrawing the fuse, Old Faithful exploded, shattering windows a quarter of a mile away. The Earl of Suffolk was killed, along with his secretary, his driver, and six other members of his team. The only trace of him found in the rubble was his silver cigarette case.

Weeks later, British newspapers noted that King George VI had posthumously awarded the George Cross—the civilian equivalent of the Victoria Cross, Britain's highest military honor—to "Charles Henry George Howard, Earl of Suffolk and Berkshire, for conspicuous bravery."

CHAPTER 7

"My God, This Is a Lovely Place to Be!"

The Exhilaration of Wartime London

FOR ALL THE FEAR, TERROR, AND DESTRUCTION OF THE BLITZ, there was an excitement, a sense of energy about living in London during this period that, in the view of many who were there, would never be equaled. In the high-voltage current of war, the threat of death seemed only to heighten the exhilaration and elation of survival. "You walk through the streets . . . and everyone you pass seems to be pulsating with life," the American magazine correspondent Quentin Reynolds wrote in his diary.

Much of wartime London's zest could be credited, too, to the presence of the European exiles, who added a splash of color and life to London's bomb-blasted streets. Throughout the war, a native Londoner never knew who might be sitting next to him on the bus or tube, in a restaurant or pub. It might be a Polish pilot just returned from a bombing raid, a Norwegian seaman rescued from his torpedoed ship, a resistance fighter smuggled out of France. For the Canadian diplomat Charles Ritchie, walking in Kensington Gardens alongside the European allies, in their strikingly varied military uniforms, was like "swimming in the full tide of history."

The displaced Europeans seemed to be everywhere. General Władysław Sikorski, the leader of the Poles, conducted business at the Rubens Hotel, opposite Buckingham Palace. The Norwegian, Dutch,

*At Buckingham Palace, King George VI and Queen Elizabeth entertain exiled
European royals and government heads, including Queen Wilhelmina
(third from left), Czech president Edvard Beneš (third from right),
and King Haakon (second from right).*

and Belgian governments operated from Stratton House, across from
the Ritz on Piccadilly. Other foreign government offices were established in houses and office buildings scattered throughout the posh
neighborhoods of Belgravia, Kensington, Mayfair, and St. James's. The
Grand Duchy of Luxembourg, the postage-stamp-sized country that
adjoined Belgium and had also been invaded and occupied by Germany, kept offices at Wilton Square in Belgravia.

Late in 1940, Charles de Gaulle and his growing Free French movement moved from their meager quarters at St. Stephen's House to a
stately four-story mansion in Carlton Gardens, overlooking St. James's
Park, that once had been the residence of Lord Palmerston, one of the
most Francophobe prime ministers in British history. De Gaulle himself lodged at the luxury Connaught Hotel, courtesy of the British
government, while his wife, Yvonne, and their mentally handicapped

twelve-year-old daughter, Anne—who had been spirited out of France in July—took up residence at a spacious country house in Shropshire, out of reach of German bombs.

Queen Wilhelmina, meanwhile, lived in a small, bomb-pitted town house in Chester Square, not far from Buckingham Palace. One of the most elegant London neighborhoods before the war, the square and its Belgravia environs had been heavily bombed in the Blitz, with many of its homes now empty. Every evening, before the Luftwaffe raids began, Wilhelmina took a small suitcase stuffed with official papers to Claridge's, where she spent the night in the hotel's large, reinforced bomb shelter and returned home the next morning after breakfast. Her own house had been patched up after suffering minor bomb damage; like the others in the square, its front was shabby and badly in need of paint. Her staff urged her to have it repainted, but such nonessential jobs required a special permit unavailable to ordinary citizens and she refused to consider it. She also resisted pressure from Dutch officials to move to larger, grander quarters more fit for a head of state. "The Queen," said Dutch prime minister Pieter Gerbrandy, "always had the idea that she did not belong in a palace while her people were in such a miserable condition at home."

Wilhelmina's fellow monarch King Haakon spent his nights with his son, Crown Prince Olav, at a country house in Berkshire, some forty-five miles west of London, to which they commuted every day by car. Haakon and Olav were close to King George VI, who had been Olav's best man at his wedding in 1929, and were frequent guests at Buckingham Palace.

After his own traumatic experience in Norway, Haakon, known as "Uncle Charles" by his nephew and the rest of the British royal family, was dismayed at what he considered the palace's lax security. George assured Haakon that he need not worry. Having practiced diligently at shooting ranges at Buckingham Palace and at Windsor, the British king said, he and the queen had become quite skilled at defending themselves. George also proudly showed Haakon a room in the palace's basement that served as a jerry-rigged bomb shelter. A former sitting room for the palace's maids, the space was furnished with overstuffed

Victorian chairs and settees and equipped with pails of sand and a hand pump for putting out fires.

Haakon was decidedly not reassured. Knowing full well how skilled the Germans were at tracking down their enemies, he asked his nephew what plans were in place to evacuate the royal family from London in case of a German invasion. He was told that a handpicked group of officers and men from the Household Cavalry, an elite military unit assigned to guard the king, stood ready day and night to defend the family against surprise attacks and to whisk them away in armored cars to safety.

Still skeptical, Haakon requested a demonstration of the evacuation plan. The bemused George complied, pressing an alarm buzzer meant to alert his rescue unit. Nothing happened. When the king dispatched an equerry to find out why, the aide returned with a report that the officer of the guard had been informed by a London police sergeant, also on duty at the palace, that no attack was in progress, "as he had heard nothing of it." Under royal orders to act as if there *were* an attack, a swarm of guardsmen rushed into the palace gardens and proceeded to whack the bushes and flower beds "in the manner of beaters at a grouse shoot rather than of men engaged in the pursuit of a dangerous enemy." King George and Queen Elizabeth burst out laughing; King Haakon was horrified. As a result of this incident, security precautions at the palace were considerably tightened.*

THE LIFESTYLES OF MOST of the Europeans who fetched up in wartime London were considerably less rarefied than those of their countries' monarchs and high government officials. While the foreign leaders lunched and dined at Claridge's or the Ritz Grill, their compatriots spent much of their time in Soho, the center of London's émigré

---

* After the war, German records revealed that, in the case of an invasion of Britain, a highly trained paratroop unit of more than one hundred men had been assigned to parachute directly onto the grounds of Buckingham Palace to seize King George VI and his family. Also high on the German "capture list" were Winston Churchill and foreign leaders such as King Haakon, Queen Wilhelmina, General Sikorski, and Edvard Beneš.

life and a haven for European expatriates since the seventeenth century. Bohemian, noisy, and inexpensive, the neighborhood was filled with French, Italian, Greek, Chinese, and other ethnic restaurants favored by the exiles. The York Minster, off Old Compton Street, was one of the best-known meeting places, attracting, among others, the Free French and lower-ranking officials from the Belgian government. Old Compton Street itself, with its wide variety of butchers, greengrocers, and patisseries, was said to be "as French as the rue St. Honoré" in Paris.

By the end of 1940, well over 100,000 continental exiles—military and civilians—had taken up residence in the British capital. As the war proceeded, thousands more joined them, mostly young men who had escaped from their nations and made their way to Britain to battle on. "Everybody's goal was the same: to get to England and join the Allied forces," noted Erik Hazelhoff Roelfzema, a Dutch law student who, at the age of twenty-three, fled his country as a deckhand aboard a rusty freighter flying the Panamanian flag. "To cross over to England you had to sacrifice all you loved . . . for this one privilege: to fight the Nazis as a free man."

Among the émigrés was Josef Korbel, the head of the broadcasting department of the Czech government in exile, who, with his wife and small daughter, Madlenka, lived on the third floor of a redbrick apartment building on Kensington Park Road. More than sixty years later, Madlenka—now known to the world as Madeleine Albright, the United States' first female secretary of state—would describe in vivid detail how, as a four-year-old during the Blitz, she curled up nightly on a bunk bed in the building's basement shelter as Luftwaffe bombs exploded in streets nearby.

In the summer and fall of 1940, the German bombing raids were hardly the only difficulty with which the Korbels and other European exiles had to cope. Like the United States and a number of other nations, Britain was gripped with fears of foreign "fifth columnists" roaming its cities and countryside in preparation for a German invasion. Many Britons shared the widespread belief that Germany's stunning victories in Norway, Denmark, Belgium, the Netherlands, and France could not be explained solely by those countries' political and military weaknesses; the triumphs must also have been due to the effectiveness of Nazi agents and sympathizers in undermining them before Germany invaded—a belief that was later found to have very little basis in fact.

Nonetheless, at the time, much of the British public regarded the foreigners descending on their shores with deep suspicion, especially those from Germany, Austria, and eastern Europe. All non-Britons were ordered to register with the police and were subject to stringent restrictions on their activities, including travel and employment. Throughout 1940, more than 20,000 "enemy aliens" from Germany, Austria, and Italy—many if not most of them Jews who had fled Nazi persecution—were taken from their homes and jobs and shut up in internment camps on the Isle of Man, off the west coast of Britain.

But as the threat of German invasion faded and the public was made aware of the shabby treatment of interned foreigners, Britons became more welcoming. Most of the internees had regained their freedom by the summer of 1941. While the British government itself offered only paltry, grudging assistance to European émigrés, many private individuals pitched in to provide aid, some opening their homes as temporary shelters for the newcomers. Others helped start canteens and organize theatrical and cultural events, football matches, dances, and English classes for European troops. The English branch of the Rothschild family turned its London mansion into a club for military officers in exile, while Olwen Vaughan, the daughter of a Liverpool clergyman, opened what became the renowned Petit Club Français in the basement of a town house near Piccadilly. A former employee of the British Film Institute and a fervent Francophile, Vaughan wanted to create a space where the Free French in London "might find a little of the spirit of the country they had been forced to leave behind." Although cramped and threadbare, the club was extraordinarily popular, among not only the French but the rest of the exile community as well. It also became a magnet for British filmmakers and, after the United States entered the war, for their American film colleagues. "Its reputation was such," said one observer, "that if during the war you heard that Orson Welles, say, or Rita Hayworth was in town, the practical thing would be to try the French club first and the Savoy second."

Although the British public was hardly known for reaching out to foreigners, many Britons temporarily packed away their traditional prejudices, not only because of sympathy for the exiles' plight but because the Europeans were unofficial citizens of Britain during that perilous time, sharing in equal measure the dangers of the Blitz and the country's severe wartime privations. "Basically [the British] were no more interested in strangers than before—but here were strangers living with them in their time of trial," the CBS correspondent Eric Sevareid observed.

Before the Blitz, many émigrés, especially those from central and eastern Europe, had felt isolated and unwelcome in Britain. "We were living in a foreign country but surrounded only by Czech people, without making friends with the English except for a very few," Madeleine Albright's mother remembered. Yet despite a "degree of separation" that continued to exist between the European exiles and the British who lived in the Korbels' apartment building, the Blitz created a momentary community of spirit, an Englishwoman who had been a neighbor of the Korbels told Albright many years later. It was, the neighbor said, "a very pleasant kind of group with a friendly warmth between the two sides; the people were very supportive of one another. They used to play great bridge games [in the basement bomb shelter] and share out their supplies."

The Europeans, in turn, were impressed with the courage, resolution, resilience, and defiant fighting spirit displayed by their hosts. Enduring the nightly bombing raids of the Blitz was made easier, one exile wrote, "because of the daily example of English stoicism, English equanimity, English humour, which lay before your eyes."

When Erik Hazelhoff Roelfzema arrived in London shortly after escaping from the Netherlands in 1941, the first thing he and another young Dutch émigré did was to hop aboard one of the city's famed red double-decker buses for a quick tour of their new home. They found that most of the bus's windows were covered with an opaque glue to protect them from bomb blasts. On one window, an official notice in verse pleaded with passengers not to scratch off the glue: "I hope you'll pardon my correction / this is here for your protection." Underneath the notice, someone had written: "I thank you for the information / But

I can't see the bloody station." The two young Dutchmen, whose own country had been intimidated and terrorized by its Nazi occupiers, were beguiled by the exchange. "It was the kind of lighthearted interplay between rulers and the ruled that no Kraut would ever understand," Hazelhoff Roelfzema later wrote.

Even the cantankerous Charles de Gaulle, never an Anglophile, had considerable praise for the Britons he encountered in his first months and years in London. "Faced with the prospect of a German attack,

A rare photo of Charles de Gaulle with his wife, Yvonne, taken at the country house outside London where Madame de Gaulle lived with the couple's mentally impaired daughter during the war.

everyone demonstrated an exemplary resolution," he recalled. "It was truly wonderful to see each Englishman behaving as though the salvation of the country depended on his own conduct."

De Gaulle, however, was far less happy with the British government and what he considered its misguided efforts to establish him as a recognizable figure in Britain. With Churchill's approval, government officials had hired Richmond Temple, a public relations consultant, to publicize the general and his cause—an initiative that infuriated de Gaulle. "I do not want to be made a film star by the press," he growled, adding that Churchill seemed to want to sell him "like a new brand of soap."

De Gaulle was particularly intent on shielding his wife and young daughter, who had Down syndrome, from the prying eyes of reporters and photographers. By all accounts, he was devoted to his daughter, lavishing on her a gentleness and tenderness that almost no one else around him ever saw or experienced. "Without Anne, perhaps I should never have done all that I have done," he later said. "She gave me so much heart and spirit."

As it turned out, the government's campaign to promote de Gaulle proved unnecessary: the British were already captivated by this lone Frenchman who had refused to acknowledge defeat and who had come to join their seemingly quixotic fight against Hitler. "We in this country always have had an affection for lost causes, and in the early days of the war, the cause of France did in fact seem lost," said Harold Nicolson, a member of Parliament and noted writer. "De Gaulle inspired us with a glow of wonder that he should be so positive that he could lead his people out of the abyss by the force of his dreams and theirs."

During his daily walks to and from Free French headquarters in Carlton Gardens, the towering Frenchman was greeted warmly by the many Londoners who recognized him. "The generous kindness which English people everywhere showed towards us was truly unimaginable," de Gaulle remembered. "Countless people came to offer their services, their time and their money. . . . When the London papers announced that Vichy was condemning me to death and confiscating my property, many jewels were left at Carlton Gardens and dozens of unknown widows sent their gold wedding rings" to help finance his movement.

No event illustrated the affinity between de Gaulle and the British people more clearly than London's commemoration of Bastille Day on July 14, 1940, less than a month after de Gaulle first arrived in the British capital. On that warm afternoon, he and the handful of Free French troops he had recruited thus far—two lines of sailors, airmen, and soldiers stretching less than a city block—marched proudly down Whitehall to the Cenotaph, a stone monument honoring those who have died in Britain's wars.

Eric Sevareid was among the scores of onlookers who far outnumbered those on parade. "I had been a spectator at a hundred displays of military might," Sevareid wrote, "and they were all essentially the same: merely a perfunctory display of organized, faceless bodies." At this one, however, "I was aware of every single face," especially that of de Gaulle, "who strode stiffly along the ranks, never opening his tightly compressed lips, glaring almost, into every pair of rigid eyes. He had the portentous air of a general surveying a great army."

For Sevareid and the other bystanders, many of whom joined the French in a rousing rendition of "La Marseillaise," the effect of the meager parade was heroic rather than comical. "You had the impulse to remove your hat and stand rigidly at attention yourself," Sevareid noted. "Every [Frenchman] there bore the conviction of consequence. There was a sense of strength in this handful that I had never felt in a demonstration by a hundred times their number."

AS WELCOMING AS THE British were to the exiles as a whole, they saved their greatest enthusiasm and affection for young Europeans in the military, particularly those wearing RAF wings. In London and other cities, bus conductors refused to take their fares, waiters would not let them pay for meals, and pub patrons bought them all the beer they could drink.

British newspapers and the BBC were assiduous in highlighting the contributions of exiled pilots to the victory in the Battle of Britain, as

Some 303 Squadron pilots with their "mother," socialite Jean Smith-Bingham.

well as to later British air campaigns. Not surprisingly, Polish fliers, especially those in 303 Squadron, attracted the most attention. "The Poles flying in the RAF are becoming the legendary heroes of this war," the *New York Times* declared in June 1941. "[They] not only are appreciated; they are pretty close to being adored." Quentin Reynolds agreed, writing in *Collier's,* "The Polish aviators are the real Glamour Boys of England now."

Among the British upper class, it became the height of chic to have a Pole or two at cocktail parties and formal dinners. In February 1941, a leading socialite, Lady Jean Smith-Bingham, "adopted" the 303 Squadron and threw a splashy dinner dance in its honor at the Dorchester

Hotel. Lavishly chronicling the affair in a two-page spread, *The Tatler,* a British society magazine, pronounced it "one of the gayest and most amusing that London has seen for many months." Smith-Bingham's social coup started a trend: within days, other women in the upper reaches of London society hastened to adopt Polish squadrons of their own.

Tadeusz Andersz, a fighter pilot attached to the recently formed Polish 315 Squadron, was at a party in London one night when an attractive blonde asked if his squadron had a mother yet. She turned out to be Virginia Cherrill, a movie actress who, among other starring roles, had appeared opposite Charlie Chaplin in *City Lights.* Once married to Cary Grant, Cherrill was now the wife of the 9th Earl of Jersey. When Andersz informed her that the squadron was still an orphan, Lady Jersey asked if she might have the honor of adopting it.

After getting permission, she threw party after party in the airmen's honor, sometimes at her London town house, sometimes at her country estate. (She always took care, Andersz recalled, "never to invite girls prettier than herself.") She attended the squadron's Christmas dinners, mailed parcels to captured Polish pilots in German stalags, and gave the others her old silk stockings to wrap around their knees to keep them warm during high-flying missions. Once when a journalist was interviewing her at her country house, they heard the roar of an airplane overhead. Looking out the window, they saw a Spitfire circling low over the house. "That is one of my Poles," Lady Jersey proudly proclaimed. "I'm their mother."

All over England, other citadels of the British class system were falling to the Poles. One 303 Squadron pilot, shot down during the Battle of Britain, parachuted onto an exclusive golf course, landing near the eighth tee. The men playing the hole insisted on carting the dazed flier off to the clubhouse for drinks. Another parachuting pilot drifted into a copse near a private tennis club in the London suburbs. Three club members observed his descent as they awaited the arrival of a fourth for their weekly doubles match. They helped extricate the Pole from the trees and, giving up on their expected fourth, asked if he played. When the young pilot said he did, he was dressed in borrowed white flannels and was soon on the court, borrowed racket in hand.

Overwhelmed by the effusive welcome, one delighted Polish pilot declared about London, "My God, this is a lovely place to be!" Other foreign troops felt the same. Throughout the war, European pilots from nearby air bases and soldiers on leave from more distant posts such as Tobruk and Tripoli swarmed into the British capital, seeking relaxation, camaraderie, excitement, and romance.

"No matter our varied origins and uncertain futures, we stood shoulder to shoulder, even if only for beer," recalled Erik Hazelhoff Roelfzema, who had a busy war after his escape from the Netherlands—first as a Dutch intelligence agent, then as an RAF pilot, and finally as a military aide to Queen Wilhelmina. "We drank together, took our girls to the same nightclubs—the Suivi, the Embassy Club, the 400. Norwegians, Hollanders, Poles, French, English, all were there—everyone packed together on those tiny dance floors." Holding each other tightly, couples swayed to such wistful hits of the day as "A Nightingale Sang in Berkeley Square," "I've Got You Under My Skin," and "I'll Be Seeing You." Like the cigarette smoke curling up to the nightclub ceilings, romance and sex hung thick in the air, and conventional morality was put away for the duration. About wartime London, the novelist Elizabeth Bowen wrote, "There was a diffused gallantry in the atmosphere . . . an unmarriedness. It came to be rumored about the country that everybody in London was in love."

Of all the Europeans, the Free French and the Poles had the greatest success in winning the company and affections of British women, who were captivated by their dash, daring, and ebullient joie de vivre. Winston Churchill's youngest daughter, Mary, and the novelist Nancy Mitford were among the many smitten by, in Mitford's words, "the Free Frogs." During the war, Nancy, who was the eldest of the famous Mitford sisters, had flings with three Frenchmen, falling deeply in love with one of them—Gaston Palewski, de Gaulle's witty, worldly, incorrigibly womanizing chief of staff. Their tempestuous, on-again, off-again relationship lasted until shortly before Mitford's death in 1973.

But it was the Poles, with their hand kissing and penchant for sending flowers, who won the greatest reputations as gallants. John Colville, one of Churchill's private secretaries, once asked a woman friend, the daughter of an earl, what it was like to serve as a WAAF driver for Polish officers. "Well," she replied, "I have to say 'Yes, sir' all day, and 'No, sir,' all night." The head of a British girls' school made headlines when she admonished the graduating class about the pitfalls of life in the outside world, ending her speech with "And remember, keep away from gin and Polish airmen."

Such admonitions were widely ignored; indeed, many young women became the pursuers of the Poles as well as the pursued. And that was true not only of the British. At a London cocktail party in her honor, Martha Gellhorn, Ernest Hemingway's talented and beautiful American wife, then covering the war for *Collier's,* ignored the rest of the guests and "devoted her entire attention to a couple of Polish pilots."

In contemporary diaries and letters and in later recollections, a number of the Poles described their wartime romances with some amazement. "As for the women," a Polish pilot wrote in his diary, "one just cannot shake them off." When he was in his eighties, one of 303 Squadron's ace fliers recalled those days with a chuckle: "I think English women should have some monument, some big monument. They were wonderful to us."

YET AS ENJOYABLE AS the Poles' experiences with British women—and the British public as a whole—were, they came at an emotional cost. The Poles (and many Britons as well) were troubled to discover that the country's upper crust—with their nation at war, their skies and cities regularly filled with fire and death—could act as if nothing had changed. Like other European exiles, the Poles had witnessed the devastation of their own country and were acutely aware that the families and friends they had been forced to leave behind were now living under German (and in the Poles' case, also Soviet) occupation. Yet here they were in England, feted and pampered by the upper echelons of society, glamorized in movies and the press, often enjoying themselves—and just as often feeling guilty for doing so.

Among many Poles, there was also the feeling that the British, as kind and friendly as most of them were, had no real understanding of them and their country. In their coverage of the Polish pilots' exploits, for example, the British press had perpetuated what the Poles considered to be inaccurate and condescending stereotypes of themselves and Poland. In many of the stories, Poles were portrayed as amusing foreigners with a funny way of speaking or as savage romantics who lived only to kill Germans. (One headline managed to include both clichés: BOMBING REICH THRILLS POLES—WE GO TONIGHT, YES?)

The Polish author Arkady Fiedler, whose book about 303 Squadron was published in Britain in 1943, observed that the best thank-you gift that Britain could give the squadron for its invaluable service would be "to get to know the Poles better. To know them honestly, intimately, through and through, putting aside prejudices and preconceptions, to know the Poles as they really are."

Other Europeans, seeking to integrate as much as possible into British society, expressed the same desire. But for many, the barriers between themselves and the reserved, restrained British were too great to surmount. "As Great Britain had always described itself as 'of Europe' but not 'in Europe,' so we were well aware that we were 'in England' but not 'of England,'" the Austrian novelist Hilde Spiel later wrote in an essay about her wartime experience in London.

Anxious to "embrace all things English," Spiel and her husband, the German author and journalist Peter de Mendelssohn, were constantly reminded in subtle ways that they did not quite belong. When they visited Mendelssohn's British publisher and his wife, they were taught, in Spiel's words, "basic principles of English life" that seemed so antithetical to their own way of living: "don't fuss; don't ask personal questions, don't touch the teapot (this was reserved for the hostess)," and, above all, adopt "understatement and a stiff upper lip."

Spiel received a similar lesson when she and her small daughter were evacuated from London during the height of the Blitz and took refuge with a family in Oxford. During lunch one Sunday, her hosts' twelve-year-old son was told by a neighbor that his beloved dog had just been hit by a car and killed. When the boy burst into tears, his parents glared at him, and his mother ordered him to control himself. At that, Spiel broke into tears herself and rushed from the table. "I cried about [the dog], about the suppressed feelings of a child, and out of homesickness for [Austria], a country in which one could sob unrestrainedly when a sorrow befell one," she wrote. When she left the room, she remembered, "no one looked at me, no one uttered a word."

For Spiel and the tens of thousands who joined her in wartime exile in Britain, that pain and that yearning for lost countries and lost lives were ever-present aches that no amount of English hospitality, however well meaning, could heal.

CHAPTER 8

"This Is London Calling"

The BBC Brings Hope
to Occupied Europe

Soon after the end of World War II, a group of Russian historians asked Field Marshal Gerd von Rundstedt, the commander of German forces on the western front, which battle he thought had been the most decisive of the war. Without hesitation, he answered: the Battle of Britain. If the Luftwaffe had crushed the RAF, he declared, Germany would have gone on to invade and defeat first the British and then the Russians. Only the United States, still militarily weak at the time, would have stood in Hitler's way.

In the short run, Britain's victory in the air had another powerful effect: it helped spark a psychological revolution among the benumbed inhabitants of occupied Europe. "Nobody ever imagined that you British could offer the magnificent resistance you are putting up against Germany," a resident of Marseille wrote to the BBC in a letter smuggled out of France in late 1940. "It makes one think that things might have been different last June if our rulers could have had this feeling."

Britain's stubborn defiance of Germany, as relayed to the Continent by the BBC's new European Service, raised the spirits of millions of people who had been traumatized by the shock, humiliation, and terror of Nazi occupation. "People of France, stop thinking that the world is with you!" German radio had broadcast to the French the day before the armistice was signed. "It is overwhelmingly ours!" The realization that such boasting was premature—that Germany might not be omnipotent after all—helped the French and others in Europe begin to shed their feelings of isolation, fatalism, helplessness, and despair. Thanks to the British and the BBC, they could "escape for a few minutes into an anti-Nazi thought world" and believe there might indeed be hope for the future. "In a world filled with poison, the BBC became the great antiseptic," noted Léon Blum, the former premier of France, who was handed over to the Germans by the Vichy government and sent to Buchenwald in 1943.

In a letter spirited out of Czechoslovakia, a girl wrote, "People who are almost too poor to buy bread have now a radio. They need it. A man told me, 'The stomach is hungry but the soul still more so. London is the only thing to feed the soul.'" A French journalist reported, "The initials BBC have quickly become part and parcel of the daily vocabulary of French citizens."

ONLY A YEAR BEFORE the Battle of Britain, the idea of the British Broadcasting Corporation serving as a beacon of liberty for the rest of Europe would have been laughable. For one thing, it had strongly supported the Chamberlain government's decisions to appease Hitler and to refrain from military commitments to other countries threatened by Germany. Founded in 1922 but already the world's oldest national broadcasting service, the BBC was a curious amalgam: although it received government funding and was ultimately answerable to Parliament, it was supposed to have editorial independence. Sir John Reith, a tall, craggy-faced Scot who was its first director general, viewed its charter differently. "Assuming that the BBC is for the people, and the Government is for the people, it follows that the BBC must be for the government," he declared. Under Reith, the BBC squelched news that Neville Chamberlain found unpalatable and relied almost entirely on official sources for its news broadcasts. It provided no analysis, no context, and no alternative points of view.

When the government pressed newspapers and the BBC to go easy on Hitler and Germany, declaring that controversial foreign policy issues should not be publicly discussed, the BBC complied. Urged by Chamberlain and his men to downplay Hitler's pressure on Czechoslovakia in 1938, the BBC minimized both Britain's unpreparedness for war and the magnitude of the threat facing the Czechs. In the aftermath of the Munich crisis, a high-level BBC official wrote a confidential memo to his superiors accusing them of embarking on a "conspiracy of silence."

Those who disagreed with appeasement, like Winston Churchill and a few other members of Parliament, were largely banned from BBC programs. In early September 1938, one of those MPs, Harold Nicolson, was told he could not broadcast a speech he had prepared because it urged British support of Czechoslovakia. In its place, the "very angry" Nicolson was forced to deliver an "innocuous" talk, but even then a radio engineer was standing by, ready to interrupt the broadcast if he so much as mentioned Czechoslovakia.

During the bungled British expedition to Norway, the BBC, having no foreign correspondents of its own, relied on government handouts that reported an unending string of successes for British troops. Leland Stowe, an American journalist covering the Norway campaign, was astonished when he heard a BBC announcer declare one night that "British expeditionary forces are pressing forward steadily from all points where they have landed in Norway. Resistance has been shattered . . . and the British and French are advancing successfully." An American photographer traveling with Stowe stared at him in amazement. "Christ, what's the matter with those mugs?" the photographer exclaimed. "Are they crazy?"

With Reith at the helm, the BBC was as out of touch with the lives of ordinary Britons as it was with events in war-battered Europe. Before the war began, it was, as one employee remembered, "an agreeable, comfortable, cultured, leisured place, remote from the world of business and struggle." To set that highbrow tone, BBC announcers, whose voices tended to have "an exquisitely bored, impeccably impeccable Oxford accent," were instructed to wear dinner jackets while at the microphone.

In a meeting with Reith in the late 1930s, the CBS broadcaster Edward R. Murrow, who had just arrived in London, made clear that he and his American network had no intention of adopting the BBC's nose-in-the-air attitude. "I want our programs to be anything but intellectual," he said. "I want them to be down to earth and comprehensible to the man in the street." With a dismissive wave of his hand, Reith replied, "Then you will drag radio down to the level of the Hyde Park Speakers' Corner." Murrow nodded: "Exactly."

The BBC's ivory-tower attitude changed soon after Britain declared war against Germany. The most immediate differences were physical: Sandbags were heaped high around Broadcasting House, the BBC headquarters, with rifle-toting sentries guarding its massive bronze front doors. The building's graceful art deco interiors were divided by steel partitions and gas-tight doors, while its trompe l'oeil murals were covered by heavy soundproofing. The seats of the concert hall were ripped out to create a giant employee dormitory, with mattresses lining the stage and floor. And its announcers no longer wore black tie to read the news. "This," remarked a BBC writer, "was the end of the boiled-shirt tradition in radio."

But dramatic as the physical shifts were, the transformation in the BBC's attitude and style was nothing less than revolutionary. After Reith was named head of the government's new Ministry of Information in early 1940, the network began an extraordinary metamorphosis that by war's end would weave it into the fabric of everyday British life, as well as make it the most trusted news source in the world.

In the newsroom, a throng of new producers and editors was hired, many of them former print reporters who brought with them a burst of energy and journalistic fervor. R. T. Clark, a classics scholar and former editorial writer for the *Manchester Guardian,* was put in charge of the domestic news service. Soon after he came on board, Clark, a cigarette dangling from his lips, signaled a seismic shift in the BBC's news policy when he announced to his staff, "Well, brothers, now that war's come, your job is to tell the truth. And if you aren't sure it is the truth, don't use it." In an internal memo, he wrote, "It seems to me that the only way to strengthen the morale of the people whose morale is worth strengthening, is to tell them the truth, and nothing but the truth, even if the truth is horrible."

Clark's philosophy was cheered not only by the new hires but also by a group of young staffers already on the payroll, dubbed the "war-mongers," who had been openly critical of their employer's manipulation of the news and its refusal to allow critics of the Chamberlain government to broadcast. Arguably the most outspoken warmonger was a tall, dark-haired dynamo named Noel Newsome, who, as head of the network's new European Service, was, in the words of the BBC historian Asa Briggs, "one of the most industrious, lively, and imaginative" of all the BBC's wartime personnel.

The son of a country doctor from Somerset, the thirty-four-year-old Newsome had studied at Oxford, where he'd received a first-class degree in modern history. After several years of working as a subeditor for the *Daily Telegraph* and foreign correspondent for the *Daily Mail,* he became foreign editor of the *Telegraph*—and immediately made clear his opposition to Chamberlain's appeasement of Germany.

Hired by the BBC two days before the war began, Newsome objected to the government's position that radio news should be used as an instrument of propaganda, the main function of which was to promote the official British point of view. Like R. T. Clark, Newsome argued for accurate reporting and vehemently protested when the government misinformed the BBC and other media about what was happening in Norway. "I cannot but resent most strongly that we were used as a blind tool," he remarked.

While urging the BBC and the government to give editorial independence to the European Service, Newsome instructed his staffers to act as if they already had it. "Noel Newsome set the style and called the tune," noted Alan Bullock, Newsome's wartime assistant, who went on to become one of Britain's most eminent postwar historians. Bullock quoted Newsome as saying, "What we have to do in this period of the war, when [Britain is] on the defensive, is to establish our credibility. If there's a disaster, we broadcast it before the Germans claim it. . . . And when the tide turns and the victories are ours, we'll be believed."

Most British policy makers, however, thought that the BBC should act as the voice of the government. In 1935, the Committee of Imperial Defence, a high-level Whitehall group that coordinated defense strategy, declared that in time of war, the government must take "effective control of broadcasting and the BBC." Curiously, one of the strongest advocates of that position was Winston Churchill himself. Even though, as a foe of appeasement, he had been kept off the air in prewar days, he became one of the most enthusiastic supporters of using the BBC for propaganda purposes when he reentered the government as Chamberlain's first lord of the admiralty in September 1939. In fact, as first lord, he was directly responsible for the falsely optimistic reports about the Norwegian campaign issued to the newspapers and BBC.

When he became prime minister, Churchill continued to oppose the idea of independence for the network. Early in his premiership, he told a subordinate that the BBC "was the enemy within the gates, doing more harm than good." On another occasion, he referred to it as "one of the major neutrals." Yet for all his grumbling, Churchill ultimately decided against government control, thanks in large part to the influence of Brendan Bracken, the prime minister's closest political adviser, who was appointed minister of information in July 1941. With Bracken, a longtime supporter of BBC independence, in overall charge, Whitehall held only a loose rein on the network. Two advisers were appointed to oversee all entertainment and news programs, but their supervision was relatively light-handed.

In the months and years to come, there would be spirited battles, to be sure, among BBC journalists, the Ministry of Information, and the Political Warfare Executive, Britain's wartime propaganda agency. Overall, though, the BBC succeeded in holding the government at arm's length for the rest of the war.

WITHIN THE NETWORK'S EUROPEAN Service, the spirit of innovation and excitement was palpable. Virtually everyone working there was a newcomer to broadcasting, caught up in this grand experiment to bring truth and hope to millions of people under Nazi domination. Britons rubbed shoulders with dispossessed Europeans. Journalists, novelists, and poets worked with actors, college professors, businessmen, philosophers, former military officers—all thrust into a world they never could have imagined in their prewar days. Alan Bullock, an Oxford graduate who had worked as a research assistant for Churchill before going to the BBC, remembered it as the time of his life. Working in the European Service, he added, was like "being a historian, living through history, in history."

When Bullock came aboard, the European Service was just two years old. Until 1938, the BBC, reflecting Britain's insularity from Europe and the rest of the world, broadcast only in English. When it began its embryonic transmissions to Europe in September 1938, its first broadcast—in French, German, and Italian—was, ironically, the text of Neville Chamberlain's speech in which he expressed horror over the thought of Britain going to war to defend Czechoslovakia.

When war broke out, the BBC's foreign operation was still relatively small, with broadcasts in only seven languages. In just a few months, it exploded to forty-five languages, half of them beamed to Europe. The larger language sections, such as French and German, broadcast up to five hours a day, including talks by and interviews of exiled heads of state and other prominent figures. But for all the sections, news was the centerpiece of the broadcasts. It was "the rock," Alan Bullock recalled. "When people are listening to you with very considerable danger and difficulty, news is what they want."

Working up to sixteen hours a day, the maverick spirits in the European Service were waging a war in which, as one observer put it, "their only weapons were wit, intelligence and a passionate conviction that they were going to win." And for two long years, they did so under the chaotic conditions of the Blitz.

Until the end of 1940, the European Service was based at Broadcasting House, which proved to be a prime landmark—and target—for the Luftwaffe's attacks on London. In mid-October 1940, a bomb crashed into the BBC headquarters, destroying the music library and several studios and killing seven staffers. Less than two months later, it was hit again. With the exception of the domestic news service, which took up space in Broadcasting House's subbasement, the BBC's major departments were evacuated to buildings in other parts of London and the country.

The European Service first took shelter in an abandoned ice rink, complete with glass roof, in the west London neighborhood of Maida Vale. Horrified at the thought of being under all that glass during a bombing raid, Noel Newsome noted that "we were packed like cattle" into tiny makeshift offices in a place "which might at any moment become a slaughterhouse." To his great relief, he and his colleagues were moved three months later to Bush House, a white sandstone behemoth near London's financial center, which would remain home to the European Service—and, from 1958, all of the BBC World Service—until 2012.[*]

Noted for its impressive exterior of pillars and arches, its soaring central hall, its marble staircases and bronze-doored elevators, Bush House had been the world's most expensive office building when it opened in the 1920s. Sadly, BBC staffers had no opportunity to enjoy the spaciousness and elegant art deco touches of its upper floors. Instead, because of the ever-present threat of bombing, they were jammed into its cramped basement, with its rabbit warren of corridors and minuscule offices.

The stuffy makeshift studios, usually shrouded in cigarette smoke, were also tiny. To improve the acoustics, canvas-covered screens were hung from the ceiling, and an oil lamp was placed by the door in case a bomb knocked out the electricity and extinguished the lights. Once, when General Bernard Montgomery was escorted into the bowels of Bush House for a broadcast, he stared in astonishment at the scruffy surroundings, then asked the BBC staffer accompanying him, "People don't work down here all the time, do they?"

Indeed they did—and most seemed to enjoy the seething, more than slightly chaotic atmosphere, which George Orwell, who worked there during the war, described as "halfway between a girls' school and a lunatic asylum." So many staffers and guest broadcasters rushed in and out that it was hard to keep track of all of them. When King Haakon arrived one day for a broadcast, the harassed receptionist asked him: "Sorry, dear—where did you say you were king of?"

Labeling themselves "the Bushmen," those working for the European Service considered themselves a breed apart from the BBC's domestic staffers, who "seemed to be awfully stuffy and odd." "It's curious how independent of the mother house we were," said the veteran BBC producer and executive Robin Scott, who got his start in the European Service. "I felt that they weren't nearly as clever at the broadcasting

---

[*] Two months after the European Service transferred to Bush House, the Maida Vale ice rink suffered a direct hit and was heavily damaged.

game as the people in Bush House, who were not only winning the war but shaping the future of broadcasting."

Each country section had its own editor and staff, usually a mix of Britons and natives of the country involved. British nationals wrote the news items and served as language supervisors to ensure that translations of the news copy were accurate. The translators and announcers, in turn, almost always came from the countries to which they were broadcasting. To make sure that a broadcaster didn't stray from the written text, the language supervisor sat in the control room with his finger on a switch that would shut off transmission—just in case, as Alan Bullock facetiously said, some announcer "suddenly shouted hurrah for Hitler."

The switch was almost never used. Nonetheless, prebroadcast conflicts often erupted between independent-minded foreign staffers and their equally forceful British supervisors, usually over the content of talks and other nonnews programs. Some of the exiled governments were allowed to develop their own programs, but they had to submit their scripts in advance. "We were very, very careful about what they were saying," Bullock noted. "What we imposed was quite different from the normal censorship [for security]. . . . We exercised a political censorship over them."

Resenting such control, some European staffers were scornful of the British, whom they thought of as "ignorant, knowing nothing about Europe," Bullock added. "But in the end, all the rows and arguments died away. They were insignificant compared to the fact that everyone shared a common purpose and that London went on broadcasting."

FOR THE HEADS OF STATE and other prominent figures of occupied Europe, the BBC provided an invaluable opportunity to reconnect with their countrymen, whose confidence in their leaders had, in some cases, been badly shaken by their abrupt departure after German invasion. That was certainly true for King Haakon, who took to the air in July 1940 when a group of Norwegian members of Parliament, pressured hard by their German occupiers, demanded that Haakon abdicate and hand over power to a German-controlled governing body. In a BBC broadcast to Norway, the king rejected that demand as resolutely as he had Hitler's ultimatum in April to cede power to Vidkun Quisling.

Assuring his people that he and his government had escaped not from fear or cowardice but to continue the struggle, Haakon declared in a deep, calm voice tinged only slightly with emotion, "The liberty and independence of the Norwegian people are to me the first commandment of the Constitution. . . . I consider I am obeying this commandment by remaining king. I say today, as I shall say all my life: All for Norway!"

The king's broadcast was heard by tens of thousands of Norwegians, and copies of his speech were surreptitiously circulated to those with no access to a radio. Faced with Haakon's refusal to abdicate, the Norwegian parliamentarians broke off negotiations with the Germans, and the king once again became the focal point of Norwegian defiance. Soon after his broadcast, the country's infant resistance movement adopted Haakon's monogram as its symbol for fighting back against the Germans. "H7" (for Haakon VII) was scribbled everywhere—on the walls of government buildings, school entrances, barns, and jail cells, even on the sheer faces of mountain cliffs. An underground chain letter, entitled "The Ten Commandments of Norwegians" and widely circulated through the country, had as its first commandment "Thou shalt obey King Haakon whom thyself has elected."

Thanks to the BBC, Queen Wilhelmina of the Netherlands had a similar influence. Erik Hazelhoff Roelfzema, the young Dutch law student, spoke for many of his countrymen when he recalled that Wilhelmina's pervasive presence "had been part of our world since the day we were born." Nonetheless, when she escaped to London in May 1940, Roelfzema's first reaction was fury at her "leaving us in the lurch. To hell with her!" He changed his mind soon afterward when he started listening to her broadcasts on Radio Orange, the BBC's Dutch program: "The Queen had been right, obviously, to leave the country when she did, in the nick of time. We recognized the wisdom of her move and regretted our violent first reaction to the news."

Like Haakon, Wilhelmina swiftly emerged as the soul of her country's resistance. But while the Norwegian king's BBC speeches were grave and dignified, Wilhelmina's were fiery, passionate, and above all, intensely human. The Dutch found it hard to believe that this was the same distant, forbidding woman who had ruled them for more than forty years. In her first broadcast, made from Buckingham Palace on the day after she arrived in England, Wilhelmina made it clear that she would never compromise with Hitler, whom she called "the arch-enemy of mankind," not to mention his "gang of criminals" or the "scoundrel" Dutchmen who collaborated with the Germans.

"Her speeches were highlights in our lives, especially when she attacked the Germans and the Dutch Nazis," recalled Henri van der Zee, a Dutch writer. A joke made the rounds in Holland that Wilhelmina's young granddaughters were forbidden to listen to her on the radio because she used such foul language when she talked about the Nazis. When German authorities confiscated the queen's palaces and other possessions in Holland in retaliation for her anti-Nazi attacks, Wilhelmina, in her next broadcast, vented her wrath in "amazingly heated swear words," according to German translators.

In a gesture of solidarity with the House of Orange, orange became the predominant national color in Holland—for clothing, flags, posters, even flowers (marigolds were ubiquitous). On June 29, 1940, the birthday of the queen's son-in-law, Prince Bernhard, thousands of residents of The Hague braved the Nazis' wrath by signing a message of congratulations to the prince and laying flowers at the statue of William of Orange in front of the queen's palace.

For occupied Czechoslovakia, the star BBC broadcaster was not its former head of state, Edvard Beneš, a chilly, austere former philosophy professor. It was fifty-four-year-old Jan Masaryk, the country's future foreign minister, who, of all the European exiles in London during the war, knew Britain best: he had spent twelve years there as head of his country's diplomatic mission before Hitler conquered Czechoslovakia. Tomáš Masaryk, Jan's father and the founder and first president of Czechoslovakia, had literally put the nation on the map in 1918, and it had been Jan's job in the 1930s and 1940s to keep it in the public eye, to establish that it was "a country," as he wryly put it, "and not a contagious disease."

Jan Masaryk

Before and during World War II, the tall, balding, nattily dressed Masaryk was one of the most popular diplomats in London. He charmed and cajoled journalists, socialites, and Foreign Office higher-ups alike. As a young man, he had spent several years in the United States, been briefly married to the daughter of a wealthy Chicago manufacturer, and spoke English with an American accent. A gifted pianist and raconteur, he was warm, witty, irreverent, and seemingly irresistible to women. "Jan," said a British friend, "had only to enter a room—and society fell at his feet." Underneath that urbane, lighthearted exterior, however, was a serious, sensitive, and highly gifted statesman—and broadcaster. Indeed, he was judged by many to be, with the exception of Churchill, the most effective Allied broadcaster of the war.

Masaryk's first broadcast to his countrymen began with these combative words: "The hour of retribution is here. The struggle to exterminate the Nazis has begun. By the name which I bear, I solemnly declare to you that we shall win the fight and that truth will prevail." Using simple, homely language that even the most uneducated peasant could understand, his subsequent broadcasts employed the same mix of hope, inspiration, and pugnacity. When referring to individual Nazi leaders, Masaryk was not averse to using crudity. Once, after Joseph Goebbels, the Reich's minister of propaganda, visited a theater in Prague, Masaryk said he hoped "one of the old servants at the National Theater had lit a scented candle in Goebbels' box to fumigate the place after he left."

Masaryk's nickname in Czechoslovakia was "Honza," which was also the name of the main character in a well-known Czech fairy tale called "The Tale of Honza." Soon after he began his wildly popular broadcasts, posters appeared on walls throughout Prague with the announcement "Hear *The Tale of Honza* tonight at 9:30." It took the Germans several months to figure out the posters' real meaning.

In 1938, Masaryk had been horrified by the British betrayal of Czechoslovakia at Munich, telling Neville Chamberlain and Foreign Secretary Lord Halifax, "If you have sacrificed my nation to preserve the peace of the world, I will be the first to applaud you. But if not, gentlemen, God help your souls." Yet despite that betrayal, Masaryk never lost his love for Britain. In his transmissions to Czechoslovakia, he repeatedly paid tribute to the courage, determination, and high ideals of the British people, telling his compatriots in one broadcast that "the English know anxiety but they don't know fear."

ALTHOUGH GERMANY BANNED LISTENING to BBC broadcasts in every country they occupied, they were relatively lenient, at least during the first months of the war, in the punishments handed out to transgressors in the western European nations—Norway, Belgium, the Netherlands, France, and Luxembourg. But when it became clear that listening to the BBC had become a national pastime in all those countries, the Germans began to crack down, outlawing the sale and use of radios and imposing increasingly heavy fines and jail sentences on those caught with them.

In the eastern European countries of Czechoslovakia and Poland, the penalties for that offense and others were far more extreme. Like the inhabitants of other Slavic nations, Poles and Czechs were regarded by the Nazis as *Untermenschen,* subhumans, who, in Hitler's view, occupied land meant for the expansion of the Aryan "master race." In both nations, listening to the BBC was punishable by death, although in Czechoslovakia, enforcement of the ban was somewhat hit-or-miss, at least early in the war. Not so in Poland, Germany's centuries-old enemy, where the campaign of terror was far more severe. None of the other occupied countries, in the words of the historian John Lukacs, "was handled [by the Germans] with such a frightful mixture of brutality, torture and truly inhuman contempt."

For the Third Reich, the eradication of the Poles and their nation, the destruction of their culture and identity, were primary goals. As part of that effort—which included the closing of all Polish schools, universities, and libraries—the country's broadcasting system was shut down and all radios were ordered confiscated. A few thousand remained hidden, however, most of them in the hands of members of Poland's widespread resistance movement, who listened to the BBC knowing they faced immediate execution if discovered. Their focus of attention was the news programs, which they transcribed and then printed in the more than 150 underground newspapers that circulated throughout the country.

Knowing that many of its listeners were risking their lives by merely turning on their radios was, for the BBC's European Service, a major challenge and heavy responsibility throughout the war. In 1940, however, its most immediate dilemma was the question of how to deal with the crazy-quilt, divided nature of broken, humiliated France.

CHAPTER 9

"An Avalanche of Vs"

The First Spark of
European Resistance

O N JUNE 18, 1940, A YOUNG BBC PRODUCER WAS TOLD TO MAKE arrangements for a broadcast to France that night by a general who had just arrived in London. For the producer, it was a ho-hum moment. "There was no great excitement over another French general, who also happened to be the under secretary for national defense," he later recalled. "Most people in Britain hadn't the faintest idea who the under secretary for national defense in France was." After meeting the chain-smoking Frenchman at the doors of Broadcasting House, the producer escorted him to the studio, both of them unaware of the imbroglio that had erupted in the British government's highest levels over the idea of Charles de Gaulle broadcasting to France.

The day before, Marshal Pétain had announced to his countrymen his plan to seek an armistice with Germany. De Gaulle, who had only just arrived in London, asked Winston Churchill if he could use the BBC to challenge Pétain's submission to the enemy. The prime minister immediately agreed—a decision that threw his War Cabinet into a tailspin. During a meeting at which Churchill was not present, members of the Cabinet concluded that "it was undesirable that General de Gaulle, [being] persona non grata to the present French government, should broadcast at the present time." When he learned about this, Churchill dispatched General Edward Spears to cajole his ministers into changing their minds.

As the War Cabinet feared, de Gaulle's speech was nothing less than a repudiation of Pétain's government and a call to rebellion. "I, General de Gaulle, now in London, appeal to all French officers and men who are at present on British soil, or may be in the future . . . to get in touch with me," he declared. "Whatever happens, the flame of French resistance must not and shall not go out." Later, de Gaulle would regard that broadcast as one of the most fateful moments of his life. "As the irrevocable words flew out upon their way," he wrote in his memoirs, "I felt within myself a life coming to an end." Speaking on the BBC, he remarked, had provided him with "a powerful means of war."

In retrospect, that turned out to be true. Yet at the time, almost no one else shared de Gaulle's belief in the broadcast's importance. Very few Frenchmen heard him that night. And, for the most part, those who did listen had no idea who he was—and no interest in following him anywhere.

Under the armistice, France was split into two zones—the German-occupied north, which included Paris, and the south, governed from Vichy by Pétain and his men. Most of the French, regardless of which part of the country in which they lived, revered the aged Pétain and looked to him as their leader, whose wisdom and firm direction would help heal the trauma of the country's collapse. "The Marshal's authority was accepted by all with more than resignation," noted the French historian Henri Michel. "He offered consolation and hope."

Despite his initial lack of an audience, de Gaulle continued broadcasting over the BBC. From July 1940 onward, he and his Free French movement were given five minutes of airtime each evening for a segment called *Honneur et Patrie*. Over the next couple of years, the general would indeed begin to rouse many in his country to resistance and rebellion. But in 1940, the BBC's most urgent task was to encourage and give hope to deeply demoralized France. Thanks to a daily half-hour program called *Les Français Parlent aux Français,* both goals were abundantly met.

Of all the BBC programs beamed to occupied Europe, *Les Français Parlent aux Français* was the most skillfully produced and performed and by far the most popular, becoming a cult favorite even among British listeners (at least among those who could understand French). It was,

wrote Asa Briggs, "a feast of radio at its original and best," staged, in
Alan Bullock's words, by "one of the wittiest, most brilliant bunch of
broadcasters there's ever been."

Virtually from the day it went on the air in July 1940, the program
took France by storm. An Englishman who lived in the French city of
Chambéry for several months after the armistice reported that it resembled a ghost town during the 9 P.M. *Les Français* broadcast. "Generally at this time of night, people are out for a stroll," he noted. "But not
now. The wireless is the cause."

Although its staff was entirely French, *Les Français* was the brainchild of a young Englishwoman named Cecilia Reeves, a talks producer
in the BBC's French section. A graduate of Cambridge's Newnham
College, Reeves had joined the BBC in 1933 as a member of its foreign
liaison department, which assisted non-British broadcasters who used
BBC facilities to transmit their stories and other material to their home
countries.

Among the foreigners was Edward R. Murrow, then in his early
thirties, who became a friend of Reeves's and other bright, talented
young staffers at the BBC. He and his British colleagues had the same
views about truth and independence in broadcasting. "We were giving in full the bad news, the hellish communiqués," said one BBC
editor, "and this meshed with Ed's desire to tell the truth even if it
was a hard and nasty truth. There was a complete meeting of minds
on that."

Reeves, who made the arrangements for Murrow's broadcasts and
inspected his copy before its transmission, had a particularly close relationship with him. She was in the studio late one night in 1937 when,
having returned from covering the German takeover of Austria, Murrow, looking "shattered and terribly fatigued," described the Nazis'
orgy of violence against Austrian Jews. After the broadcast, he asked
her to come back with him to his apartment, where, over tumblers of
whiskey, he talked to her until dawn about the atrocities he had seen. "I
still have a picture of the horror . . . this hideous picture, and of the
agony with which he told it," she said.

After the war began, Reeves was sent to the BBC's Paris office. She
returned to London just before France fell and was assigned the job of

putting together a team of Frenchmen to broadcast to their newly occupied homeland. In doing so, she drew on her experience with Murrow and his broadcasting style, which was considerably more informal and colloquial than that of BBC announcers. Pictures in the air were what Murrow wanted. Throwing out the rigid, traditional rules of news writing and broadcasting, he was calm and conversational in his broadcasts, sometimes telling his stories in the first person singular, just a friend chatting with other friends.

Before the war, Reeves had also been impressed with CBS's innovative European news roundups, which featured reporting and analysis by Murrow and other CBS correspondents. She thought that a similar format would be perfect for broadcasts to France, given the French passion for discussion and argument.

To oversee the program, she chose a charismatic French theater director named Michel Saint-Denis, who had come to Britain in 1930 after founding and leading an avant-garde theater group in France. By 1939, Saint-Denis was regarded as one of the most innovative stage directors in London, working closely with such theatrical luminaries as John Gielgud, Laurence Olivier, and Alec Guinness. His 1937 version of Chekhov's *Three Sisters,* starring Gielgud, Peggy Ashcroft, and Michael Redgrave, is still regarded as a landmark production of twentieth-century British theater. After the war, Saint-Denis would help found two major drama schools—the Old Vic Theatre Centre in London and Juilliard in New York. Arguably, however, his greatest achievement was his wartime work at the BBC.

ALTHOUGH ALREADY FORTY-TWO YEARS old when war broke out, Saint-Denis returned to France from London and rejoined his old World War I regiment. Attached to the British Expeditionary Force as a liaison officer and interpreter, he was evacuated at Dunkirk. He came back to Britain infuriated by France's capitulation to Germany and determined to do all he could to help liberate his homeland. That, however, did not mean joining de Gaulle's Free French forces: like many Frenchmen in London, Saint-Denis was put off by what he viewed as the general's haughty arrogance and autocratic style of leadership. He was about to enlist in the British army when Reeves persuaded him he could help France more by working at the BBC.

Despite his lack of broadcast experience, Saint-Denis instinctively understood radio's potential as a powerful weapon in this war. According to a BBC colleague, he treated "the mike as an old friend from the beginning" and trained the six men he hired for the *Français* team— a cartoonist, an actor, an artist, a poet, and two journalists—to do the same. To protect their families in France from German retribution, several members of the team, including Saint-Denis (who had a wife, a mistress, and children by both women living there), adopted pseudonyms: the name he chose, "Jacques Duchesne," had been the nom de guerre of the spokesman of the working-class supporters of the French Revolution.

Agreeing with Cecilia Reeves that unadorned propaganda wasn't going to convince anyone, Saint-Denis turned politics into entertainment, using humor, drama, and music in imaginative, sophisticated ways to discuss and analyze the political and military events of the day. "The French frequently were not scrupulous in following the official directives," one BBC executive recalled. "If they had, [their broadcasts] would have sounded like official directives. . . . Instead, they struck a note that the listeners understood instinctively." Another BBC official remarked of Saint-Denis, "With a message to give and enough theatrical experience to invent original ways of giving it, half an hour's propaganda became more exciting in his hands than any other radio program I have ever heard."

Every evening from July 1940 to October 1944, the shirtsleeved members of the Saint-Denis group gathered around microphones in a small, stuffy underground studio at Bush House. Occasionally the muffled crump of a nearby bomb blast could be heard in the studio, but *Les Français* never stopped broadcasting. At the beginning of each program, Saint-Denis spoke to his audience in the guise of a fictional working-class Frenchman; as this earthy, shrewd, patriotic character, he tried to bolster the spirits of his compatriots. The program's daily news segment reported Allied setbacks and losses as well as victories, following the BBC's credo that telling the truth would do more to win over its listeners than lying to them.

Once a week, in perhaps the most popular feature of *Les Français,* Saint-Denis and two other members of the team transformed themselves into *"les trois amis."* The *amis* were supposedly a group of old French pals in England who got together in various locales—pubs, restaurants, parks, a cliff in Dover (all of them complete with appropriate sound effects)—to chat about current events, prominently including the latest idiocies of the Germans. In a typical *bon mot,* one of the friends justified his decision to leave France by observing that he "would rather see the English in their country than the Germans in ours."

That segment—and the entire program—appealed to the cultural identity of the French at a time when it was being undermined. Listening to *"les trois amis"* was like "eavesdropping in a French café," the English writer Raymond Mortimer remarked. "We found our spirits greatly improved by their gaiety." A letter to the *Les Français* team spirited out of France declared that "the very soul of French wit has fled to London." Another smuggled letter said, "If only you could see us listening to your broadcasts. We only live for that."

On occasion, though, *Les Français*'s mood was considerably more serious. That was the case on October 21, 1940, when Churchill appeared on the program for his first broadcast to German-controlled France. The prime minister enlisted Saint-Denis's help in preparing for the broadcast. All that day, with the noise of German bombers overhead, the Frenchman worked with Churchill in his underground War Rooms beneath Whitehall, translating his speech into colloquial French.

"I want to be understood as I am," Churchill emphasized to Saint-Denis, "not as you are, not even as the French language is. Don't make it sound too correct." When the translation was finished, Churchill read it aloud in his idiosyncratic French, which Saint-Denis realized was somewhat calculated, as Churchill himself admitted. "If I spoke perfect French," he said, sipping from a glass of brandy, the French "wouldn't like it very much."

Saint-Denis, who introduced Churchill to *Les Français* listeners, stayed with him throughout the broadcast, which took place in a tiny closetlike space in the War Rooms. Because the makeshift studio was so small, the British leader made the speech, one of his most eloquent during the war, with Saint-Denis sitting on his lap. He opened it with the words *"C'est moi, Churchill, qui vous parle"* ("It is I, Churchill, who speaks to you").

Reiterating his fierce belief in France's resurrection, Churchill declared, "Never will I believe that the soul of France is dead. Never will I believe that her place amongst the great nations of the world has been lost for ever." Liberation, like the next day's morning, would surely come, he said. Until then, he advised the French to "sleep to gather strength for the morning. . . . Brightly will it shine on the brave and the true, kindly upon all who suffer for the cause, glorious upon the tombs of heroes. Thus will shine the dawn. *Vive la France!*"

France's response was electric. "Every word you said was like a drop of blood in a life-saving transfusion," the French painter Paul Maze, a friend of Churchill's, wrote him. On the day of the broadcast, a teacher in a lycée near Paris gave his students an unusual assignment. "We are going to hear the leader of the Allied forces tonight," he said. "The broadcast will be badly jammed, so will each of you take down every sentence he can hear properly? We will piece it together tomorrow." The next day, the class succeeded in reconstructing the speech in full.

Considering the overwhelmingly favorable reaction, it may not have been a coincidence that just a week after Churchill's address, the Vichy government, which had been somewhat more lenient in its curbs on the public than the German regime in northern France, imposed a harsher ban on listening to foreign broadcasts, quadrupling the penalties to a ten-thousand-franc fine and two years in prison.

One of the greatest headaches facing the *Les Français* team and the rest of the BBC's French section was how to deal with Marshal Pétain and Vichy. Cecilia Reeves discovered how ticklish those topics were when she oversaw a broadcast critical of Pétain in the summer of 1940. Afterward, she told Ed Murrow that "half the French in London attacked me for attacking Pétain; the other half for paying him any mind, or not attacking him enough." Murrow advised her to back off for the moment. "Pétain is necessary now for France," he said. "You can attack his policy. But to attack him personally would be disastrous."

When it came to the Vichy leader, the *Les Français* broadcasters were as divided as the rest of their countrymen. One of them was so passion-ately opposed to Pétain that he challenged a pro-Pétain colleague at the BBC to a duel. Saint-Denis, by contrast, had great esteem for the mar-shal and argued that condemning him too precipitously would alienate millions of Frenchmen. Unlike de Gaulle, who had attacked Pétain from the start as "the father of our defeat," Saint-Denis refused to crit-icize him in the first months of his broadcasts. His caution, as well as his earlier refusal to join the Free French, infuriated de Gaulle, who re-peatedly tried—and failed—to force the BBC to take *Les Français* off the air. The two men continued their political and personal wrangling throughout the war, yet in France, their listeners, unaware of the deep rifts and rivalries bedeviling the French community in London, thought of them as close allies.

Thanks to Vichy's continuing cooperation with the enemy and Germany's escalating oppression of the French (which *Les Français* re-ported and commented on), the public mood in France slowly began to shift. On October 24, 1940, Pétain met with Hitler in Montoire, a small town in central France. Soon afterward, he announced that he and his government had embarked on a policy of official collaboration with Germany; from then on, the Nazis could rely on the help of French authorities in both zones to carry out their policies, which would soon include the roundup and deportation of French Jews.

After the Montoire meeting, Pétain's popularity gradually declined throughout the country, with a growing number of the French (al-though still a small minority) expressing outright opposition to him and his government. Among them was Michel Saint-Denis, who, nine months after the fall of France, finally decided to take Pétain on. "To-night," he told his listeners on March 19, 1941, "I must speak to you about the poor Marechal . . . who let himself be influenced by the par-asites surrounding him and who took a position in favor of the enemy." It was time, he said, to confront both Vichy and the Germans.

SAINT-DENIS WAS NOT THE first BBC broadcaster to encourage occu-pied Europeans to throw off their passivity. Two months earlier, Victor de Laveleye, the organizer of BBC programs to Belgium, had urged his compatriots to demonstrate their resistance to German rule by scrawling the letter *V* on the walls of buildings throughout the country. De Laveleye, who had served as Belgian minister of justice before coming to London, told his listeners that the letter would serve as a symbol to unite and rally their sharply divided nation. (Belgium's population in the north spoke Flemish, a variant of Dutch, and had close cultural and religious ties to the Netherlands, while Belgians in the south spoke French and were intimately linked to France.) As de Laveleye noted, V was the first letter of both the French word *"victoire"* and the Flemish word *"vrijheid"* ("freedom"), not to mention the English word "victory."

Belgians accepted de Laveleye's challenge with gusto, chalking Vs on walls, doors, pavements, and telegraph and telephone poles. So did a growing number of the French, many of whom learned about the V campaign by listening to the BBC's Belgian service. Although de Laveleye's initiative was meant for Belgium, it spread across France in a matter of days. In both countries, chalk sales skyrocketed. A letter to the BBC from Normandy noted "a multitude of little Vs everywhere." A correspondent from Argentière in the French Alps reported "an avalanche of Vs, even on vehicles and on the roads." A Marseille resident remarked that his city was so inundated with Vs that there was "not a single empty space."

In London, *Les Français* devoted a special program to the V sign, which included a catchy song urging the program's listeners to adopt what had become the alphabet's most famous letter:

> *Il ne faut pas*
> *Désespérer*
> *On les aura!*
> *Il ne faut pas*
> *Vous arrêter*
> *De résister!*
> *N'oubliez pas*
> *La lettre V*
> *Écrivez la!*

Chantonnez la!
V!V!V!V!V!*

By early spring of 1941, the BBC's entire European Service had taken up the V campaign, with similar results. Not long afterward, someone realized that the first four notes of Beethoven's Fifth Symphony sounded like Morse code for the letter V—dot-dot-dot-dash. Those notes, played on a kettle drum, became the European Service's call sign, and they, too, spread like wildfire throughout occupied Europe. People hummed and whistled them. Restaurant diners tapped them with spoons on their glasses and cups. Train engineers blasted them on their whistles. Schoolteachers called their students to order by clapping their hands with the V rhythm. When a British bomber used its landing lights to flash the letter *V* over Paris, its crew watched the city light up with Vs beamed back from cars and apartment windows.

The V campaign was, as the historian Tom Hickman noted, "the first pan-European gesture of resistance," helping Europeans shed their sense of helplessness and come together in thumbing their noses at the Germans "who paraded in their streets, packed their restaurants, and were billeted in their homes." Exposed repeatedly to the V sign, a German soldier, according to Victor de Laveleye, could no longer doubt that he was "surrounded by an immense crowd of citizens eagerly awaiting his first moment of weakness, watching for his first failure."

As it turned out, Europeans were not the only ones caught up in V madness. In the neutral United States, jewelers sold V brooches and earrings, and Elizabeth Arden featured V for Victory lipsticks. Wendell Willkie, the 1940 Republican presidential candidate, sported a V tiepin. But the most celebrated champion of the V symbol was Winston Churchill, whose raised right hand, with his index and middle fingers

---

*   Don't despair,
Don't stop resisting,
We will beat them!
Don't forget the letter V,
Write it!
Sing it!
V! V! V! V! V!

A German poster in occupied Holland tries to appropriate the BBC's highly successful V campaign by insisting that "V Means That Germany Is Winning on All Fronts."

pointed up in the shape of a V, became his signature gesture for the rest of his life. In a BBC speech to the Continent in the summer of 1941, Churchill called the V sign "a symbol of the unconquerable will of the people of the occupied territories and a portent of the fate awaiting Nazi tyranny. So long as the people of Europe continue to refuse all collaboration with the invader, it is sure that his cause will perish and that Europe will be liberated."

The extraordinary success of the V campaign was sourly acknowledged by none other than Joseph Goebbels, Nazi Germany's master propagandist, who noted "the intellectual invasion of the Continent by the British radio, an invasion of which the letter V was the symbol." Incredibly, in their attempt to neutralize the campaign, Goebbels and the Germans appropriated the letter V for their own use. German-controlled radio stations in Norway, the Netherlands, and other cap-

*Winston Churchill
and his signature
"V" gesture.*

tive countries began using the first bars of Beethoven's Fifth Symphony
to open their programs. In Amsterdam, a thirty-foot banner, bearing
the words V FOR VICTORY WHICH GERMANY IS WINNING IN ALL EUROPE ON
ALL FRONTS, hung from one of Queen Wilhelmina's palaces. Huge V
streamers adorned the main hotels of Oslo, and the Germans draped a
gigantic V sign on Paris's Eiffel Tower. In Prague, one of the main
thoroughfares was renamed Victory Street, with large Vs painted down
its center; in Poland, Vs were printed on the front pages of German-
run newspapers.

The BBC's Noel Newsome, who broadcast to Europe three times a
week in the guise of an anonymous British "man in the street," made
savage fun of the Germans' effort to adopt the V sign as their own.
"Soon, perhaps," Newsome remarked to his listeners, "the Germans
will be forced to pretend that the letters RAF stand for the Luftwaffe

and that [de Gaulle's Free French symbol] the Cross of Lorraine is a
new type of swastika."

BY EARLY 1941, THANKS in no small part to the BBC, much of captive
Europe was finally emerging from the lethargy of defeat. When de
Gaulle, in a BBC broadcast, urged the French to stay home for one
designated hour on January 1, 1941, as a sign of protest against German
occupation, hundreds of thousands complied. Passive though it was,
the protest was France's first organized anti-German demonstration.
"The [French] underground movement was built up by the BBC," the
resistance leader André Philip said later in the war. "We needed help
from outside, and the BBC gave that help."

In Norway, meanwhile, teachers, clergymen, actors, and theater directors all staged mass protests against German control. In the Netherlands, thousands of university professors and students stayed away
from classes to protest Nazi persecution of Dutch Jews. When the Germans there began to round up Jews in early 1941, "virtually the entire
working population of Amsterdam and other nearby cities," the Dutch
historian Louis de Jong noted, went on strike. The two-day action,
which German police put down by force, was, de Jong believed, "the
first and only anti-pogrom strike in history."

Buoyed by the success of the V campaign and other signs of nascent
resistance, Noel Newsome and Douglas Ritchie, Newsome's deputy in
the European Service, wanted to go even further, using the BBC to
spark open warfare against the Germans. As early as July 1940, Newsome told European Service staffers that they should encourage their
listeners to "desire an anti-Nazi revolution rather than to fear it"—
a campaign that would end in a "great uprising of the peoples against
a morally and spiritually bankrupt tyranny whose actual material
strength is waning."

In calling for aggressive action, Newsome was following the lead of
Frederick Ogilvie, John Reith's successor as BBC director general,
who, the month before, had urged the BBC to adopt the motto "Every
patriot a saboteur." As Ogilvie saw it, "a civilian population which is

not actively hindering the enemy are in effect traitors to the common cause."

British intelligence officials swiftly squelched Newsome and Ogilvie's quixotic idea, underscoring the absurdity of promoting a European revolt against Hitler just weeks after his stunningly successful blitzkrieg. Newsome was quite wrong in his blithe assumption that Germany's strength was diminishing, they said. How could Europeans be expected to rise up when the Reich so totally dominated the Continent and almost no one expected Britain to survive?

With the British still holding out a year later, Newsome and Ritchie decided it was time to revive the concept of a BBC-led revolution on the Continent. Ritchie, who, like Newsome, hosted a weekly broadcast to English-speaking listeners in occupied Europe, told his audience that they were "the unknown soldiers. Millions of you . . . A great silent army, waiting and watching. The V is your sign." In a memo to BBC higher-ups, he was even more forceful: "When the British Government gives the word, the BBC will cause riots and destruction in every city in Europe." Under Ritchie's direction, an unofficial committee was set up to "encourage, develop, and coordinate British broadcasts to enemy-occupied countries about action against the Germans."

Others in the BBC and the government strongly disagreed with what they viewed as the European Service's meddling. For one thing, Newsome and Ritchie were encroaching on the functions of the Special Operations Executive, a new government agency set up to promote resistance activities in occupied Europe. For another, the idea of a mass European rebellion in the near future continued to be hopelessly out of touch with reality.

Already, the Germans had begun to crack down on the mild defiance sparked by the V campaign. In the French town of Moulins, German authorities imposed "severe punishments" on town residents in retribution for the epidemic of Vs and other anti-German inscriptions daubed on walls and other surfaces. On April 29, 1941, the German radio in Paris broadcast a chilling warning to those in France inclined to continue the campaign: "Silly people, you can take up your chalks again and put in full this time, the word you began so well—'Victims'—in blood red letters." A few days later, the BBC received a letter from Vichy France that noted, "Many speak of revolt . . . but what can we do? We have no weapons, although it is true we still have chalk."

Such communications reinforced the anxiety of many at the BBC about "giving any kind of false lead to people to start firing guns . . . until the time was right." As one staffer put it, "We were not the ones who'd be put up against a wall if anything went wrong."

CHAPTER 10

Spying on the Nazis

Cracking Enigma and Other
European Intelligence Coups

THE GENTEEL ST. ERMIN'S HOTEL, TUCKED AWAY FROM LONDON'S bustle on a quiet side street in Westminster, seemed an unlikely spot for the hatching of plots to incite violence and revolution. A Victorian redbrick dowager of a building, St. Ermin's was known for its tree-lined courtyard and a spectacular soaring lobby featuring a rococo plaster ceiling and curved balcony. There, at the turn of the century, guests could enjoy afternoon tea while listening to the soothing strains of a small string orchestra.

Forty years later, as France fell and Hitler prepared to hurl his Luftwaffe thunderbolts at Britain, a handful of government officials gathered in a room on the fourth floor of St. Ermin's to try to figure out a way to fight back. Clearly, the British could not return to the Continent anytime soon: their army was too meager, their arms almost non-existent, and the ally they coveted—the rich and powerful United States—showed no signs of wanting to join the conflict. In the desperate summer of 1940, Britain's skimpy offensive arsenal contained just two weapons: the Royal Navy's blockade of Germany, begun in 1939, and a nascent bombing effort against the Reich by the RAF. The plotters at St. Ermin's added a third: a campaign to undermine the enemy from inside the occupied countries of Europe.

Thus was born a new top secret government agency, the innocuously named Special Operations Executive, which took over three floors of St. Ermin's in a matter of weeks. SOE's assignment was to foment sabotage, subversion, and resistance in captive Europe, a goal that its creators hoped would disrupt and eventually help destroy the German war machine. Winston Churchill, an enthusiastic champion of the idea, dubbed the SOE "the Ministry of Ungentlemanly Warfare" and instructed its first chief to "set Europe ablaze."

Before it could attempt to do so, however, SOE had to deal with the unpleasant reality that, aside from Churchill and a few other leading political figures, virtually no one in high British government circles wanted it to exist, much less succeed. That was particularly true of the top men in Britain's vaunted Secret Intelligence Service (MI6), who loathed the thought of a new clandestine government agency independent of their control, especially one whose goals and methods were so fundamentally in conflict with those of their own.

MI6, which was responsible for collecting military, economic, and political intelligence from countries outside the British Empire, prided itself on its secrecy and discretion; the idea of a sister agency drawing unwelcome attention to itself (and to MI6 operations) through sabotage and other public acts of violence was anathema. The intelligence historian Nigel West aptly summed up the key difference between the two organizations in his description of how each would react if one of its operatives witnessed enemy troops crossing a bridge: an MI6 spy would observe the troops and estimate their number while an SOE agent would blow up the bridge to prevent the enemy from getting across.

As SOE officials would soon discover, MI6 was a dangerous foe to have. Globally renowned, it enjoyed a sterling reputation as an all-seeing, all-knowing spy organization. Churchill considered the British intelligence service "the finest in the world." So, interestingly, did Hitler and other Nazi higher-ups, including SS head Heinrich Himmler and Himmler's murderous deputy Reinhard Heydrich, known as "the blond beast," who used MI6 as a model when he created the SS's fearsome intelligence and security operations: the SD and Gestapo. Heydrich even signed some of his letters and memos with the letter "C," the initial used by MI6 chiefs in their official correspondence.

Yet for all his admiration of the British intelligence service, Heydrich, like his Nazi colleagues, actually had very little substantive knowledge of how it worked. His romantic views of its omniscience and omnipotence came from reading the prewar British spy novels that filled the Gestapo's reference library. And as it turned out, all those awestruck notions were as fanciful as the fiction on which they were based.

FROM THE LATE NINETEENTH century on, British novelists found that one of the surest routes to fame and fortune was to write about the fictitious exploits of British secret agents in continental Europe. When the spy novel genre began in the 1890s, the enemy was usually French, but as German military power exploded in the early twentieth century, he became almost exclusively German. The heroes, however, remained remarkably the same.

With few exceptions, they were well-born English gentlemen who belonged to the best London clubs, rode to hounds, and were connoisseurs of gourmet food and vintage wines. Yet all those paragons were willing to forsake their good life for a time to spy for their country in the face of great danger and appalling odds. The protagonist of Robert Erskine Childers's wildly popular novel *The Riddle of the Sands,* written in 1902, was typical: he was, according to Childers, "a young man of condition and fashion, who knows the right people, belongs to the right clubs, and has a safe, possibly a brilliant future in the Foreign Office."

For decades, the adventures of these amateur patrician spies not only attracted millions of readers from around the world but also inspired a generation of young Englishmen to follow in their footsteps. An SOE agent noted after the war that "practically every [SOE] officer I met at home and abroad was, like me, imagining himself as Richard Hannay or Sandy Arbuthnot," two British agents who were the fictional creations of John Buchan, whose famed adventure novels included *The Thirty-nine Steps.*

Reflecting Britons' instinctive distrust of foreigners, the enemy agents in those books tended to be stereotypes as well—unshaven, badly dressed, shifty, and duplicitous. The novels' overall moral seemed to be that getting involved with foreign countries and individuals was a dangerous undertaking and that Britain—"the supreme country in the world," as one fictional British agent described his homeland—should do its best to stay well clear of them.

Given the xenophobia of British spy novels, it's a wonder that foreigners like Himmler and Heydrich were so taken with them. Heydrich, whose addiction began during a post–World War I stint in the German navy, was convinced that the success of the British Empire was due to the brilliance of MI6 and that every upstanding Englishman was "ready to aid the Secret Intelligence Service, regarding it as his obvious duty. . . . The SS has adopted as its idea this English view of intelligence work as a matter for gentlemen." In an attempt to emulate the British, he focused on recruiting for his own operations young, well-educated Germans from good families.

Even Hitler joined in the chorus of praise for MI6. Speaking to Nazi intelligence officials early in the war, the Führer remarked that "the British Secret Service has a great tradition. Germany possesses nothing comparable to it. Therefore, each [German] success means the building up of such a tradition and requires even greater determination. . . . The cunning and perfidy of the British Secret Service is known to the world."

Such Nazi paeans, however, could not have been further from the truth. Starved of government funds after World War I, MI6 in the late 1930s was underfinanced, understaffed, and woefully short of both talent and technology. In 1935, two years after Hitler came to power in Germany, the then-current "C"—Admiral Hugh Sinclair—despairingly remarked that his agency's entire annual budget equaled the cost of maintaining one British destroyer for a year. Although Sinclair borrowed money from wealthy relatives to keep MI6 afloat, it was never enough; at the time of the Munich Conference in 1938, he could not afford to buy wireless transmission sets for the few agents he had in Europe to let them communicate directly with London.

MI6 did indeed tend to recruit well-born young men as its intelligence officers; preferring "breeding over intellect," it shied away from those who had attended college, seeking instead "minds untainted by the solvent force of a university education." Many of its operatives in the interwar period were former military officers with substantial private incomes. The early education of "these metropolitan young gentlemen . . . had been expensive rather than profound," observed Hugh Trevor-Roper, the noted historian, who, after a stint as an Oxford don, worked for MI6 during World War II. "[They] didn't have much use for ideas." They were, he added, "by and large pretty stupid—and some of them very stupid."

In one area, however, the real intelligence officers had much in common with their fictional counterparts: both groups operated with an air of casual, gentlemanly amateurism. When a new MI6 operative named Leslie Nicholson asked in 1930 about the kind of training he would receive, he was told that "there was no need for expert knowledge." When Nicholson persisted in seeking "tips on how to be a spy," the MI6 station chief in Vienna responded, "You'll just have to work it out for yourself. I think everyone has his own methods, and I can't think of anything I can tell you."

As it happened, Stewart Menzies, who was named "C" in 1939 and who ran MI6 throughout the war, had never himself experienced what it was like to be a secret agent, trained or otherwise. The grandson of an enormously wealthy whiskey baron from Scotland, Menzies had attended Eton, then joined the British Army's prestigious Life Guards Regiment, many of whose officers were aristocrats. During the Great War, he fought with distinction in France and was awarded a Distinguished Service Order and Military Cross. After being gassed, he signed up with army intelligence, and when the war ended, he shifted to MI6 headquarters in London, where he remained for the rest of his career.

A charter member of the clubby, upper-class "old-boy network" that had dominated British society for generations, Menzies, like many if not most of his contemporaries, was conservative in his social and political attitudes and wary of foreigners. He once said that "only people with foreign names commit treason," obviously unaware that someone with the very English name of Kim Philby, who was without question one of the "old boys," was committing treason within Menzies's own agency.

Tall and slender, with thinning blond hair, the fifty-year-old Menzies belonged to the Beaufort Hunt, the select fox-hunting group sponsored by the Duke of Beaufort, and White's, the most exclusive men's club in London, where no bottle of nonvintage wine was ever served and no woman was ever allowed to enter. It was in White's bar that he did much of his recruiting for MI6, focusing mostly on young men from families in his own cloistered milieu.

An amiable social butterfly, Menzies was generally regarded as an intelligence lightweight by senior MI6 officers, who noted his lack of practical experience in the field and his propensity for procrastination. "He was not a very strong man, and not a very intelligent one," recalled Victor Cavendish-Bentinck, wartime chairman of the Joint Intelligence Committee, a Cabinet-level group that oversaw all of Britain's intelligence operations.

The person who really ran MI6, in the opinion of many, was Menzies's deputy, Claude Dansey, a shadowy figure well versed in stealth and deceit, who was, as the writer Ben Macintyre put it, "a most unpleasant man and a most experienced spy." An anomaly in MI6's uppercrust world, Dansey had not gone to Eton or served in one of the army's posh regiments. Instead, he had spent much of his early career as a military intelligence officer in Africa, where he had run spy networks that gathered information on and helped put down rebellious native groups. During World War I, Dansey had worked for British intelligence in London, where his myriad duties included rounding up suspect aliens and engaging in counterespionage in Britain and western Europe.

Short, balding, and bespectacled, Dansey had a luxuriant white mustache and, in Ben Macintyre's words, the sharp, penetrating "eyes of a hyperactive ferret." He was witty, spiteful, and widely disliked. "Everyone was scared of him," said the journalist and author Malcolm Muggeridge, who, like Hugh Trevor-Roper, worked for MI6 during the war. "He was the only real professional there. The others at the top were all second-rate men with second-rate minds." Trevor-Roper was far more jaundiced in his view of Dansey, describing him as "an utter shit; corrupt, incompetent, but with a certain low cunning."

As dissimilar as they were, Menzies and Dansey shared at least one common trait: an obsessive devotion to clandestine behavior that bemused Muggeridge, Trevor-Roper, and the other young outsiders brought into MI6's inner sanctum during the war. According to Muggeridge, MI6's motto seemed to be that "nothing should ever be done simply if there are devious ways of doing it." Like small boys playing secret agent, the agency's old hands often used code names when they were not needed, communicated in code when writing innocuous messages, and left those messages in out-of-the-way places such as a potting shed rather than posting them in an ordinary mailbox.

Muggeridge realized he had been infected with that same obsession for secrecy when he found himself tiptoeing past Menzies's hushed offices on the fourth floor of MI6 headquarters at 54 Broadway, just off Parliament Square. "Secrecy," Muggeridge recalled, "is as essential to Intelligence as vestments and incense to a Mass . . . and must at all costs be maintained, quite irrespective of whether or not it serves any purpose."

While the use of pointless code names and other forms of clandestinity struck newcomers like Muggeridge as richly comic, MI6's fixation with secrecy served an important purpose for those in its upper reaches, one that had nothing to do with protecting the safety and security of Britain. Its usefulness was far more personal: it helped shield those in power from the scrutiny of Parliament, the British public, and the rest of the government. The novelist John le Carré, who worked briefly for MI6 shortly after World War II, noted how devoted the agency had been to "the conspiracies of self-protection, of using the skirts of official secrecy in order to protect incompetence, of gross class privilege, of amazing credulity."

In the years immediately preceding the war, MI6, as it happened, had a considerable amount of incompetence to protect.

ON A DARK, DREARY November day in 1939, a car pulled up to the Backus Café in the small Dutch town of Venlo, just across the Meuse River from Germany. Two middle-aged Englishmen—sporting trim gray mustaches, monocles, and bowlers—got out. They were MI6 officers, come to Venlo on a top secret mission that, if successful, might well lead to the overthrow of Hitler and the end of the two-month-old war.

Ever since the Führer had come to power, the neutral Netherlands, situated as it was next to Nazi Germany, had served as one of the main spy centers of Europe, playing reluctant host to countless intelligence agents from all over the globe. Like most of their foreign counterparts, the two mustachioed MI6 officers—Sigismund Payne Best and Richard Stevens—were based at The Hague, which British intelligence used as its unofficial European headquarters.

For such an important post, MI6 had been astonishingly inept in its selection of station chiefs, who operated there under the flimsy cover of passport control officers. In 1936, station chief H. E. Dalton had killed himself after the revelation that he had embezzled fees paid by Jewish refugees for British visas. Then came the discovery that two Dutch operatives working for Dalton's successor had been recruited as double agents by the Abwehr, Germany's military intelligence agency.

Realizing that one of its most important bases in Europe had been penetrated by the Germans, MI6 chose not to take the logical step of closing down the office and starting over. Instead it created a second, unofficial intelligence service throughout Europe that would exist alongside the old operation but have no connection with it. Known as the Z Organization, it was run by Claude Dansey, who chose as its agents a varied collection of amateurs: businessmen, industrialists, journalists, politicians, and other British subjects who either lived in Europe or were frequent travelers there.

The man Dansey picked for The Hague was Sigismund Best, the owner of a Dutch pharmaceutical and chemical company, who passed on to Dansey mostly worthless items of intelligence. At the same time, he claimed significant amounts of money as expenses for the thirteen agents he alleged he was running, nine of whom turned out to be fictitious.

When war broke out in September 1939, Best was ordered by London to join forces with Richard Stevens, MI6's latest station chief in The Hague, thereby destroying the whole point of having an alternative intelligence operation. A newcomer to the agency, Stevens, who had earlier served in the British Army in India, was as unimpressive in his job as Best had been in his. Before going to The Hague, Stevens had "never been a spy, much less a spymaster," he later acknowledged. "I was totally lacking in experience and felt I was altogether the wrong sort of man for such work." When he told MI6 higher-ups of his fears, they assured him that The Hague was largely an administrative post— a gross misstatement, as Stevens would find out a few months later when he and Best were thrown into a situation for which neither was remotely prepared.

In October 1939, the month after Germany invaded Poland, Neville Chamberlain quietly let it be known that the British government would consider making peace with Germany if Hitler were deposed. At the same time, Stevens and Best received word that a dissident military faction in the Reich was plotting to get rid of Hitler and open peace negotiations with Britain. After several clandestine sessions with men reportedly from this rebel group, the MI6 officers, with the backing of Chamberlain and the Foreign Office, agreed to a November 9 meeting at the Backus Café in Venlo to meet the German general leading the resistance.

But when Stevens and Best entered the café, they discovered to their shock that there was no general and that the officers with whom they'd met were in fact SS intelligence agents; one of them was Walter Schellenberg, Reinhard Heydrich's deputy and another longtime British spy novel addict and MI6 fan. After shooting to death a young Dutch military intelligence agent who had accompanied the British operatives, Schellenberg and his colleagues bundled Stevens and Best into a car and sped over the border into Germany.

The intelligence officers' abduction, one of the most embarrassing episodes in MI6 history, became an even greater disaster thanks to their behavior after the kidnapping. Ignoring London's directive that captured agents should reveal only the names and addresses of their cover businesses or other jobs, Stevens and Best, without being subjected to physical torture, collaborated fully with the Germans. Stevens was caught with a complete list of Dutch agents in his pocket; he also apparently handed over the names of all MI6 station chiefs in western and central Europe, along with the identities of their foreign operatives. In addition, he and Best provided extensive information about MI6's hierarchy in London, including the names of department heads and the location of their offices in the Broadway headquarters.

As a result of the Venlo disaster, the MI6 network in western Europe was in ruins by the time of the 1940 German blitzkrieg. Yet despite this fiasco, Menzies and Dansey managed to retain their jobs, thanks in large part to the providential arrival in London of the exile European governments and their intelligence services.

In exchange for providing financial, communications, and transportation support to the secret services of Czechoslovakia, Norway, Poland, Belgium, the Netherlands, and de Gaulle's Free French, MI6 gained control of most of their operations. The foreign services, in turn, provided virtually all of the wartime intelligence the British received about German activities in occupied Europe.

Though hardly a genius as spymaster, Menzies was brilliant at protecting and promoting himself and his agency in the often brutal infighting in Whitehall. One of his major weapons in that bureaucratic war was never to reveal MI6's sources to anyone, not even Churchill. In that way, he and Dansey could claim sole credit for any successes that came their way.

As it happened, the European intelligence agencies with which MI6 was now aligned scored many coups. Yet almost nobody outside MI6 knew it. An exception was David Bruce, head of the London branch of the Office of Strategic Services (OSS), the United States' newly formed espionage and sabotage agency. Bruce, who arrived in London in 1942, noted in his diary that MI6's intelligence capabilities were "lamentably weak. Most of the reports they send us are duplicates of those already received by us from European secret intelligence services."

The first foreign intelligence group to arrive in London was the Czechs'. Just before Hitler occupied all of Czechoslovakia in March 1939, František Moravec, the head of that country's highly respected spy agency, and ten of his top officers, along with dozens of boxes of files, escaped to Britain. Moravec's arrival was a particular windfall because he brought with him the reports of one of the top Allied intelligence sources of World War II, a disaffected German Abwehr officer named Paul Thümmel. Code-named A54, Thümmel, the chief of the Abwehr station in Prague, provided the Czechs—and indirectly the British—with remarkably accurate information about German military plans for more than two years. Thanks to Thümmel, for example, MI6 learned beforehand of German plans to invade France through the Ardennes in 1940 and to conquer Yugoslavia and Greece in the spring of 1941. (The Ardennes intelligence coup—and Britain's and France's failure to do anything about it—prove that, however good intelligence might be, it is of little or no use unless action is taken as a result.) In the fall of 1940, Thümmel also reported that Hitler had abandoned his plans, at least temporarily, for Operation Sea Lion, the proposed invasion of Britain.

The Norwegian intelligence service, meanwhile, passed on to MI6 reports from hundreds of coast watchers in Norway, who monitored the movements of German submarines and warships. In 1941, one of them informed London that he'd spotted four German warships in a fjord in central Norway—information that led to the sinking of the battleship *Bismarck* and the crippling of the heavy cruiser *Prinz Eugen*. Hundreds more ordinary Norwegians reported on fortifications, airfields, camps, and German troop movements.

In the early years of the war, however, France was MI6's main focus. As the occupied country closest to Britain, it was Hitler's springboard—the country from which the Luftwaffe bombed British cities and the German navy dispatched submarines to sink British merchant shipping. It would also serve as the launching point for any invasion of Britain. As a result, intelligence about the movements and disposition there of German troops, ships, submarines, barges, and aircraft was of vital importance to the British government.

To gather such information, MI6 was able to draw on a cornucopia of sources. One was de Gaulle's fledgling intelligence service, headed by André Dewavrin, a young army officer and former professor at Saint-Cyr, France's foremost military academy, who agreed to dispatch secret agents to France under British control. The first to be sent was Gilbert Renault, a French film producer who was working on a movie about Christopher Columbus when France fell. He escaped to London and joined de Gaulle's Free French; six weeks later, he landed secretly on the coast of Brittany to begin what turned out to be a phenomenally successful two-year stint as a spy.

Renault's first job was to collect and send back to Britain detailed, up-to-date maps of France. (As with Norway, the only maps the British had of France in the 1940 campaign were those they'd collected from travel agencies.) Once he had done that, Renault, although a complete novice at intelligence, put together a far-flung spy network, called the Confrérie de Notre Dame, that eventually covered much of occupied France and Belgium. The information it provided led to such military successes as the 1942 British commando raids on the northwest French ports of Bruneval and Saint-Nazaire.

When Renault, whose code name was Colonel Rémy, finally left France in June 1942, he took with him copies of plans for Germany's defense installations on the Normandy coast, stolen by one of his agents. Those blueprints later proved to be an invaluable resource for the British and American planners of the D-Day invasion.

While Renault and other Free French agents were vital assets for MI6, so, too, were a number of Frenchmen who at one time or another had been allied with Pétain's Vichy government. Some of MI6's closest wartime links in France were with the prewar French intelligence services, many of whose members worked for Vichy. Soon after France's capitulation to Germany, a group of anti-German officers in the French army formed an underground intelligence organization called Organisation de Résistance de l'Armée (ORA) that accepted Pétain as French head of state but conspired to end the German occupation of France.

Indeed, among the most important Allied spy rings in France was one organized by Georges Loustaunau-Lacau, a colonel who had earlier served as Pétain's top military aide. When the colonel was arrested by the Vichy police, his former secretary—a petite, elegant thirty-year-old mother of two named Marie-Madeleine Fourcade—became the leader of the network, called Alliance, which included hundreds of men demobilized from the French army, navy, and air force. At its height, Fourcade's operation (known as Noah's Ark by the Gestapo because its members used animal code names) was active throughout all of France and numbered more than three thousand agents, some five

*Marie-Madeleine
Fourcade*

hundred of whom were arrested, tortured, and executed by the Germans over the course of the war. Fourcade herself was arrested twice but escaped both times, once by stripping naked and forcing her slender body through the bars of a Gestapo jail cell.

Sure that the British would never accept the idea of a woman as head of a major intelligence network, Fourcade kept her identity secret until the end of 1941, when she was smuggled into neutral Spain in a diplomatic bag for a meeting with her MI6 handler. She needn't have worried about how the meeting would go. Her handler was bedazzled by her "Nefertiti-like beauty and charm," and although the irascible Claude Dansey grumbled that "letting women run anything was against his principles," he couldn't argue with Fourcade's accomplishments. MI6 continued to supply money, wireless sets, and other equipment to her Alliance network, which repaid that largesse with a flood of top-level intelligence, including information about German coastal fortifications and German troop movements prior to D-Day.

As good as the French were, however, it was the Poles who provided the lion's share of British—and Allied—intelligence during the
war. In 2005, the British government acknowledged that nearly 50 percent of the secret information obtained by the Allies from wartime Europe had come from Polish sources. "The Poles had the best special
services in Europe," said Douglas Dodds-Parker, a British intelligence
official who worked with them during the war. Actually, the Poles
were even better than that, according to the deputy chief of American
military intelligence, who argued in 1942 that "they have the best intelligence in the world. Its value for us is beyond compare."

The Poles were longtime masters of covert activity, having been
occupied and partitioned for more than a century by three powerful
neighbors: Russia, Germany, and Austro-Hungary. "With generations
of clandestine action behind them," Dodds-Parker noted, "they had
educated the rest of us."

From the day Poland regained its status as a sovereign country in
1918, it had given top priority to spying and code breaking, specifically
aimed at its two chief historic enemies, Germany and Russia. In the
words of a former chief of Polish intelligence, "If you live trapped between the two wheels of a grindstone, you have to learn how to keep
from being crushed." In 1939, Polish leaders were unable to prevent
that from happening, but before escaping to the West, they did leave in
place sophisticated intelligence and resistance networks.

From a large town house in the fashionable west London neighborhood of Knightsbridge, Polish intelligence officials maintained close
radio contact with a widespread network of agents inside Poland, who
provided a flood of information, including statistics about industrial
production and the deployment of German military and naval forces.
In return, the British gave the Poles, along with the Czechs, a high degree of autonomy. Unlike the other exile intelligence services in London, the two eastern European countries were allowed to operate their
own training establishments, codes, ciphers, and radio networks without MI6 control, with the proviso that they pass on all intelligence rel-
evant to the Allied war effort.

Years after the war, the Polish historian Jan Ciechanowski estimated
that as many as 16,000 Poles—most of them members of the Home

Army, the country's highly organized resistance movement—were involved in the gathering of military and economic information inside Poland. "No place in Poland, where there was anything of great significance, could be kept from the prying eyes of Home Army intelligence," Ciechanowski said. In fact, the Home Army's network was even more far-flung than that, with contacts throughout Austria and Germany, including outposts in Cologne, Bremen, and Berlin. Much of the information from the Reich came from Poles who had been forced to work in German factories as slave laborers.

The Poles' extraordinary talent for spying won admiration, albeit grudging, from German intelligence officials as well. After two days of poring over captured Polish intelligence documents in 1939, Walter Schellenberg wrote in his diary that "the amount of information, especially concerning Germany's production of armaments, is quite astonishing." Later, he dourly noted, "One always has to be prepared for unpleasant surprises with the Poles."

Besides its spies in Poland, Austria, and Germany, Polish intelligence boasted agents in Scandinavia, the Baltic States, Switzerland, Italy, Belgium, the Balkans, and North Africa. Among its most successful operatives was Halina Szymańska, whose husband had been the Polish military attaché in prewar Berlin. While there, the couple had become friends with Admiral Wilhelm Canaris, the head of the Abwehr. One of the most enigmatic figures of the war, Canaris, who had grown increasingly disenchanted with Hitler, was playing a double game: while counterintelligence operatives in his agency ruthlessly tracked down Allied spies and saboteurs, he encouraged other Abwehr colleagues to pass on intelligence to MI6. After Poland's defeat, Canaris arranged for Szymańska's escape to Switzerland, where he put her in touch with Hans Bernd Gisevius, who was in charge of Abwehr operations there. For more than two years, Szymańska acted as a conduit between Gisevius, Polish intelligence, and MI6, providing information about high-level Nazi decision making, including German plans to invade Russia in 1941.

In France, too, the Poles organized and ran several important intelligence networks. The seeds of the French operations were planted by several Polish officers who remained behind in the unoccupied zone after Allied forces were evacuated from France in June 1940. Their first network, called F-1, established an escape route for marooned Polish and other Allied troops that led from Toulouse, in southwest France, to Britain. But F-1's major effort was to gather intelligence about German aircraft, weapons, and troop movements. Its organizers recruited dozens of additional agents, many of whom came from the large communities of Polish immigrants scattered throughout France. A sizable number of these émigrés were industrial workers, able to provide detailed reports about the output and location of factories producing armaments and other items of interest to the Allies.

By early 1941, the original Polish network had split into several new cells. The most important of these operated out of a rented room in the heart of occupied Paris. It was organized and run by an intense, adventurous Polish air force officer named Roman Garby-Czerniawski, one of F-1's original organizers. A fighter pilot before the war, the French-speaking Garby-Czerniawski had been recruited by Polish military intelligence in late 1939. He was, said a colleague, "a man who lives and thinks spying."

Garby-Czerniawski's goal in establishing himself in Paris was to provide London with as detailed a picture as possible of German forces and installations in occupied France. He christened his operation the Interallié network, declaring that "the boss will be a Pole, the agents mostly French, and all working for the Allies." Hundreds of full- and part-time operatives gathered information for him, among them railway workers, fishermen, policemen, and housewives.

The agents' reports were sent to various "cutouts" in Paris—a restroom attendant at the iconic La Pallette restaurant, a teacher at a Berlitz language school, and a concierge at an apartment building, among others—who then passed them on to Garby-Czerniawski. After collating and typing the reports (some of them up to four hundred pages long and containing maps and diagrams), he gave them to a Polish courier, who boarded a train to Bordeaux, in unoccupied France, and secreted them in a hiding place in the first-class restroom. Once the train reached Bordeaux, another Polish operative would retrieve the reports, which were then relayed to London. Eventually Garby-Czerniawski's network acquired several wireless sets and was able to transmit directly to London, providing such a huge volume of information that those on the other end were hard pressed to keep up with it.

Stewart Menzies and MI6, meanwhile, happily took credit for the rich mother lode of information they were receiving from Polish and other European spies. But all that intelligence, highly valuable as it was, paled in comparison with what one historian called "the most important intelligence triumph of this or any other war." Just before the Battle of Britain began, British cryptographers at Bletchley Park, the country's code-breaking center, succeeded in cracking the Luftwaffe version of Germany's fiendishly complex Enigma cipher. Ultra—the name given to the information obtained from Enigma—proved critical in winning the Battle of the Atlantic and the campaigns in North Africa and Normandy, as well as the Allied victory as a whole.

"It was thanks to Ultra that we won the war," Winston Churchill told King George VI. Actually, according to the noted intelligence historian Christopher Andrew, Churchill was overstating it a bit. "Intelligence did not decide the outcome of the war," Andrew observed. "The Red Army and the U.S. did. But the successes of Allied intelligence undoubtedly shortened it—and saved millions of lives."

What almost no one, including Churchill, knew was that Britain's code-breaking success had been due in large part to the French and, above all, to the Poles. The Ultra operation "would never have gotten off the ground if we had not learned from the Poles, in the nick of time, the details both of the . . . Enigma machine and of the operating procedures that were in use," wrote Gordon Welchman, one of Ultra's top cryptographers.

THE STORY OF THIS top secret Allied collaboration began in July 1939, when the leading code breakers of Britain and France were invited by Polish military intelligence to a meeting at a camouflaged, heavily guarded concrete bunker in a forest near Warsaw. Once inside this newly built transmitting station and cipher center, they were shown a small black device resembling a typewriter, with keys that rotated a cluster of three-inch wheels—an exact replica of Germany's astonishingly sophisticated Enigma machine. With it, unbeknownst to the

British and French, the Poles had been reading much of Germany's military and political communications for more than six years—a feat that, in the Germans' estimation, should have taken 900 million years to accomplish.

The visitors responded to the Poles' disclosure with stunned silence. Alfred Dillwyn "Dilly" Knox, an eccentric former classics scholar from Cambridge and Britain's top cryptographer, was particularly upset. The son of an Anglican bishop, the tall, bespectacled Knox was, in the words of a colleague, "a bit of a character, to put it mildly." He combined a keen intellect with an absentmindedness so extreme that he forgot to invite two of his three brothers, one of whom was the Catholic theologian Ronald Knox, to his wedding. Among his closest friends were John Maynard Keynes and other prominent members of the Bloomsbury Group, including Lytton Strachey, E. M. Forster, and Leonard Woolf.

Knox's early training in deciphering ancient Egyptian papyrus fragments had been instrumental in his emergence as a matchless code breaker, beginning in World War I, when he had cracked the cipher used by the German naval commander in chief. For more than a year, he and his colleagues in Britain's underfunded code-breaking agency, called the Government Code and Cypher School (GC&CS), had been studying Germany's latest cipher system, Enigma, but had gotten nowhere in their attempts to decipher it. He found it impossible to believe that the Poles had beaten them to it.

According to Alastair Denniston, who headed GC&CS and was also at the session in Poland, Knox sat in "stony silence" as he and the other British and French participants were briefed on the Poles' success. "It was only when we got back into a car to drive away that he suddenly let himself go and, assuming that no one understood any English, raged and raved that they were lying to us," Denniston said. "The whole thing was a fraud, he kept on repeating—they never worked it out—they pinched [the machine] years ago . . . they must have bought it or pinched it."

The temperamental Knox apparently didn't know—or had forgotten—that Poland had given top priority to intelligence gathering and code breaking since regaining its independence. After World

Marian Rejewski

War II, another leading British cryptographer acknowledged that he and some of his fellow code breakers had been "very slow to admit that the Poles might have anything to teach us."

Dilly Knox, however, was not among them. The day after his outburst, he calmed down considerably when he met Marian Rejewski, the thirty-three-year-old mathematician who had been the first to crack Enigma, and the two colleagues who worked with him. The three young Poles explained the intricacies of the machine and their novel technique for breaking ciphers, called mechanical combination theory, to Knox, who, in Rejewski's words, "grasped everything very quickly, almost as quick as lightning."

Denniston had once remarked about Knox, "He can't stand it when someone else knows more than him." But the British code breaker made an exception when it came to Rejewski. An assistant to Knox later recalled that "Marian and Dilly struck up a bond right away— a true meeting of the minds."

Knox, Denniston reported, soon "became his own bright self and won the hearts and admiration" of the Poles. Even though alcohol was banned at the Polish cipher center, a few bottles of beer were found, and everyone drank to the Polish triumph. When he left the center that day, Knox chanted, *"Nous marchons ensemble"* ("We are traveling together"). On his return to Britain, he sent Rejewski and his colleagues a thank-you note in Polish: *"Serdeznie dziękuję za współpracę i cierpliw ość"* ("My sincere thanks for your cooperation and patience"). Accompanying the note were three silk scarves depicting a horse winning The Derby—Knox's graceful acknowledgment that the Poles had come in first in the Enigma race.

Yet that exploit did not belong solely to the Poles. If it hadn't been for the efforts of a short, stout French military intelligence officer named Gustave Bertrand, they might never have solved Enigma's complexities. In 1933, Bertrand, the head of French radio intelligence, had approached his Polish counterparts with an intriguing story and offer. He told them he had paid a substantial amount of money to an official in the German military cipher department for top secret documents relating to Enigma, including instructions for operating the machine and four diagrams of its construction.

Bertrand's superiors in France had had no interest in the documents, declaring that even with them, Enigma could not be broken. He next approached MI6, which also dismissed the idea. When he contacted the Poles, however, they accepted the material, according to Bertrand, as if it had been "manna in the desert."

The documents were turned over to three new recruits in the Polish cipher bureau, all in their twenties. The standout of the three was Rejewski, a twenty-eight-year-old mathematical genius who had just returned from a year of graduate study at Germany's University of Göttingen, an international mecca for mathematicians.

Armed with the documents, Rejewski and his colleagues built their own Enigma machine, as well as what they called a "bomba," an electromechanical device that allowed them to scan all the possible permutations of the Enigma code at high speeds. (The "bomba" was named after a popular Polish ice cream dessert that the mathematicians were eating when they came up with the idea.)

By early 1938, the Poles were able to decrypt about three-quarters of the Enigma intercepts. The Germans, however, began adding even more complexity to their machine, introducing two new rotors and making significant changes in their methods of enciphering. Hampered by a lack of money and other resources and realizing that war was drawing near, the Poles decided to share what they had accomplished with the British and French. Not long after the visit of Dilly Knox and the others to the forest outside Warsaw and only days before Germany invaded Poland, the Poles sent replicas of Enigma to Britain and France, along with detailed information on how to use it.

Knox and his team went immediately to work on the "Polish treasure trove," as he called it. In the past, the GC&CS had recruited academics from various disciplines for cryptography work, but, like Poland, it had begun to focus on mathematicians, notably including Gordon Welchman and Alan Turing. After thoroughly examining the design and details of Enigma and the Polish "bomba," the shy, absent-minded Turing took what he had learned and built a far more powerful and accurate decoding machine, which he called "the bombe."

In May 1940, just days after Churchill came to power, Bletchley Park code breakers used the bombe to begin cracking the Luftwaffe's version of Enigma; months later, they did the same to the Enigma codes of the German navy and army. The information gleaned from these decrypts was "of almost unbelievably high quality," the British historian M.R.D. Foot wrote. "Operation instructions from Hitler . . . to his supreme commanders were now and again read by his enemies even before they had been gotten into the hands of their addressees."

Since MI6 oversaw Bletchley Park, Stewart Menzies had the daily pleasure of presenting Churchill with an old buff-colored leather box containing the latest priceless information gleaned from Enigma—a box to which Churchill had the only key. For both men, it was a high point of their day. "As 'C' quickly saw, he would never have to fear criticism or cuts in his budget as long as he could drop in on the prime minister at breakfast time with some tasty item of Intelligence," Malcolm Muggeridge noted. About Menzies, Victor Cavendish-Bentinck, the chairman of the Joint Intelligence Committee, remarked, "He would not have held the job for more than a year if it had not been for Bletchley."

And if it hadn't been for a quirk of fate, the young Polish cryptographers whose early work had led to Ultra might well have been working at Bletchley Park alongside Knox, Turing, and the others to produce what Churchill called "golden eggs" of intelligence.

After the fall of Poland, Marian Rejewski had escaped to Romania with his two colleagues, leaving his wife and two young children behind in Warsaw. In the Romanian capital of Bucharest, the three Poles contacted the British embassy, only to be told by a harried diplomat that staffers there were too busy at the moment to deal with them. The mathematicians then went to the French embassy, where they were warmly welcomed and, within a day or two, given travel documents to travel to France.

When Dilly Knox and Alastair Denniston learned of the Poles' escape, they asked the French to send them on to Bletchley Park. "The experience of these men may shorten our task by months," Denniston told the French. His request, however, was rejected by Gustave Bertrand, as was a proposal by the Poles to invite British cryptographers to Paris. Bertrand, who was fiercely anti-German, didn't much like the British, either. He was particularly irritated that they had, as he put it, "profited gratuitously from a Franco-Polish friendship of eight years' duration [that was] sustained by mutual trust." Even though the Poles would have much preferred to work at Bletchley Park, Bertrand was determined to keep them in his own country.

Until the fall of France, the Polish cryptographers worked with him and his code-breaking team at the French military's radio intelligence and deciphering center, housed in a handsome château about twenty-five miles northeast of Paris. Despite Bertrand's frostiness toward the British, his operation cooperated closely with Bletchley Park, daily exchanging decrypts as well as other information and ideas.

When the Germans marched into Paris, Bertrand evacuated his code breakers—not to a safe location outside France but, in an audacious and breathtakingly risky move, to the south of the country, where they set up shop again at a secluded château in the countryside of Provence.

This was now an underground operation: Pétain and the higher-ups in his collaborationist regime knew nothing about it.

For the next sixteen months, Bertrand's team was faced with the daily threat of detection; although Provence was in unoccupied France, the Vichy government allowed German agents, in the guise of armistice commissioners, to move about freely in its territory. The code breakers were given a certain amount of protection by anti-German intelligence officials at Vichy, who tipped Bertrand off about German agents or overzealous Vichy police officers who might be roaming around in the area. Nonetheless, the team remained on constant alert, watching for vans or cars with circular aerials on their roofs—a telltale sign of radio-direction-finding equipment inside.

The cryptographers rarely left the château, whose ground-floor windows were barred and kept shut, making working conditions distinctly unpleasant in the hot, sultry summer of 1940. As a further precaution, three cars were ready, day and night, to whisk the team and their equipment away in case of a sudden German or Vichy police raid. Yet for all their difficulties, the French and Polish code breakers never lost contact with Bletchley Park, providing the British with a constant stream of decrypts about the movements, locations, and equipment of the Reich's air, ground, and naval forces in France and other occupied countries.

Throughout the war, the British and their allies were never free from the worry that the Germans would realize that their Enigma ciphers were being read. But despite repeated indications that the British had advance knowledge of many of their military plans, Reich officials refused to acknowledge that their vaunted machine could possibly have yielded its secrets. Ironically, they preferred to believe that agents of the all-powerful MI6 had somehow obtained information about German plans and tactics on their own and passed them on to Churchill and his government.

CHAPTER 11

"Mad Hatter's Tea Party"

SOE and Its Struggle
to Set Europe Ablaze

W HILE STEWART MENZIES AND MI6 WERE BASKING IN THE RE-
flected glow of Ultra's success, their archenemy in Whitehall was still
struggling to get off the ground. As it turned out, the grandiose vision
painted by SOE's supporters of its future accomplishments was, to put
it mildly, premature.

"We shall aid and stir the people of every conquered country to
resistance and revolution," Churchill had proclaimed in a meeting with
the leaders of the European governments in exile. "Hitler will find no
peace, no rest, no halting place, no parley." Hugh Dalton—who, as
head of the Ministry of Economic Warfare, was in overall charge of
SOE—had promised that by the end of 1940, "the slave lands which
Germany had overrun" would rise up in rebellion, causing Nazi occupation to "dissolve like snow in the spring."

But when the spring of 1941 arrived, the Germans were still firmly
ensconced in Europe, and only a handful of SOE operatives had actually been dispatched there. From its beginning, the new agency had
been faced with an overwhelming array of problems, not least of which
was the fact that most of the officials setting it up hadn't the slightest
idea what they were doing.

With a few prominent exceptions, SOE simply did not have the
kind of leadership one would expect for such a daring, innovative, upstart organization. Hugh Dalton, a major figure in the Labour Party,

had made no secret of his loathing for Britain's landed gentry and aristocracy; it seemed to follow, therefore, that he would avoid the well-bred, well-connected types who populated MI6. But such was not the case. Like Menzies, Dalton, whose father had been chaplain to Queen Victoria and a tutor to the future King George V, had been educated at Eton. That old school tie won out in his recruitment of SOE's central staff, most of whose members came from the informal social networks in London composed of ex-Etonians and the products of other elite public schools.

Unlike MI6, which drew heavily from the military, SOE recruited largely from legal, banking, and other business circles. But as their schooling indicated, the staffs of the two agencies were similar in the sheltered, privileged lives they led. As a result of their insular backgrounds, they knew almost nothing about the real world in Britain or anywhere else. And as citizens of an island nation that had not been invaded or occupied in more than eight hundred years, they had no idea how ruthless an occupier could be.

"Only a country which had withstood foreign invasion really knew what war was," observed the French journalist Eve Curie. Enduring repeated bombing raids, as the British had, was not the same as living with the Germans. As dreadful as those aerial attacks were, the bombers came and went. There was no intimate daily contact with an enemy that, unlike the British, "was not prepared to play the gentleman," in the words of the military historian John Keegan. The Germans, Keegan added, had no compunction about "breaking all laws and conventions against those who challenged them."

"Fear never abated," recalled one Frenchman of his country's more than four years of occupation. "Fear for oneself; fear of being denounced; fear of being followed without knowing it; fear that it will be 'them,' when at dawn one hears or thinks one hears a door slam shut or someone coming up the stairs. Fears, too, for one's family. Fear, finally, of being afraid and of not being able to surmount it."

THE MISGUIDED OPTIMISM ABOUT SOE's potential for immediate success also stemmed from the experience of several SOE officials who had been in Poland during the German invasion as members of a British military mission. The key figure in the group was Brigadier General Colin Gubbins, SOE's first director of operations and training. An anomaly in the agency because of his military background, Gubbins differed from the majority of his colleagues in other striking ways: he possessed an original and daring mind, was widely read and traveled, and was fluent in two foreign languages—French and German. He had fought against guerrilla fighters in the Irish independence movement and then joined Allied troops battling Bolshevik forces in Soviet Russia in the years immediately following World War I. In both encounters, he had been impressed by his foes' swift, sudden attacks and quick retreats. Shortly before World War II began, he had written a pamphlet about such guerrilla tactics, urging the British military establishment to study and learn from them—to no avail.

As a firsthand witness to the 1939 campaign in Poland, Gubbins had admired the courage with which the Polish army had fought against overwhelming odds. But he was even more impressed by the Poles' later determination to fight back while under German control. Even before the country fell, Polish officials had already laid the groundwork for widespread armed resistance. Some nine hundred Poles trained in guerrilla warfare were left in place, and dynamite, grenades, rifles, and pistols were stored in three hundred underground bunkers throughout the country.

More than any other country occupied by Germany, Poland rejected collaboration. Its Home Army—the largest, most sophisticated, and best-organized resistance movement in all of Europe—made it clear it expected all Poles to defy the Germans in every possible way, from noncooperation to outright sabotage.

When their occupiers shut down government institutions such as courts and the national legislature, the Poles re-created them as part of a remarkable underground society. They did the same with schools and cultural institutions, also banned by the Germans. Throughout the country, orchestras and chamber quartets performed behind closed doors, as did troupes of professional and amateur actors. Clandestine classes, meanwhile, were held for more than a million children and young adults. Like other major Polish institutions of higher education, Poland's most esteemed university—Jagiellonian University in

Kraków—offered underground study in all its departments. More than eight hundred students, including Karol Wojtyła, the future Pope John Paul II, attended the Jagiellonian classes.

Colin Gubbins used Poland as his model for the type and scale of resistance that he wanted to foment throughout Europe. The problem, as he soon discovered, was that Poland was unique in its determination to rebel. Shortly after the war, Gubbins told an audience that whereas the shock of German occupation had stunned the peoples of western Europe, "only the Poles, toughened by centuries of oppression, were spiritually uncrushed."

When other European countries had been invaded and occupied, their citizens, in contrast to the Poles, had not known how to react, much less fight back. As one Frenchman noted, "The French have no experience of clandestine life; they do not even know how to be silent or how to hide." In the occupation's early days, the Germans also did not treat western Europeans, whom they regarded as having Aryan blood, as savagely as they did the Poles and other Slavs. While all Poles lived in constant fear of arrest, torture, and death, enemy forces in the Netherlands, Norway, Belgium, France, and Luxembourg were generally well disciplined and more or less polite to those countries' non-Jewish populations, as long as they did nothing to defy German rule.

For most captive Europeans, the paramount aim was simple survival. Notwithstanding their surface politeness, the Germans were ruthless in their occupation policies, sending most of Europe's foodstuffs and energy sources, particularly coal, to the Reich and inflicting great privation on the people under their control. Focused as they were on acquiring the daily necessities of life and anxious to protect their families and themselves from perilous confrontations with the enemy, it's not surprising that the majority of Europeans did not put resistance high on their personal agendas.

Years after the war, Stanley Hoffmann, a Harvard professor who lived in occupied France as a child, had some gentle but pointed advice for historians and others who criticized the failure of much of occupied Europe to stand up to the Germans: "For [people] who have never experienced sudden, total defeat and the almost overnight disappearance of their political elites; who have never lived under foreign occupation; who do not know what Nazi pressure meant; who have never had to worry first and last about food and physical survival . . . the warning must be heeded: do not judge too harshly."

It's important to note, too, that while the vast majority of Europeans were never active resisters during the war, neither were they active collaborators. Most were antagonistic toward their captors, displaying their animosity primarily through silence and social ostracism. In Paris, one observer remarked, "people pass by the Germans without seeing them. They are surrounded by silence . . . in the trains, in the metro, in the street. Each [Parisian] keeps his thoughts to himself. And yet one senses the hostility."

By mid-1941, a growing number of Europeans had overcome their initial sense of shock and helplessness and had begun to embrace more overt signs of passive resistance, as demonstrated by the success of the BBC's V campaign. Still, incidents of active rebellion were extremely rare. Until the people of Europe were willing to take direct action themselves, SOE could accomplish very little.

ACTUALLY, IT WAS PROBABLY just as well that widespread underground activities were not yet taking shape in the early years of the war. If they had been, SOE would have been unable to provide the resources—both human and material—necessary to keep any such flicker of insurgency alive.

Another of the new agency's myriad difficulties was the fact that the officials who had created it had not provided it with the means needed to carry out its mission. It was forced to rely on other government agencies, most of which were opposed to its very existence, for such essentials as communications and transport for its operatives.

MI6, not surprisingly, remained SOE's fiercest opponent. The two competed bitterly throughout the war in their efforts to recruit agents from among European troops in Britain, as well as from the thousands of young foreigners who fetched up there each year. They also fought each other for the use of scarce transportation resources, particularly RAF aircraft, to ferry their operatives (and, in SOE's case, weapons, ammunition, and other equipment) to Europe. The RAF preferred not to use its planes for anything but its bombing campaign against Germany, but when it reluctantly did allocate them for other uses, it tended to assign them to MI6, sharing the intelligence agency's belief that SOE was "an ungentlemanly body it was better to keep clear of."

Air Chief Marshal Sir Charles Portal, the head of the British air staff, bluntly made that point when he wrote to a senior SOE official in early 1941, "I think that the dropping of men dressed in civilian clothes for the purpose of attempting to kill members of the opposing forces is not an operation with which the Royal Air Force should be associated. I think you will agree that there is a vast difference in ethics between the time-honoured operation of the dropping of a spy from the air and this entirely new scheme for dropping what one can only call assassins." (Interestingly, Portal's public school fastidiousness about the ethics of killing did not seem to carry over to Bomber Command's area bombing of German cities later in the war, which resulted in the deaths of hundreds of thousands of civilians.)

Adding to SOE's woes was its creators' decision that it must use MI6's signals and coding systems to communicate with its agents. Furthermore, any intelligence that SOE received had to be sent straight to Claude Dansey, which ensured that MI6 always knew far more about SOE and its operations than SOE knew about MI6. Dansey and Stewart Menzies, both highly skilled bureaucratic infighters, used that knowledge in their ceaseless struggle to bring the upstart new agency under their control or, failing that, to kill it outright.

For the remainder of the war, SOE and MI6, known by SOE staffers as "the bastards of Broadway," engaged in "full-scale and dangerous brawls the like of which Whitehall bureaucracy had rarely if ever seen before," as one historian put it. Malcolm Muggeridge, a bemused witness to the bureaucratic mayhem, later wrote, "Though SOE and MI6 were nominally on the same side in the war, they were, generally speaking, more abhorrent to one another than the [Germans] were to either of them."

NOT UNTIL FEBRUARY 15, 1941, eight months after SOE's creation, did it infiltrate its first agents into occupied Europe. The target country, fittingly, was Poland. After months of wrangling with MI6 and the RAF, SOE was finally able to obtain the services of a single aircraft— a slow, cumbersome, obsolescent Whitley bomber—to carry three Polish parachutists and several crates of weapons more than nine hundred miles to a location not far from Warsaw. The fourteen-hour round-trip journey was, as one SOE official noted, an "extraordinary feat of navigation and endurance, flying blind across occupied Europe in the depths of winter, finding the parachute zone and making it back" to Britain.

Hugh Dalton was elated by the mission's success in dropping the men and weapons, as was the Polish government in exile. Dalton and the Poles began planning for a series of regular flights to supply agents, weapons, ammunition, and other equipment to the Polish Home Army as it mounted sabotage operations against the Germans and prepared for an open revolt later in the war.

In their enthusiasm, however, neither Dalton nor the Poles realized that the flight had actually been a Pyrrhic victory, highlighting the enormous logistical difficulties of air operations to Poland. To reach that country, British bombers had had to cross hundreds of miles of German-occupied territory, braving antiaircraft fire and enemy fighter planes along the entire route. Another significant obstacle was the weather. Even if skies were clear in Britain and over the North Sea when the bombers took off, there were often clouds or rain over Poland, making drops or landings highly problematic. And with the flights taking up to fourteen hours, the bombers' fuel reserves were so limited that even the slightest navigational mistake could be disastrous.

Colin Gubbins and his planning and operations staff were faced with a catch-22 situation: the Poles were by far the most active resisters in Europe, but the logistics of supplying them were considered too difficult for frequent flights. In the future, political considerations would also play a major role in the eventual decision by British policy makers to concentrate on SOE missions closer to home, particularly in France and the Low Countries and later in Yugoslavia.

While Gubbins understood and reluctantly agreed with that reasoning, he could not bring himself "to discourage the Poles from planning operations on the greatest scale," Gubbins's biographer wrote. As a result, the Home Army continued its preparations for a mass uprising against the Germans, unaware that the British had abandoned any idea of helping them. The stage was set for disaster.

THOUGH SOE MAY HAVE been laggardly in carrying out its mandate in the first year of its existence, it was not at all slow in expanding its bureaucratic empire. In October 1940, the rapidly growing staff moved from St. Ermin's Hotel to a large modern office building at 64 Baker Street, just down the road from Sherlock Holmes's fictional lodgings. Within a month, that location was full, and five neighboring buildings were commandeered to house all of SOE's country sections, which soon included not only western and eastern Europe but also the Balkans and the Middle East.

In Scotland and the north of England, several large country estates were requisitioned for the training of agents. Country sections also maintained a number of apartments in west London neighborhoods such as South Kensington and Marylebone, where agents were interviewed, briefed, and sometimes housed without ever knowing where SOE's main offices were.

As with MI6, a dense aura of secrecy enveloped SOE's operations. Almost no one outside the organization, except for a relatively few high government officials, even knew it existed. No version of its name ever appeared in the London phone directory, and its heavily guarded buildings were adorned with brass plaques bearing vague cover names, such as "Inter Services Research Bureau." There were no nameplates on the doors of its offices. Although most of the European country sections were in the same building, they had almost no contact with one another, thanks to an SOE rule forbidding any sharing of information between sections.

In this new, unwieldy fiefdom, there was a topsy-turvy atmosphere from the very start. With no prior knowledge or understanding of clandestine operations, most SOE staffers were making things up as they went along. "Of course, it was amateurish," a staffer said. "We were all amateurs doing our best. There was nobody there who was a professional anything." When one SOE officer was asked by a colleague why he had adopted "White Rabbit" as a code name, the officer replied, "I work for a fucking mad hatter's tea party. Can you think of a better reason?"

Other than Colin Gubbins's pamphlet, SOE had no guidelines to follow in setting up its training in the arts of "ungentlemanly warfare." As Lord Selborne, who replaced Hugh Dalton as SOE's ministerial overseer in 1942, acknowledged, "Underground warfare was an unknown art in England. . . . There were no textbooks for newcomers, no old hands to initiate one into the experiences of the last war. . . . Lessons had to be learned in the hard school of practice."

Because of the need for secrecy, SOE employed no advertising or other public means of searching for potential agents. Instead, staffers combed the ranks of émigrés and made discreet approaches to British friends and acquaintances for the names of people fluent in at least one foreign language who might be interested in "special foreign service." "Entry into SOE was so largely a matter of accident that there was nothing which deserved the name of a recruiting system," observed the historian M.R.D. Foot. When potential agents were approached, they were told virtually nothing about what the job would entail. "Interviewers," said one SOE officer, "were under orders not to disclose the function of SOE or even its name."

Francis Cammaerts, the Cambridge-educated, French-speaking son of an eminent Belgian art historian and poet, recalled that at one point during his interview with SOE, an official vaguely suggested that he might be able to use his French in North Africa. "The idea that I would go into occupied France didn't even occur to me," he said. "I didn't know that anyone was doing that sort of thing." Cammaerts, who would become one of SOE's top agents in France, added, "The name SOE was never mentioned. I never heard those initials until after the war." Indeed, many recruits believed they were joining MI6.

When it came to the selection and training of agents, those headed for Poland and Czechoslovakia were handled far differently than those assigned to Norway, the Low Countries, and France. Because both the Poles and Czechs had functioning resistance movements that predated SOE's creation, the agency exerted much less control over their operations, confining itself to providing transport and communications facilities, along with parachute, weapons, and other specialized training. The Polish and Czech governments in exile were responsible for selecting the SOE agents from their countries, most of whom were already highly trained veterans of their military branches.

In the case of Norway, SOE was responsible for choosing the agents, all of whom were Norwegian; in that country's small, intimate, and closed society, foreigners, even those who spoke Norwegian well, would have been dangerously obvious. Conducted in the mountains of Scotland, the Norwegians' training was paramilitary in nature, focusing on commando work and on learning to survive for weeks, if not months, in harsh, remote, wintry terrain akin to that of the agents' homeland. "Even hardened Norwegians found survival tough in those conditions," noted an observer. "Few British agents would have made it."

Agents selected to work in Holland and Belgium were also, for the most part, natives of their countries. By contrast, the agents sent to France were an astonishingly diverse group, coming from a wide variety of nations and social backgrounds that ranged from pimps to princesses. They included businessmen, teachers, journalists, a *Vogue* fashion artist, a traveling acrobat, the former head of a Paris fashion house, and a receptionist from a hotel in London's West End. Besides speaking French, the one thing that almost all of them had in common was that, like much of the staff that selected them, they were rank amateurs in the business of war.

The agents who made it through the first stages of screening were sent to various remote country houses (each country section had its own house) for rigorous physical training. If they passed that hurdle, they were dispatched to another country estate for military training, which involved learning how to use pistols, grenades, and other small arms. At that point, most of them still had no idea what potentially lay in store for them. Only if they made it to the third round of training, which taught them the basics of survival in an enemy-controlled country, were they told they had been selected for drops into occupied Europe.

As SOE prepared these operatives to lead clandestine lives, their instructors were handicapped by knowing almost nothing about the current conditions in occupied Europe. There also was little information about German methods of detecting and suppressing individuals who opposed them. As a result, SOE staffers were just guessing "at the sorts of things they were instructing us on," Francis Cammaerts said. "They were trying to teach us something that they themselves didn't know."

Cammaerts—who was later caught by the Gestapo in France, only to be rescued by a female SOE colleague just hours before he was to be executed—made clear after the war that the gentlemanly British were unequipped to prepare agents for the real-life savagery of the enemy. During his final round of training, he, like other prospective operatives, had been pulled out of bed in the middle of the night and interrogated by SOE instructors in Gestapo uniforms. In retrospect, he said, "that was pretty useless. You couldn't have them doing [what the Gestapo did], such as pulling out fingernails." His faux German interrogators, Cammaerts added, were "very severe and asked really hard questions—but that was not what it was like. Those likely to know exactly what happened under such circumstances were hardly likely to have survived."

SOE spent a great deal of time and energy trying to make sure that agents could pass muster with the false identities they were given before they were dispatched to Europe. They were provided with forged identity cards and other documents, as well as the kinds of clothing and other personal effects worn and used by residents of the countries to which they were sent. Above all, they were told, they must never draw attention to themselves. Instead, they should do everything possible to blend into the background.

Not infrequently, though, SOE itself was responsible for making such unobtrusiveness impossible. One agent, newly arrived in France, was on a train when a Frenchwoman turned to him and said, "Oh, I love your shoes. My brother bought shoes exactly the same as yours in London." The agent later remarked, "Can you imagine how I felt? I'd barely been in France a couple of hours and I'd already been noticed because my shoes were conspicuous, because at that time you couldn't buy leather in France."

For an agent, the key element in keeping a low profile was an impeccable command of the language of the country to which he or she was sent. Yet a number of the operatives dispatched to France were known for their mangling of the language or for speaking it fluently with a strong foreign accent.

While such blunders and oversights were due in part to SOE's amateurishness, they were also attributable to an increasingly anxious, frenetic atmosphere within the agency. Having little or nothing to show for its first year of operation, SOE was under intense pressure from Whitehall, especially from its foes in MI6, to justify its existence. In its hurry to produce results, it was not as thorough and careful as it should have been in the screening and preparation of agents. Indeed, it sometimes sent operatives into action who should never have been dispatched at all.

One startling example was an agent headed for Belgium, who, during his training, had been rated as "absolutely appalling" by SOE security officials. They noted his "fondness for drink," irresponsibility with money, and a tendency to pick up "the most awful women." These negatives were ignored by SOE's Belgian section, which sent the man to Brussels anyway. Within days, he had picked up a peroxided blonde and taken her to a prominent hotel that was "absolutely stiff with the Gestapo." His arrest soon followed, which, according to an SOE official, resulted in the arrest and execution of eighteen members of the Belgian resistance.

Another red-flag agent was an exotic twenty-six-year-old WAAF officer named Noor Inayat Khan, whose father came from Indian Muslim nobility and whose mother, an American, hailed from Albuquerque, New Mexico. Born in Moscow, Khan had grown up in London and Paris, where she had studied piano at the Paris Conservatory under the famed composer and teacher Nadia Boulanger.

Maurice Buckmaster, the head of SOE's F [French] Section, described Khan as "a sensitive, somewhat dreamy girl" with a penchant for absentmindedness. In a fitness report, one of her instructors wrote that she "tends to give far too much information. Came here without the foggiest idea what she was being trained for." Two SOE colleagues who trained with her urged Buckmaster not to send her to France. The officer in charge of instructing her and other would-be operatives in survival tactics wrote that she was "temperamentally unsuitable" to be an agent and would be a major security risk in the field. He based that judgment in part on a mock interrogation of Khan by a Bristol police superintendent, whose force worked with SOE. After the interrogation, the superintendent informed the agency that "if this girl's an agent, I'm Winston Churchill." Despite all the warnings, Khan was dispatched to France as a wireless operator—the most difficult and dangerous job an agent could be assigned. She was later caught up in one of SOE's biggest fiascoes of the war.

Years after the conflict, Major Hermann Giskes, an Abwehr officer in wartime Holland and arguably SOE's most successful German adversary, had some harsh words for "the amateurish way the British had gone about organizing SOE. They built it up too hastily and set in motion a machinery which was too vast, too ambitious, for the means at their disposal. SOE needed professional intelligence officers at its head, as we had in the Abwehr, and in particular it needed men adept at subversive warfare. Instead they sent us infants, keen and willing, but quite unfitted for that kind of combat."

For the first three years of SOE operations, Giskes's unsparing analysis was undoubtedly correct, at least in regard to France, Belgium, and Holland. Later the agency began to learn from its mistakes and ended up having a significant impact on the war in those countries. But its early seat-of-the-pants approach came at an appalling cost: the deaths of dozens of agents, not to mention the hundreds of Europeans who lost their lives by working with often unqualified and ill-prepared SOE officers in resisting the enemy.

CHAPTER 12

Factions, Feuds, and Infighting

The Shock of Exile

ON AN ICY MORNING IN MARCH 1941, MORE THAN FOUR HUNdred British Army commandos, accompanied by several dozen SOE-trained Norwegians, landed on two remote Norwegian islands high above the Arctic Circle. In a matter of hours, they had forced the surrender of the islands' small German garrisons, destroyed German and Norwegian ships in the harbors, and blown up four fish-oil factories.

The inhabitants of the snow-covered islands, in the rugged Lofoten chain off the coast of northern Norway, were overjoyed. They turned out en masse to welcome the raiders and guide them to their targets. According to one observer, many of the islanders were "almost ready to fight each other as to who should answer the British officers' questions." Several hours later, as the commandos prepared to return to England, more than a hundred young Lofoten residents, most of them fishermen, insisted on going along. The majority wanted to join the elite Norwegian SOE team that had participated in the raid. Officially called Norwegian Independent Company No. 1, it was commonly known as the Linge Company, after its tough, aggressive commander, Martin Linge, a former actor from Oslo who had fought in the 1940 battle for Norway.

Militarily, Operation Claymore, as the raid had been christened, was of no real significance. The starkly beautiful Lofotens were not strategically important, the German presence there was almost nil, and the destruction of fish-oil factories was hardly a major coup. Nonetheless, the British touted Claymore as a triumph, "a classic example of a perfectly executed commando raid." It was left unmentioned that the British desperately needed a victory of some kind, regardless of size or significance.

The spring of 1941 was one of Britain's lowest points in the war. Although it had survived the Battle of Britain, German bombs still rained down on its cities. Merchant shipping losses in the Atlantic had risen to astronomical proportions, and starvation for British civilians loomed as a distinct possibility. The British Army, meanwhile, had suffered one disaster after another. In the course of those dire months, Germany had conquered Yugoslavia and overpowered Greece, routing British forces there and on the island of Crete. In the Middle East, the early British triumphs over the Italians in Libya had turned to dust when German troops under General Erwin Rommel had rushed to the Italians' rescue. In only ten days, the Germans had regained almost all the ground that the British had captured.

That melancholy string of defeats led to an upsurge of parliamentary criticism of Winston Churchill and his government. Acknowledging a sense of "discouragement and disheartenment in the country," Churchill wanted action—any action—against the Germans to prove to the world that Britain was not defeated. "We are in that awful period when everything is going wrong, and those in authority feel they have to do something," Alexander Cadogan, the permanent undersecretary of the Foreign Office, noted grimly in his diary.

Churchill had been the guiding light of the Lofoten raid. With his innate love of adventure and danger, he had been drawn all his life to such daring, flamboyant, unconventional enterprises. What's more, Operation Claymore required only a relatively small attack force, with little or no risk of serious casualties.

In its self-congratulations over the raid's success, the British government deemed it the "perfect example of Allied collaboration." In fact, no such collaboration had existed: prior to the operation, the British had consulted neither the Norwegian government in exile nor the leaders of Norway's nascent underground army, known as Milorg, which had been slowly and cautiously expanding since its founding shortly after Norway's defeat. Milorg, made up of small, informal groups of young Norwegians who had fought against the Germans in 1940, had few weapons, little security, and no formal military training.

Although many Norwegians had engaged in acts of civil disobedience against their occupiers, most shied away from taking part in sabotage or other forms of direct resistance. To do so, they believed, would be suicidal. Because of Hitler's unshakable belief that Britain planned to invade Norway at some point during the war (a fear that Churchill encouraged), the Germans had turned the country into an armed fortress, defended by formidable coastal batteries, warships, submarines, and aircraft. More than 300,000 highly trained, well-equipped Wehrmacht troops—one German for every ten Norwegians—were stationed there.

Acutely aware of its members' lack of preparation and experience in clandestine activities, the Milorg leadership had only one aim: the gradual buildup of a secret army to take part in the liberation of Norway and the rest of Europe. That was not good enough, however, for a steady flow of impatient young Norwegians like Martin Linge, who, from the beginning of the occupation, had escaped to Britain by boat across the North Sea. Anxious to strike back hard against Germany, a good number of them had volunteered for the special SOE unit of Norwegian volunteers headed by Linge but operated under overall British command.

Both Milorg and the Norwegian government in exile were upset when informed of the Lofoten raid. At the very least, Norwegian officials told the British, they should have been consulted about an operation staged on their country's soil and carried out by more than fifty of their own citizens. What angered them most, however, were the immediate German reprisals against the islands' inhabitants. Several dozen homes were destroyed and more than seventy residents arrested and sent to a concentration camp. For its part, Milorg contended that the raiders' destruction of factories and fishing trawlers had been far more damaging to the islanders' livelihoods than to the German war effort.

Unmoved by the Norwegians' complaints, SOE dismissed them out of hand. The agency's Norwegian section contemptuously dubbed Milorg "a military Sunday school" and declared that sabotage was essential "to cause Germans in Norway as much trouble as possible and to force them to keep large garrisons there." The Norwegians, SOE made clear, would have no say in the matter.

To prove the point, British commandos and members of Company Linge staged another, much larger raid on the Lofotens and nearby coastal towns in late December 1941, nine months after the first operation. Nearly 15,000 tons of shipping were sunk, German installations and gun emplacements were destroyed, and 150 Germans were killed, with 98 taken prisoner.

Assuring the local population that this time they had come to stay, the raiders received another warm welcome. Residents took the commandos into their homes, held public demonstrations of support, and helped identify local collaborators. But the joy over their sudden liberation vanished as quickly as it had appeared. The day after the raid, German planes bombed the Lofotens, and British intelligence warned that German troops were massing in northern Norway, apparently with the intention of staging a counterattack. After receiving that news, the expedition's commander ordered an immediate evacuation of British and Norwegian personnel.

As the raiders marched back to their ships, the islanders cursed and spat at their erstwhile saviors, who were fleeing, as the Norwegians saw it, without so much as a fight. As an SOE report later noted, the local population was furious that the British once again had scored a major propaganda triumph with few casualties of their own, while Norwegian citizens were left facing "the horrors of German reprisals." The vengeance was swift to arrive. SS squads descended on the Lofotens, destroying homes and businesses and dispatching hundreds of people to concentration camps, many of them relatives of the young men who had earlier escaped to Britain.

Once again, the Norwegian government in exile erupted. This time, its outrage was shared by more than twenty Linge Company members, whose leader, Martin Linge, had been one of the mission's few Allied fatalities. Demoralized by Linge's death and the latest reprisals against their countrymen, they declared that unless they received prior authorization from the Norwegian government, they would refuse to take part in any future operations. Faced with such insubordination, the British government realized it could no longer brush aside Norway's disaffection.*

NORWEGIAN OFFICIALS WERE HARDLY alone in feeling resentment toward the British. Accustomed to exercising power and command within their own countries, all the European governments in exile had a profoundly difficult time adjusting to their dependence on the nation that had given them refuge. For each government, its relationship with Britain, unequal as it was, was essential to its survival. For Britain, with its myriad problems and responsibilities, the relationship was just one among many.

Still trying to cope with the humiliation and trauma of defeat, the exile governments were also entangled in bitter, explosive battles within their own ranks. "Political émigrés are strange people," Josef Korbel, Madeleine Albright's father, later reflected. "Uprooted from their national environment and deprived of a political base, they struggle among themselves for power."

Even before the war, the political scene in most of Europe had been highly fractious, reflecting each country's social, economic, and religious divisions. European prewar governments, for the most part, were coalitions stitched together from several political parties with often widely divergent views. Such groupings tended to be fragile and short-lived, enduring frequent crises and behind-the-scenes machinations for as long as they existed.

These homegrown pressures and strains were exacerbated by the twin shocks of defeat and exile. "Intrigues flourished like toadstools in London's hothouse atmosphere," Erik Hazelhoff Roelfzema observed, "carried out by men with many personal and political scores to settle." The backbiting and finger-pointing were especially evident in internal debates about which officials and political parties were most to blame for their countries' defeat by the Germans.

---

* Unknown to SOE, one of the key aims of the Lofoten raids was to seize Enigma machines and operating manuals from captured German ships. Those materials would play a role in Bletchley Park's later success in cracking the German naval Enigma settings.

For young Europeans who had managed to escape from German rule, the quarrels and animosities of their government leaders were a source of anger and disillusionment. Speaking of the Dutch government in exile, Hazelhoff Roelfzema remarked, "They lived in a world of jobs and salaries, promotions and raises, which to us escapees, after fifteen months of occupation, was as illusory as the cell and firing squad were to them. . . . They were unaware that reality had left them behind."

British officials, for their part, grew increasingly impatient with their European guests' factions, feuds, and infighting. Few Britons were as empathetic as Mary Churchill, the young daughter of the prime minister, who later noted, "For the British, life had become very simple. We meant to fight; we thought we would win; but we would fight anyway. We were spared the [Europeans'] agonies of divided loyalties and complicated issues."

INTERESTINGLY, DESPITE THE NORWEGIANS' unhappiness over the raids on the Lofotens, their liaison with the British turned out to be the smoothest of all the wartime Anglo-European relationships. Thanks to its abundant resources, Norway was one of the few occupied nations able to pay its own way during the conflict. That was due largely to the income from its merchant fleet, which transported nearly 60 percent of Britain's oil and half of its foodstuffs, thus also playing an invaluable role in that country's survival.

Unlike other governments in exile, the Norwegians also made things easier for the British by not bringing a complicated political agenda to the table. Their main postwar goal, the liberation and independence of their country, never caused a problem for the strategic interests of Britain or the two other powerful nations that would soon join the alliance, the Soviet Union and United States.

In almost every other way, the Norwegians were low-maintenance guests. Although their officials always spoke up when they thought their interests were at stake, they committed themselves from the start to working closely with their British counterparts. In mid-1942, the cooperation paid off, with a resolution of the Lofoten furor. Agreeing to end all raids on Norway's territory without its government's approval, British officials announced that from then on, they would work with the Norwegians to create a partnership among SOE, the Norwegian high command, and Milorg. "In time, all realized that it was impossible to run two independent paramilitary underground movements side by side," acknowledged the head of SOE's Norwegian section. "Inevitably, it would lead to . . . the two cutting each other's throats." The new collaboration proved, in the words of one historian, to be "remarkably successful."

In February 1943, that cooperation led to arguably the most dramatic and daring Allied sabotage coup of the war—the partial destruction of the Norsk Hydro electrochemical plant in Norway responsible for producing the heavy water used in making a nuclear bomb. After the French had spirited away all the existing heavy water from the Norsk Hydro plant in March 1940, the Germans, following their occupation of Norway, had greatly stepped up production there. In late 1942, workers at Norsk Hydro sent word to the Allies that their German masters were on the verge of sending large quantities of heavy water to the Reich.

On Winston Churchill's orders, a small band of SOE-trained members of Linge Company were dropped onto one of the harshest, coldest, and bleakest terrains in Norway—the Hardanger mountain plateau near Norsk Hydro. Battling heavy snow, fierce winds, and below-zero temperatures, the Norwegians slogged their way to the factory—a seven-story building perched like a medieval castle halfway up a steep mountainside, whose apparent sole access was a heavily guarded seventy-five-foot suspension bridge spanning a gorge six hundred feet below.

The saboteurs chose a different route—rappelling down the sheer cliff to the gorge, crossing it, then climbing up the other side. Evading the factory's many German guards, they overpowered two Norwegian watchmen and slipped inside, placing timer-operated explosive charges and fuses around the heavy water tanks. By the time the explosives detonated, the Norwegians had vanished, returning the same way they'd come. None was caught.

Although the explosion caused massive destruction and the loss of some 500 kilograms of heavy water, the Germans had the factory back in operation within a few months. Finally, an Allied air raid and another, smaller SOE sabotage operation ended Germany's heavy water production altogether.

As it turned out, the Reich had never made much of an effort to produce a nuclear bomb—a fact that the Allies discovered only after the war. Certainly, the repeated attempts to deny heavy water to the Germans helped to discourage them in that endeavor.

THE SUCCESS OF THE Norsk Hydro mission contributed to the strengthening of ties between Norway and Britain, as did King Haakon's popularity with the British people. The Norwegian monarch's influence was enhanced by his intimate relationship with King George VI, who once told his daughter, the future Queen Elizabeth II, how much the "unshakable courage and resolution" shown by his uncle had "supported and uplifted him during those heavy days." As the rallying point for his own people, Haakon also proved to be an essential link between those living under German domination in Norway and their compatriots in London.

As was true in every occupied country, a huge psychological divide existed between those who had left Norway at the time of its defeat and those who had been given no such alternative. Since 1940, many Norwegians at home had been vocal in their criticism of the government officials now in London for the deplorable condition of the nation's defenses that had helped lead to the German victory.

At one point, considerable pressure was put on Haakon to push the government aside and take over leadership of the country. He emphatically rejected that idea, just as he had spurned the earlier proposal from Norwegian collaborationists that he abdicate. Both suggestions, he noted, were blatant violations of the Norwegian constitution. Proclaiming his solidarity with the current government, Haakon declared, "We are all in the same boat. . . . Mutual trust is essential for Norway's struggle for freedom."

When Haakon turned seventy in August 1942, his subjects in Norway joined their compatriots in London in massive celebrations of the man who had emerged as their nation's most important unifying figure. In cities and towns all over Norway, tens of thousands of people carrying flowers and wearing badges emblazoned with "H7" (Haakon VII) marched to honor their king. In London, more than five thousand Norwegians, including top government officials, paraded past Haakon and his son, all of them headed for a huge birthday party for the monarch at the Royal Albert Hall. It was the largest gathering of Norwegians to take place during the war.

LIKE KING HAAKON, Queen Wilhelmina of the Netherlands also became a major player in her nation's wartime affairs. Highly respected by British officials and the British public, she, too, acted as a bridge between members of her government and her countrymen back home. But unlike Haakon, the combative Wilhelmina was no peacemaker, trying to ease and remain above fractious exile politics. Instead, elbows out, she charged right into the fray.

Wilhelmina's struggle with her government ministers began almost as soon as they all had arrived in London in May 1940. Several members of the Cabinet, including Prime Minister Dirk Jan de Geer, had not wanted to come to the British capital at all. A fervent pacifist, de Geer was convinced that Germany would win the war, and he initially wanted the Dutch government to approach Hitler to seek a compromise peace. After losing that struggle, he argued that the government should move from London, which he feared would be invaded by the Germans or destroyed by bombs, to the Dutch East Indies, more than seven thousand miles away.

Only a couple of ministers were opposed to leaving Britain. They were joined by the queen, who was appalled and outraged by the defeatism she saw in a majority of her Cabinet. Wilhelmina was determined to fight on in London. If Germany invaded, she planned to try to cross the Atlantic to join her daughter, Princess Juliana, in Canada. But if that should prove impossible, she had already ordered her private secretary to shoot her before the Germans could capture her. She told de Geer she would not go to the East Indies, that her health would not permit such a long, arduous journey. In an audacious and unprecedented move, she also informed the prime minister that she had lost all confidence in him. He promptly offered his resignation, which she just as promptly accepted.

Such a display of queenly displeasure would never have been successful at home in the Netherlands, where she had no real authority and the Cabinet and parliament ruled. But in London, there was no parliament. The Cabinet now had to take her views into account: any action it wanted to take required her approval. If she withheld her signature, there was no other government body that could overrule her.

For Wilhelmina, exile meant power, and she took full advantage of it. Having resigned as prime minister, de Geer expected that he would be allowed to stay on as minister of finance. The queen informed him otherwise, then named as prime minister the only member of the Cabinet she thought shared her fierce hatred of the Nazis and determination to fight them to the end. He was Pieter Gerbrandy, the minister of justice, who had recently entered politics after teaching law at the University of Amsterdam.

Outwardly, Gerbrandy was hardly prepossessing. Standing only four feet, eight inches, he had a luxuriant mustache that, as one friend put it, "dropped incongruously from his small, round face like the whiskers of a walrus." His command of English left much to be desired: at his first meeting with Winston Churchill, he put out his hand and said, "Goodbye." The amused Churchill, who became quite fond of the man he called "Cherry Brandy," replied, "Sir, I wish that all political meetings were as short and to the point." But as Wilhelmina knew and Churchill soon realized, Gerbrandy was no figure of fun. Fearless and tough, he believed that the war should be prosecuted with the utmost vigor, a conviction he backed up with all the resources of his country, including its merchant fleet and the riches of the East Indies.

Having helped keep the Netherlands in the war, Wilhelmina now sought to transform her own life. Thanks to her move to London, the doors to her hated "cage" had finally sprung open, and she was no longer cut off from the real world. In the British capital, she had been given what she had always yearned for: the chance "to meet people as they really were, not dressed up for a visit to the palace."

Although she led an active social and official life in London, the queen's main focus was on her people back home. She insisted on meeting every Dutch citizen who escaped to England, often inviting them to tea at her small house in Chester Square. The escapees, known as *Engelandvaarders* ("England farers"), told Wilhelmina how important her fiery BBC broadcasts had become to her compatriots and how she had emerged as the prime symbol of hope and freedom in the Netherlands.

"For the Queen, there was only one good Dutchman, the *Engelandvaarder,* the man who had risked his life to come and fight for freedom," a Dutch writer observed. Erik Hazelhoff Roelfzema, perhaps the best-known and most defiant *Engelandvaarder,* noted, "The simplest sailor from Rotterdam commanded more of her attention than the highest functionary in the government-in-exile."

Roelfzema had met Wilhelmina shortly after his escape from Holland in June 1941. Tall, blond, and good-looking, the former law student had been a rebel for much of his life against the rigidly structured society of Holland and the staid, conservative lifestyle of most of his countrymen. Shortly before the war began, he had traveled across the United States by hitchhiking and hopping freight trains, then written a bestselling book about his experiences.

When he arrived in London, he and Peter Tazelaar, another escapee, had ricocheted around various departments of the Dutch government in exile, trying to find a way to get into the fight. They were soon "suffocating in the porridge of bureaucracy," Roelfzema said. "Whether it was me or my dirty clothes, the subjects I raised or the intensity of my arguments, I made everyone visibly nervous. . . . If I used words like 'Occupation' or 'secret contacts,' they recoiled as if I showed signs of advanced leprosy." When the Dutch war minister told Roelfzema that he was "really too busy" to see him, the frustrated young escapee smashed his fist on the official's desk and upset his teacup. "Greatly alarmed, he hastily terminated the audience," Roelfzema recalled, "by calling in the MPs and having me thrown out of his office."

His and Tazelaar's reception by the queen, however, was an altogether different experience: "Instead of unbalanced adventurers, we were suddenly treated as exceptional people. After . . . these last few weeks of humiliation by our own countrymen in London, we hardly knew any more how to accept respect, let alone admiration."

Wilhelmina was charmed by the two young men's dash and daring, by their thumbing their noses at Dutch officialdom, but, above all, by their determination to defy and defeat the Nazis. With her blessing, the Dutch government's intelligence chief installed them in a mews house behind her Chester Square home and sanctioned their plan to establish better connections between the Dutch resistance and the English and Dutch intelligence services. Over the next few months, they and a couple of associates made several trips to deliver radio equipment, agents, and light arms to Holland; one mission, in which Tazelaar was put ashore to contact the underground, was particularly audacious.

Erik Hazelhoff Roelfzema (right) and Peter Tazelaar (left)

Early one morning in November 1941, a fishing boat quietly dropped anchor off a beach near The Hague. Tazelaar swam to shore, zipped off his wet suit—and revealed white tie and tails underneath. Removing a small bottle of Hennessy XO cognac from his pocket, he took a swig and sprinkled a few drops over his elegant evening clothes. Only then did he stroll nonchalantly past a luxury seaside hotel crawling with German officers and hop on a tram—just another tipsy young Dutchman on his way home from a long night of partying. (Tazelaar's exploit inspired the opening scene in the James Bond film *Goldfinger,* in which Bond goes ashore wearing a tuxedo under his wet suit.)

Tazelaar and Roelfzema, as it happened, were less intimidated by the Germans than they were by their queen. When he first met Wilhelmina, Roelfzema recalled being totally dumbstruck. For his entire life, the remote, aloof monarch of the Netherlands had been "the focal point of my existence," as she had been for other Dutch citizens. The fact that she was also a human being, he said, had never entered his mind. She, in turn, was initially shy and awkward with him, trying to reach out but having little experience in doing so. Ever since her childhood, Wilhelmina later noted, she had been afraid that "people would laugh at me if I showed too much feeling for them." That feeling never quite left her, reinforcing her reserved demeanor toward others. But in London, she made a determined effort to unbend a little, especially with young *Engelandvaarders* such as Roelfzema.

As the months passed, he and the queen became closer, engaging in several discussions about wartime conditions in Holland. "I got the impression that she enjoyed our informal, democratic relationship," Roelfzema observed, "and liked to experiment with the ways of common folks." During one conversation, she took out a pack of English cigarettes and asked him if he'd like to smoke. Roelfzema was astounded. "Everyone in Holland knew of the Queen's fierce antagonism to smoking," he said. "Cigarettes were not even allowed inside any palace where she lived." But she continued to hold out the pack, and he finally took a cigarette. As he did so, he realized the significance of her action: "That was the past; she had done with it. She knew how to behave like ordinary people now, and she would no longer inflict her personal preferences on them."

———

WHILE WILHELMINA AND HAAKON came into their own during their London exile, the opposite was true for Czechoslovakia's Edvard Beneš, whose sojourn in the British capital was, at least for the first two years of the war, an exercise in frustration and humiliation.

President of Czechoslovakia at the time of the Munich agreement, Beneš had resigned under German pressure five days after the pact was signed. He had traveled first to Britain and then to the United States, where he had taught sociology at the University of Chicago. When war approached in the summer of 1939, he had returned to London, arguing that because Germany had taken over the whole country a few months before, he should be reinstated as Czechoslovakia's rightful leader.* While the Czech government that had succeeded his regime was initially legal, it had forfeited any legitimacy, he contended, by becoming a docile front for German rule.

Beneš and his associates, including Jan Masaryk, the former Czech minister to Britain, asked the British government to recognize them as their nation's government in exile. Neville Chamberlain and his subordinates were appalled at the thought. Not only did they refuse the request, they told Beneš they would not grant him political asylum unless he promised to refrain from all political activity while in Britain.

The government then did its best to forget about the country it had betrayed and about Beneš, who was tucked away, out of sight and certainly out of mind, in a small redbrick house in suburban London. "The men of Munich had to find a scapegoat for what they had done, and Dr. Beneš was the obvious choice," said Robert Bruce Lockhart, a former British diplomat and journalist who served as an unofficial liaison between Beneš and Whitehall. The government's response, Lockhart added, was "a tragic illustration of the dislike that men feel for those whom they have wronged." Having ordered Beneš not to fight at

---

* In its takeover of Czechoslovakia, the Germans cut it in two. The western two-thirds of the country was incorporated into the Reich as the Protectorate of Bohemia and Moravia. Unlike the protectorate, which was under direct German occupation, the Slovakian region was allowed to secede, becoming a Nazi satellite state. Its government was composed of Slovaks who did what the Germans told them to do.

the time of Munich, British officials now blamed him for giving in to Hitler too easily.

For a veteran statesman like Beneš, this cold-shoulder attitude was both a shock and a personal affront. Along with Tomáš Masaryk, he had been instrumental in the creation of Czechoslovakia after World War I, convincing the victorious Allied powers to grant it independence. He then had helped transform it into the most industrialized, democratic, and prosperous state in eastern Europe.

The man who once had lived in an ornate fifteenth-century palace in Prague was now confined to a cramped bungalow in the London suburb of Putney. On the rare occasions when he and his associates were invited to official diplomatic dinners and other gatherings, they were given the least important seats and stood last in every receiving line. On Sunday evenings, when the BBC played the national anthems of the Allied nations whose governments were in London, the Czech anthem was omitted.

When Winston Churchill succeeded Chamberlain in May 1940, Beneš and the Czechs in London had high hopes that, as an outspoken opponent of Munich, the new prime minister would help them. The following month, Churchill did acknowledge Beneš and his ministers as the provisional government in exile, but he failed to grant the full recognition that would give them equal status with the other European exile governments.

Thus the Czechs' humiliation continued. At the first inter-Allied conference, Czechoslovakia was ranked last because of its government's provisional status. On the country's Independence Day in 1940, the only Allied government officials to attend a reception at the Czech government offices were Robert Bruce Lockhart and the Norwegian chargé d'affaires.

As the provisional government embarked on a long, arduous struggle for full recognition, the reserved, austere Beneš stayed in the background, giving way to Jan Masaryk, now the provisional foreign minister, who launched a masterful campaign to publicize the Czech cause. In his relentless lobbying of British officials, several of whom were his close friends, Masaryk cajoled, wheedled, and argued. He

Czech president Edvard Beneš speaking with Czech airmen in Britain.

claimed that Beneš's resignation as president had been invalid because it had been coerced by the Nazis. He contended that the Munich agreement was illegal because it had been signed without Czech approval. Pointing out that his country's pilots and troops were now fighting under British command, he sardonically asked if the deaths of several Czech fliers during the Battle of Britain should be considered as provisional as their government.

When Churchill inspected Czech troops at their training camp near London in April 1941, Masaryk took full advantage of his visit. Knowing that Churchill was in the depths of gloom over recent British military setbacks, he suggested that the soldiers learn several British patriotic songs before the prime minister came. After the inspection, as Churchill stepped into his car to return to London, the troops broke into a rousing rendition of "Rule Britannia." That stirring paean to British imperial might had its desired effect: Churchill, his eyes welling with tears, left his car and sang along. That day, when Beneš again brought up the matter of full recognition, Churchill declared, "This must be put right. I'll see to it." Three months later, with the strong endorsement of Foreign Secretary Anthony Eden,* Britain formally

---

* Eden replaced Lord Halifax as foreign secretary in early 1941.

recognized Beneš and his ministers as Czechoslovakia's official government in exile. In August 1942, the British government withdrew its signature from the Munich Agreement, thus declaring it invalid.

Although Beneš had finally got what he wanted, he never lost what associates called his "Munich complex." Haunted by the British and French betrayal of his country and by his own humiliation, he became increasingly obsessed with gaining prestige and influence for himself and Czechoslovakia—aims that were understandable but that would nonetheless lead to short-term tragedy and a disastrous long-term future for his nation.

FOR CHARLES DE GAULLE, prestige and influence were key goals, too. More important, however, was independence. Even though de Gaulle and his men could never have survived without British support, financial or otherwise, de Gaulle was determined not to give in to his hosts' various wishes and demands. His unofficial motto, in the words of one observer, was "Extreme weakness requires extreme intransigence." Two months after he arrived in Britain, he declared, "I am no man's subordinate. I have one mission and one mission only, that of carrying on the struggle for my country's liberation."

As difficult as the British found the Czechs and other European exiles on occasion, none infuriated them more than the Free French and their haughty leader. Throughout the war, de Gaulle's headquarters was the scene of constant, often violent intrigues, rivalries, and power struggles. United only by their allegiance to de Gaulle, his followers in London came from all points of the French political spectrum, reflecting the deep political and social divisions that had plagued France for generations.

"All the French émigrés are at loggerheads," Harold Nicolson, then a junior minister in Churchill's government, wrote in his diary in September 1940. "All of them come to see me and say how ghastly everyone else is." As one of de Gaulle's colleagues put it, "One had to be a little mad to be Free French."

With his aloof, autocratic manner, de Gaulle only added to the polarization of the tens of thousands of French exiles in London—the comparative few who had escaped there after France's fall and a much larger group that had been there since well before the war. The prewar French community, which contained numerous bankers, industrialists, and merchants, tended to be pro-Pétain and Vichy. But even those who wanted their country to stay in the fight had little faith or confidence in this obscure general, who, though never having been elected to anything, insisted he was the sole leader of undefeated France. "We were constantly being surprised by the ill will, the distrust that he aroused among the most outstanding members of the French community in London," remarked one British official. "In our country it was not with the British but chiefly with the French that he had trouble."

Several leading French political figures, meanwhile, decided to go to the United States rather than put up with the prickly general. They included Jean Monnet, a top economist and diplomat who had worked to promote economic cooperation between Britain and France before the French defeat. Shortly after arriving in London in the summer of 1940, Monnet left for Washington, where he became a key economic adviser to President Roosevelt.

Even de Gaulle's most loyal supporters were put off at times by his notorious rudeness and arrogance. Not infrequently, a subordinate recalled, those who wanted to join the Free French ranks "were received and interviewed in such a way that they came out with their confidence shattered." One French naval officer was so disillusioned by his icy reception at Carlton Gardens that he returned to France and eventually became a top leader of the resistance.

Nonetheless, for all the political Sturm und Drang swirling around de Gaulle, young Frenchmen continued to enlist in the Free French military, with more than 7,500 in uniform by the end of August 1940. The movement acquired even greater momentum when three French colonies in Equatorial Africa—Chad, the Cameroons, and the French Congo—abandoned Vichy and joined de Gaulle. Although thinly populated and lacking in natural resources, those colonies provided him with a territorial base outside Britain—the first step on the long and extraordinarily difficult road to becoming an independent government entity.

Winston Churchill took notice. Unlike many in his administration, the prime minister remained a stalwart champion of de Gaulle and his followers throughout most of 1940. The prime minister was particularly grateful to de Gaulle for his subdued public reaction when, under Churchill's orders, the British destroyed much of the French fleet in the North African port of Mers-el-Kebir to keep it out of German hands. More than 1,200 French sailors were killed in that July 3 attack. Although privately sharing the shock and outrage of his countrymen, de Gaulle told the French in a BBC broadcast that while he decried the assault, he understood the need for it.

Shortly after de Gaulle acquired his foothold in Equatorial Africa, Churchill demonstrated his support for him by decreeing that it was time to bring the Free French into the war. The move was inspired by a telegram in midsummer 1940 from the British consul general in Dakar, the Vichy-held capital of French West Africa. The consul argued that a British–Free French show of force in Dakar might well prompt an anti-Vichy uprising by French troops stationed there.

For Churchill, it was a tempting idea. He wanted—and needed—a successful Allied military offensive as soon as possible, and this one might be relatively easy to pull off. Moreover, if it succeeded, it would ensure that Germany would be denied Dakar, with its strong fortress and important naval base. A port city on the far west coast of the continent, Dakar was the closest point in Africa to the Americas. In the view of the nervous Franklin Roosevelt and his top military leaders, it was like "a loaded pistol pointing across the Atlantic"—a possible staging area for the transportation of German troops to the east coast of Brazil and then northward to the Panama Canal.

As Churchill envisioned the operation, the Royal Navy would transport British and Free French troops to Dakar, where de Gaulle would be installed "to rally the French in West Africa to his cause." But when he proposed the idea to de Gaulle in August 1940, the general was initially reluctant, noting the lack of any concrete evidence that French officers and troops in Dakar were in fact inclined to support him. He finally succumbed to Churchill's insistent cajoling, but with one caveat: if his men met any opposition there, "he would not consider going on with the operation."

Once de Gaulle had signed on, however, he was given no say in the operations's planning, which turned out to be as botched, in the words of one historian, as "the worst muddles of the Norwegian campaign" five months earlier. The British had acquired almost no intelligence about Dakar, its coastal defenses, or the strength of the Vichy forces there. The expedition's commanders had no experience in working with the troops assigned to them, who in turn had been given no training in landing operations. The great flotilla of ships envisioned by Churchill was whittled down to two old battleships, four cruisers, and a few destroyers and transports.

There were also several major security leaks. French officers had been overheard offering toasts "to Dakar" in several restaurants in London and the expedition's embarkation point at Liverpool. British intelligence officers had talked openly about the operation's destination when gathering information about Dakar at London travel agencies; so had Liverpool dockworkers when loading the expedition's ships. Assault landing craft had been trucked across England and then loaded onto transports, with no effort to disguise them.

The mission's chance of success, hardly auspicious to begin with, became even slimmer when, a few days after the troops had sailed from Liverpool on August 31, five Vichy warships steamed from the southern French port of Toulon to Dakar, undetected and unchallenged by the British fleet at Gibraltar. When Churchill and his military chiefs were informed of this latest difficulty, they wanted to cancel the undertaking at once, but de Gaulle and the expedition's British commanders, now close to their target, strongly objected. The War Office reluctantly gave them permission to proceed.

As the expedition neared Dakar on September 23, de Gaulle broadcast an appeal to its military forces and other inhabitants to rise up against Vichy and join the Free French cause. In response, the shore batteries of the fortress and the guns of the warships in the harbor opened up on the Anglo-French fleet, seriously damaging two cruisers. Less than forty-eight hours later, after it had become abundantly clear that the French at Dakar had no intention of switching sides, de Gaulle and the British naval commander scrubbed the mission.

Bungled from beginning to end, the Dakar expedition proved to be yet another humiliating military fiasco. It was ridiculed by Vichy and

German propaganda and lambasted by the British press: the *Daily Mirror* declared that it marked "the lowest depths of imbecility to which we have sunk." Yet although British officials were almost entirely to blame for the failure, much of the condemnation was aimed at de Gaulle and the Free French, largely because of their indiscreet security breaches before the mission was launched. In fact, the breaches, both Free French and British, played no part in what happened at Dakar; Vichy officials did not know about the expedition until it approached the port. That, however, made no difference to de Gaulle's many critics in Whitehall and elsewhere.

Churchill, however, remained steadfast. In response to calls from several British lawmakers to cut off ties with the Free French, he declared to the House of Commons that his government had "no intention whatever of abandoning the cause of General de Gaulle until it is merged, as merged it will be, in the larger cause of France." De Gaulle, for his part, refrained from casting any public blame on the British. Thanks to his restraint and Churchill's vigorous support, the attacks died down, and the furor eventually faded away.

Nonetheless, the collapse of de Gaulle and Churchill's first joint military venture had highly damaging long-term consequences. De Gaulle was devastated. Both he and his movement desperately needed a success to prove themselves to their critics; instead, this highly public failure only reinforced the naysayers' skepticism. It was a profound personal humiliation for the proud, thin-skinned general, and some of those around him feared he might try to kill himself. "After Dakar, he was never entirely happy again," recalled one of his top lieutenants.

De Gaulle's detractors in the British government, meanwhile, claimed that the Free French's indiscreet toasts about Dakar proved that they couldn't be trusted with secret information. The security lapse became a pretext for not informing de Gaulle and his men about future military operations on French territory. The failure of the mission also gave added impetus to the efforts of those in Whitehall who were still eager to establish closer ties with Vichy.

Indeed, secret discussions between Britain and Vichy had begun just a few months after the fall of France—a fact revealed by the American newspaper correspondent Helen Kirkpatrick in late 1940. The talks were sanctioned by Churchill, who, for all his efforts to further de Gaulle's cause, was unwilling to give up hope of persuading Vichy to abandon its subservience to Germany and transfer its military forces and empire to the Allies.

De Gaulle was deeply upset, of course, when he learned of the talks. It was obvious to him that his efforts to establish himself and the Free French as a political as well as military entity had failed, at least for now. He warned Churchill and his government that the discussions were bound to collapse, as they ultimately did.

General Edward Spears, Churchill's liaison to de Gaulle, took note of how "the intolerable strain of constantly recurring rebuffs and disappointments" was worsening the general's already formidable temper and sharpening his suspicions of the British. "During those days, he was like a man who had been skinned alive," another observer remarked.

"I do not think I shall ever get on with *les Anglais,*" de Gaulle stormed to Spears. "You are all the same, exclusively concentrated upon your own interests and business, quite insensitive to the requirements of others. . . . Do you think I am interested in England winning the war? I am not. I am only interested in France's victory." When a shocked Spears replied, "They are all the same," de Gaulle shot back, "Not at all."

By the end of 1940, de Gaulle's close relationship with Churchill was beginning to fray—a deterioration that accelerated throughout the difficult days of 1941. The British prime minister knew perfectly well why de Gaulle behaved as he did: "He felt it was essential . . . that, although an exile, dependent upon our protection . . . he be rude to the British, to prove to French eyes that he was not a British puppet." Churchill's insights into de Gaulle's personality, however, did not make it any easier for him to bear the Frenchman's escalating outbursts.

He finally reached the end of his patience in the summer of 1941 when de Gaulle gave an interview to an American newspaper correspondent in which, for the first time, his complaints about Britain included a personal disparagement of Churchill himself. Deeply hurt by de Gaulle's seeming lack of appreciation for all he had done for him and his cause, the prime minister erupted in rage, writing to Anthony Eden,

"He has clearly gone off his head." He ordered members of his Cabinet to cut off all relations with de Gaulle and the Free French and barred them from making BBC broadcasts. "De Gaulle's attitude is deplorable and his pronouncements, private and public, are intolerable," John Colville, one of Churchill's private secretaries, wrote in his diary. "The PM is sick to death of him."

After De Gaulle claimed he had been misquoted in the interview, Churchill calmed down and rescinded his bans against the general and his supporters. But neither leader ever fully forgave the other, and their conflicts would take on increasingly operatic dimensions until the war's end.

AS THE CONFLICT BETWEEN the British and de Gaulle deepened, the other exile groups paid close attention. Since their escape to Britain, the various European governments, with their separate and unique interests, had contended with one another for their host country's favor. But as the war ground on, they also began to see the advantages of forging tighter bonds. Haunted by their countries' prewar powerlessness and by the failure of neutrality, a number of European officials in London set out to explore the idea of gaining greater security and strength for their small nations through a possible European union. "A genuine feeling of solidarity developed between the governments and their heads of state," Queen Wilhelmina recalled.

The Europeans' need for greater unity was underscored in 1941 by the addition of two powerful countries—the United States and the Soviet Union—to the antifascist alliance. With those titans now committed, the early closeness between Great Britain and occupied Europe gave way to great power politics.

PART TWO

RULE OF THE TITANS

"Rich and Poor Relations"

The European Allies'
Fading Importance

For MOST OF WORLD WAR II, THE UNITED STATES MAINTAINED two embassies in London. The U.S. embassy in Britain took up half an acre on Grosvenor Square and numbered some seven hundred employees. Presided over by Ambassador John Gilbert Winant, this frantically humming nerve center of U.S.-British relations was the United States' largest and most important diplomatic mission in the world. It required an around-the-clock staff of twenty-four telephone operators to handle the more than six thousand calls that streamed in each day.

The second embassy—located in an apartment in Berkeley Square, just a few blocks from Winant's bustling fiefdom—served the European governments in exile. The apartment's master bedroom, festooned with a large wall map of Europe, doubled as the ambassador's office, while the remaining six staffers operated out of the drawing room and smaller bedrooms. The embassy's reception area consisted of a narrow wooden bench in the foyer, where European leaders and other visitors sat while awaiting their appointments.

At first glance, Anthony J. Drexel Biddle, Jr., the U.S. ambassador to occupied Europe, appeared as undistinguished as the space his mission occupied. A wealthy socialite descended from two of Philadelphia's oldest, best-connected families, Biddle was, as *Life* magazine quaintly put it, "financially and socially beyond reproach."

Before embarking on his ambassadorial career, the tall, lean, fortyish Biddle had been noted primarily for his frequent appearance on the annual list of best-dressed American men and for his membership on multiple corporation boards and in more than twenty private clubs. Charming and debonair, he had a penchant for calling every man he met, no matter how casual an acquaintance, "old sport" or "old boy." A British official once tartly remarked, "One rather expects him to leap into the air and begin a Fred Astaire sort of dance. Very disconcerting."

Biddle's first diplomatic nomination—as U.S. minister to Norway in 1935—came as a quid pro quo for his sizable donation to FDR's first presidential campaign. At the time, the appointment was widely scorned. His "previous career," wrote one journalist, "included nothing that could have been charitably regarded as the slightest clue to a brilliant future in diplomacy."

In fact, to the surprise of virtually everyone, Biddle proved himself an adept player of the diplomatic game. Although a millionaire many times over, he had a democratic, informal style that captivated the egalitarian Norwegians—and in particular King Haakon, who became a close friend of his. For all his jaunty ebullience, Biddle took the job seriously and worked hard to master its demands, dispatching reports to the State Department that were "notable for their authoritative accuracy and sound analysis."

In 1937, he was named ambassador to Poland. When the Germans invaded two years later, he and his staff made a hair-raising escape across the country by car, frequently stopping to jump into roadside ditches whenever Luftwaffe planes appeared overhead. After following the newly constituted Polish government in exile to France, he was assigned by President Roosevelt to accompany French officials when they fled from Paris to Tours and then to Bordeaux in June 1940. "In five years," an observer wrote, Biddle "had evolved from a society man playing at diplomacy to an agent of the U.S. in one of the most tragic and delicate situations in history." No American official knew better than he what this savage new kind of warfare was like or how much the countries to which he was assigned as ambassador had suffered as a result of it.

He also knew that naming one man as envoy to multiple governments in exile was an absurd idea; as hard as he worked and as much as he tried, he could not adequately serve their needs and assuage their concerns. He realized, too, what his appointment signaled to the European officials with whom he dealt in London: the unimportance of their countries in the eyes of the Roosevelt administration.

Even more worrying to the governments in exile, Winston Churchill, who used to be their most outspoken champion, would soon come to share the U.S. government's view. Although well aware of the debt he owed the Europeans for their contributions to Britain's survival, Churchill needed the two Allied newcomers considerably more: the Soviet Union, to lift the main burden of fighting the Germans from Britain's shoulders, and the United States, to provide the manpower and industrial strength necessary for an invasion of western Europe and final victory.

When Germany launched an unexpected lightning attack on the Soviet Union in June 1941, London was still struggling to recover from the worst German bombing raid it had yet experienced. As bad as the others were, none had come close to the destructiveness of the May 6 firestorm, which had done catastrophic damage to many of London's landmarks, including Westminster Abbey and Parliament, and killed 1,436 Londoners—the highest daily death toll recorded in all the city's history. The end to such carnage from the air seemed nowhere in sight. And with British forces on the defensive everywhere, there was also little or no hope of ultimate triumph in the war.

Small wonder, then, that Churchill was overjoyed when he learned that Germany had marched into the Soviet Union on June 22, 1941, along a vast front stretching from the Black Sea to the Baltic. Without consulting any of his European or Commonwealth allies, the prime minister went on the radio to promise unconditional support for Joseph Stalin and his country, even though the Soviets had been locked in a quasi alliance with Germany since August 1939, supplying the Reich with oil, grain, cotton, iron ore, and other raw materials crucial to its war effort.

As much as he despised what he called Stalin's "wicked regime," Churchill saw its reluctant addition to the Allied ranks as a miracle of deliverance for himself and Britain, allowing both to catch their breath and regroup. America's entry into the war after the December 7, 1941,

Japanese attack on Pearl Harbor made the deliverance complete. The Soviet Union and United States had done everything they could to stay out of the conflict, but once they had been catapulted into it, Allied victory, in Churchill's view, was certain.

Yet for most of 1942, that outlook seemed highly improbable. The Soviets were constantly on the verge of defeat as the Germans swept toward Moscow, while the United States' entry into the war was accompanied by one crushing rout after another. The shock of losing much of the U.S. fleet at Pearl Harbor was followed by Japanese conquests of Guam, Wake Island, and the Philippines. For the British, the situation was even worse. Vanquished the year before by the Germans in France, Greece, and Crete, they now lost their empire in the Far East and the Pacific to the Japanese. On Christmas Day 1941, Hong Kong fell, followed by Singapore, Burma, and Malaya. The surrender of Singapore, previously regarded as an invincible British bulwark in the Far East, was a particular shock. Britons couldn't understand how Singapore's 85,000-man garrison could have given up so readily. Speaking in the House of Commons, an emotional Churchill called it "the greatest disaster in British arms which our history records."

Singapore, unfortunately, was hardly the last of a seemingly endless string of British military calamities that unspooled in the winter and spring of 1942. In North Africa, General Rommel bottled up a new British offensive in Libya, recapturing towns and cities the British had just taken. In June, after holding out against a long siege, the port of Tobruk, a key British bastion in eastern Libya, capitulated, with more than 30,000 troops surrendering to a considerably smaller German force. A far greater strategic defeat than the loss of Singapore, the capture of Tobruk cleared the way for a German advance toward Cairo and the Suez Canal, threatening the entire British presence in the Middle East.

Ever since April 1940, with the crucial exception of the Battle of Britain, the British had suffered one humiliation after another at the hands of their enemies. As one setback followed another in 1942, the mood in Britain grew progressively fractious and sour. Among the public and in Parliament, there was widespread grumbling about the government's handling of the war. One member of Parliament went so far as to suggest sacking all the British generals and replacing them with Polish, Czech, and Free French military officers "until we can produce trained commanders of our own."

In January and again in July, Churchill faced votes of censure in the House of Commons over his direction of the war. Although he won both handily, the attacks on his leadership took a severe psychological toll. "Papa is at a very low ebb," his daughter Mary wrote in her diary in early 1942. "He is not too well physically, and he is worn down by the continuous crushing pressure of events."

Reeling from the relentless battering, Churchill was not inclined to argue with Roosevelt and Stalin about their backhanded attitude toward the smaller European allies, whom both clearly regarded as Lilliputians. In January 1942, Roosevelt and Churchill stage-managed the signing in Washington of an agreement by the United Nations, as the president dubbed the twenty-six nations then in the alliance, all of them pledging their full resources to the fight.* "The United Nations constitute an association of independent peoples of equal dignity and equal importance," Roosevelt declared. Yet only the Soviet Union and China, which the president had anointed as another major ally, were consulted in advance about the drafting of the document, and only the Soviet and Chinese ambassadors received formal invitations to the White House signing ceremony with Roosevelt and Churchill. The ambassadors of the other Allied countries were merely informed that they could drop by, at their convenience, to sign the declaration. Jan Ciechanowski, the Polish ambassador to Washington, noted that "if the concept of the United Nations could still be regarded as an international family concern, it was one definitely composed of rich and poor relations."

Throughout the war, Roosevelt had a disconcerting habit of talking about the fates of smaller nations as if they were his alone to decide. In a 1942 meeting with Soviet foreign minister Vyacheslav Molotov, for example, the president remarked that the Soviet Union needed a northern port that was not icebound in the winter and suggested that

---

* In addition to the United States, Britain, the Soviet Union, and China, the alliance consisted largely of countries from occupied Europe, the British Commonwealth, and Central and South America.

the USSR annex the Norwegian coastal town of Narvik. The startled Molotov rejected the idea, noting that his country did not "have any territorial or other claims against Norway."

A few months later, in a White House chat with Oliver Lyttelton, the British minister of production, FDR mentioned the divisions between Belgium's two main ethnic groups: the Dutch-speaking Flemish and the French-speaking Walloons. After declaring that the Flemish and Walloons "can't live together," he proposed that "after the war, we should make two states, one known as Wallonia and one as Flamingia, and we should amalgamate Luxembourg with Flamingia." Incredulous at the thought of forcing a European ally to partition itself, Lyttelton later wrote of Roosevelt, "He allowed his thoughts and conversations to flit across the tumultuous and troubled [world] scene with a lightness and inconsequence which were truly frightening in one wielding so much power."

When Lyttelton reported Roosevelt's comments to Anthony Eden, the British foreign secretary said he was sure the president was joking. But when Eden himself visited the White House in March 1943, Roosevelt reintroduced the proposal. "He seemed to see himself disposing of the fate of many lands, allied no less than enemy," Eden remarked. "I poured water, I hope politely, on [the idea], and the President did not revert to the subject again."

FDR's attitude toward the countries of occupied Europe and the other small allies revealed some of the contradictions in his immensely complex personality. He sincerely believed that the United States' mission after World War II would be to help build a freer, more just world. Yet he also felt that the United States, the Soviet Union, and Britain had the right to dictate to the less powerful states, not only during the war but afterward as well.

Although in public Roosevelt and Churchill continued throughout the war to espouse equal rights and freedoms for all nations, the occupied European countries, like the other smaller allies, were excluded from any significant role in war planning. The military staffs of the governments in exile, for example, were barred from participating in meetings of the U.S.-British Combined Chiefs of Staff, which was responsible for the planning of future Allied military operations, including campaigns that would take place on the Europeans' own territory. The Netherlands, which lost the Dutch East Indies to the Japanese in March 1942, was excluded, much to the Dutch government's indignation, from all high-level Allied decision making regarding operations against the Japanese in the southwest Pacific.

Yet while all the exiled Europeans in London were diminished by the addition of the Americans and Soviets to the Allied cause, three national groups—the Free French, Poles, and Czechs—saw their wartime efforts and the futures of their countries dramatically and immediately affected.

FROM THE START, FDR felt nothing but disdain for Charles de Gaulle and France. He had little understanding of the complexity of the situation in that defeated, traumatized country and scant sympathy for its citizens. All he knew or cared about was that it had failed the Allied cause. By capitulating to Germany, he believed, France had lost its place among the Western powers. "There is no France," he declared, insisting that it would not really exist again until after its liberation.

As for de Gaulle himself, Roosevelt considered him insignificant and absurd, a British puppet with grandiose ambitions. From the beginning, the president was scornful of this general, "who had escaped from an army of beaten men but who talked of 'indefeasible rights,' 'long-standing splendor,' and 'immortal France.'" FDR was "convinced," a U.S. official wrote, "that de Gaulle's ambitions were a threat to Allied harmony and a menace to French democracy. Accordingly, he made up his mind—and once it was made up, it was never changed—that the U.S. would make no concession which would help de Gaulle to achieve his ambitions."

Unlike Britain, the United States had formally recognized Vichy as the legitimate government of France almost immediately after the country's capitulation to Germany. In a sign of regard for Pétain's regime, Roosevelt sent a good friend—Admiral William Leahy, a former chief of naval operations—to Vichy as ambassador. As the war progressed and Pétain's government stepped up its collaboration with Nazi Germany, not to mention the repression of its own citizens, the administration's close ties to Vichy came under increasingly severe criticism in the United States and elsewhere.

FDR was unapologetic. He believed, just as Churchill did, that he could persuade Vichy to keep French North Africa and the remainder of the French fleet out of German hands and perhaps come over to the Allied side. To that end, the United States vigorously wooed Vichy by, among other things, dispatching food and critically needed supplies to North Africa. Though it accepted the American gifts, Vichy showed no sign of complying with the U.S. government's wishes, just as it had ignored British overtures. Instead of discouraging Roosevelt, the rebuff only served to redouble his determination to win Vichy over.

THE CONFLICT BETWEEN BRITAIN'S support of de Gaulle and FDR's intense antipathy toward him first surfaced in mid-1942, after the two new Western Allies began planning their first joint offensive against Germany. U.S. generals had pushed for an invasion of the Continent, but the British protested that Anglo-British forces were not ready for such a high-risk campaign. The Allies finally compromised on a British alternative proposal: an amphibious invasion of North Africa, to take place in November.

Convinced that the United States was popular with Vichy, Roosevelt argued that Vichy troops in North Africa would put up little or no resistance to the landings as long as U.S. soldiers took the lead, British forces were in the background, and Free French troops were nowhere to be seen. By early 1942, de Gaulle had acquired an army of more than 50,000 men, an air force of more than a thousand pilots and crew, and several dozen ships. None of that mattered to Roosevelt, who told Churchill that the general and his followers "must be given no role in the liberation and governance of North Africa and France." The president also insisted that de Gaulle be kept ignorant of all planning for the invasion.

Churchill was faced with a highly painful dilemma. As contentious as his relationship with de Gaulle had become, he had made a solemn pledge to support the general in June 1940, and he shrank from having to go back on his word. He also sharply disagreed with FDR's belief that France had lost its status as a great nation. When he was told of a quip by Roosevelt about de Gaulle's supposed belief that he was a lineal descendant of Joan of Arc, Churchill didn't laugh. The idea didn't seem at all absurd to him. The prime minister sadly noted that "France without an Army is not France. De Gaulle is the spirit of that Army. Perhaps, the last survivor of a warrior race."

All the same, Churchill considered himself Roosevelt's lieutenant and vowed to his staff that "nothing must stand in the way of his friendship for the President on which so much depended." The British ended up handing over all initiative for the invasion to the Americans. But just a few days before it was to take place, Churchill begged Roosevelt to at least be allowed to tell de Gaulle about it: "You will remember that I have . . . recognized him as the Leader of free Frenchmen. I am confident his military honour can be trusted." When FDR rejected his request, Churchill replied, "I am still sorry about de Gaulle. But we are ready to accept your view."

On November 8, 1942, more than 30,000 American and British troops poured onto the beaches of North Africa. Churchill would later acknowledge "the gravity of the affront offered to de Gaulle" by himself and Roosevelt. Nonetheless, de Gaulle went on the BBC that night to endorse the invasion: "French leaders, soldiers, sailors, airmen, civil servants, French settlers in North Africa, arise! Help our Allies! Join them without reservation. Fighting France adjures you to do so. . . . Ignore the traitors who try to persuade you that the Allies want to take our empire for themselves. The great moment has come."

Privately, though, de Gaulle raged against the exclusion of himself and his forces. In a bitter note to U.S. secretary of state Cordell Hull, he said he particularly resented "the ungracious attitude of the United States government toward the only Frenchmen who are carrying on the war side by side with the Allies." But he saved his sharpest barbs for the man who had so warmly welcomed him to Britain seventeen months before. "I don't understand you!" he exclaimed at a meeting with Churchill. "You have fought since the first day. One could even say that you personally symbolize this war. Yet you allow yourself to be towed along by the United States, whose soldiers have never even seen a German. It is up to you to take over the moral direction of this war. Public opinion in Europe will be behind you." De Gaulle would later observe in his memoir that "these words made a profound impression on Mr. Churchill, and he wavered perceptibly."

Yet as much as he might inwardly agree with the truth of what the general had said, Churchill was well aware of his own powerlessness to do anything about it. In joining forces with the Soviets and Americans, he had found himself in much the same position as the Europeans: he was fast becoming a junior partner to Stalin and Roosevelt, just as the small nations' leaders were subordinate to him.

WHILE ALLIED LEADERS QUARRELED over North Africa and France, the French people, at long last, had begun to stir. Initially, the idea of overt resistance in France had seemed considerably less likely than in other occupied countries, in large part because its own government was actively cooperating with Germany. By granting the Vichy regime the right to function on its own as long as French police and government administrators did what Hitler wanted, the Führer had ensured that he could control France with a minimal German presence. As one historian put it, "There was no other occupied country during World War II which contributed more to the initial efficiency of Nazi rule than France."

French officials were hardly alone in collaborating with the enemy. Many of the country's wealthiest and most noted citizens—industrialists, aristocrats, writers, movie stars, dress designers—socialized with their occupiers throughout the war and benefited economically and in other ways from their presence. So did thousands of others. According to the historian Julian Jackson, an estimated 220,000 French citizens could be classified as wartime collaborators.

The vast majority of the French, however, did not follow their lead. Although most never showed any interest in open resistance, they tended to be strongly pro-British and anti-German—attitudes confirmed by public opinion reports over the course of the conflict. As early as August 1940, a German army memo sourly noted that the "exemplary, amiable and helpful behavior of German soldiers towards [the French] population has aroused little sympathy."

In general, the French showed their hostility by ignoring their occupiers and refusing personal contact with them. "I lower my head so that you don't see my eyes, to deny you the joy of an exchanged glance," one Frenchman wrote about the Germans in February 1943. "You are in the middle of us, like an object, in a circle of silence and ice." Though emotionally gratifying, such snubs involved little personal risk to those making them. The idea of more active resistance was uncharted, far more dangerous territory.

In the first few months after France's fall, there were isolated acts of rebellion around the country: shots fired at German patrols, enemy posters and car tires slashed; laughter and jeers in movie theaters whenever Hitler appeared in newsreels. On November 11, 1940—Armistice Day—thousands of French students gathered at the Place de l'Étoile in Paris to sing "La Marseillaise" and protest against German occupation. It was the first large anti-German demonstration in France, and the enemy was determined to make it the last. German police and troops charged into the crowd and shot several students—a response that shocked the country and discouraged any future acts of mass defiance.

Nonetheless, unnoticed by almost everyone, the embers of rebellion began to smolder. The Scottish writer Janet Teissier du Cros, who lived in France throughout the war, put it another way; French resistance "grew as naturally as a mushroom grows among dead leaves," she wrote. "In its beginnings it was no organized movement. In town, village or countryside, those in whom it burned soon came to know which of their neighbors shared their views; and with no clear notion how their feelings could be translated into action, they gathered, at first simply for moral support. Heroism crept upon many of them like a thief in the night."

Gradually, these small clusters of would-be resisters joined forces with equally disorganized groups, creating fragmented movements around the country. Few of them had much, if any, knowledge of the others. What they did have in common was the first step most of them took as embryo resistance organizations: the publication of clandestine newspapers meant to counteract German propaganda and provide the French public with accurate information about what was occurring in the war and their country.

Underground newspapers were central to the existence of resistance movements in every occupied nation, but they were particularly important in France, a country that places an extremely high value on the spoken and written word. According to the National Library of France, more than one thousand underground publications were published during the country's occupation.

Like the BBC broadcasts to France, the newspapers were aimed at replacing despair and a feeling of helplessness with hope and a spirit of defiance. "We made it clear that there was an active Resistance at work, one that was growing from day to day," an underground editor remarked after the war. "[Our membership] was invisible to our readers. . . . The only sign it could give at that stage was our two-page printed sheet." The newspapers in themselves were tangible proof that Frenchmen were fighting back. Producing and distributing the papers—leaving them in post offices and on trains, slipping them through mail slots—involved considerable risk and ended up serving as the seedbed and training ground for more overt and dangerous kinds of rebellion.

As they set about retrieving their self-respect, these early resisters also developed a sense of community that had been lacking in French society for generations, if not centuries. Transcending traditional social and economic barriers, people from all classes and walks of life—journalists, teachers, railwaymen, shopkeepers, students, stevedores, engineers, clerks, and farmers—banded together in what one resister called "our passionate love for our country." Even the aristocracy was represented: Jean, Philippe, and Pierre de Vomécourt—three wealthy brothers from Lorraine who also happened to be barons—became important liaisons between SOE and resistance members.

"In our war, the soul found redemption," observed Emmanuel d'Astier de La Vigerie, a journalist, intellectual, and founder of Libération-Sud, one of the major resistance movements in the south of France. "We were in a revolt which, for so many of us, adrift in a sterile society, opened the door to a lost fraternity. The motive was different for each of us, but we were all, with others unknown, living in the same . . . state of exaltation."

In another challenge to the status quo, resistance movements welcomed Jews and others who were regarded as outsiders in French society. "The proportion of Jews in the Resistance was greater than that for the French population as a whole," a French historian noted. "[The underground] remained an alternative society that had taken in Jews on an equal basis and offered them a chance to act without changing any part of their identity."

Just as iconoclastic was the crucial role played by women, not only in France but in virtually every other occupied country of Europe. They acted as couriers, collected intelligence, transported arms, escorted downed Allied pilots to safety, hid insurgents in their homes, and even led armed bands of resistance fighters against German targets. One U.S. intelligence agent called women "the lifeblood of the Resistance."

In France, as in other countries, women's success as insurgents was due in large part to the Germans' stereotypical view of them. Coming from a traditional, conservative society, the enemy saw women chiefly in their conventional domestic roles as wives and mothers and, at least early in the war, rarely suspected them of being spies and saboteurs. "Women have such an innocent look, you know," noted Andrée de Jongh, a young Belgian woman who was arguably the most fearless and best-known resistance heroine of the war. "They look so terribly harmless. Germans are not accustomed to women with opinions of their own."

For more than a year after France's fall, the country's burgeoning resistance movements focused on making contacts, building up their membership, publishing newspapers, gathering intelligence, creating false identity papers and other documents, and weighing the possibility of future paramilitary actions. At that point, there was little thought of direct confrontation with the enemy. "Above all, no isolated violent action," a movement newspaper cautioned its readers. "The moment has not yet come."

After Germany's invasion of the Soviet Union, however, French communists decided differently. Earlier, the French Communist Party had followed Stalin's lead in cozying up to the Germans, remaining aloof when its own country was vanquished by the Reich. But in July 1941, the Soviet Comintern ordered the French communists to launch an armed struggle against munitions factories and German troops in France, in hopes of weakening the Reich's campaign in Russia. They obeyed the order with alacrity.

Their first strike came on August 21, 1941, with the fatal shooting of a young German naval cadet in a Paris subway station. The Vichy government, in an attempt to appease the Nazi authorities, ordered the execution of six French communists who had had nothing to do with the ambush. Rather than halting the communist attacks, the reprisal was followed by more assassinations: on October 20, a high-level German official was killed in Nantes, followed by another in Bordeaux. In retribution, ninety-seven additional French hostages were shot. Over the next seven months, more than four hundred French citizens would lose their lives as a result of German vengeance.

The French people, already restive over growing food and fuel shortages, were infuriated by the wanton killing of their compatriots. By 1942, words such as "hate" and "rage" were commonly used in reports about French attitudes toward their occupiers. In June of that year, a meeting of several dozen German intelligence officers in Paris was told that "ninety-nine percent of the French population are openly hostile to us. The French despise the Germans. They will not even forgive us for treating them so decently."

Fierce enmity, however, did not immediately translate into widespread direct resistance; as long as Germany seemed unbeatable, the idea of rebellion seemed quixotic in the extreme. In November 1942, that myth of invincibility finally began to crumble, thanks to Britain's victory at the Egyptian port of El Alamein, its first battlefield triumph, followed a few days later by the Allied invasion of North Africa.

Adding to the cascade of events was the German takeover of Vichy-controlled France on November 11, 1942, just three days after the attack on North Africa. The whole of France was now under enemy domination, with the severe repression in the north spreading to the comparatively more relaxed south.

The takeover also thrust Vichy's open collaboration with Germany into sharper relief. In the view of an increasing number of the French, Vichy's cooperation in the hostage shootings and its leading role in the roundup and deportation of French Jews to death camps had turned Pétain and his officials into nothing less than Hitler's henchmen.

But the biggest impetus to the resistance's growth was the Reich's decision in 1942 to draft hundreds of thousands of French citizens to work as forced labor in its factories. It did so because of a major miscalculation by the Nazis: when they had invaded the Soviet Union, they had expected the campaign to last no more than six weeks. Yet a year later, the battle was still grinding on, with millions of German troops fighting—and dying—on the eastern front. So many men had been drafted into the Wehrmacht that Hitler found himself with a severe shortage of factory workers to produce the tanks, planes, artillery, submarines, and other matériel he so desperately needed. He decided to fill the gap with conscripted workers from all over Europe.

In the late spring of 1942, the Führer demanded that some 350,000 French citizens be assigned to the Nazi industrial war effort. Vichy prime minister Pierre Laval called on the French to volunteer, but when few responded, he issued an order requiring all Frenchmen between the ages of eighteen and fifty and all unmarried women between twenty-one and thirty-five to give two years of service to German war work. The *service du travail obligatoire* (or STO, as it was commonly called) was in effect a national draft of slave labor, imposed by the French government itself. Of the more than one million French citizens shipped to Germany over the course of the war, more than two hundred thousand never returned.

Until the work draft, the lives of most of the French had not been deeply affected by the German repression. The STO, however, hit home in the most literal way: virtually every family had a loved one in danger of being rounded up. For many, enduring the occupation was now no longer an option; it was time to work to end it. Clandestine newspapers called on all French citizens to refuse to obey the order. Worker strikes and protests multiplied. More important, tens of thousands of men left their homes and went underground. The lightly populated, heavily wooded French countryside, along with mountainous regions in the east and south of the country, became favorite hiding places; in those out-of-the-way places, members of newly formed quasi-guerrilla groups, called maquis, lived off the land and began to plot sabotage and subversion.

From that point on, the resistance began to count as a real force in France, but it still had a number of major weaknesses. The various movements throughout the country were working independently and sometimes at cross-purposes. They had no money, few arms, and little sense of discipline or direction. And until early 1942, they had virtually no connection with Charles de Gaulle and the Free French in London.

Although neither realized it at first, de Gaulle and the resistance needed each other to achieve their common goal: the liberation of their country. Providentially, there was at least one man who understood that point. He was Jean Moulin, a short, stocky, boyishly handsome French civil servant who walked, uninvited and unannounced, into de Gaulle's London headquarters one day in late October 1941.

Jean Moulin

The greatest figure in France's wartime resistance, Moulin, more than any other person, would be responsible for bringing together the wide array of fragmented movements and welding them into a cohesive, relatively disciplined body. In the process, he would also give de Gaulle the legitimacy he needed to expand his own crusade and transform himself from an Allied underling into the acknowledged leader of France.

UNTIL MOULIN APPEARED AT Carlton Gardens, de Gaulle and his men knew almost nothing about France's growing insurgency, despite their having dispatched some twenty intelligence agents to their homeland over the previous eighteen months. The agents' assignment had been to collect information on German activities, and they apparently rarely, if ever, crossed paths with the underground. "We knew that men of good will, dispersed here and there in France, were ready to engage in violent action against the Germans," André Dewavrin, de Gaulle's intelligence chief, wrote in January 1941. "But we had absolutely no idea how to get in contact with them, and as a consequence, how to organize them."

In fact, the Free French never saw the home front as a priority until Moulin arrived in London. In de Gaulle's initial BBC broadcast in June 1940, he had focused on recruiting Frenchmen outside France—those who had escaped from there or were living in North Africa and other French possessions. "The general appeared to have little faith in the possibilities of either a secret army at home or effective work by paramilitary forces," Dewavrin said.

That all changed when de Gaulle met Moulin. Like others before him, the general, who could be so rude and icy to others, fell under his visitor's spell. With his commanding presence and fierce integrity, Moulin displayed "the sort of natural authority and experience which his past history has given him," said one British official who met him in London.

When the war broke out, the forty-year-old Moulin had been serving as prefect (or governor) of the department of Eure-et-Loire, a region of northwest France whose capital is the city of Chartres. Unlike his fellow administrators, most of whom willingly collaborated with the Germans, Moulin refused to accept Nazi rule; just days after the occupation began, he was arrested by the Gestapo and tortured after refusing to follow their orders. Fearing he might yield to the pressure, he tried to commit suicide by cutting his throat, but he was found and hospitalized and eventually recovered. His wounds, however, left him with a husky voice and badly scarred throat, which, when in public, he kept swathed with a scarf.

Although Moulin lost his job because of his insubordination, he retained his freedom. For the rest of what would turn out to be a short life, he worked to make the French resistance a force to be reckoned with. For much of the next year, he made clandestine trips around the country, making contacts with and collecting information about resistance groups. His main focus was on three large movements in Vichy France, which he called "the main organizations of resistance to the invader."

In a handwritten report that he presented to de Gaulle after smuggling himself out of France, Moulin laid out in detail the movements' achievements, goals, and potential for sabotage and military action against the enemy. "There is a rising tide among thousands of young Frenchmen who want to take part in war again," he told de Gaulle. "They want someone to tell them that they are already in the front line in France. This must be combined with some promise of organization and direction." If properly directed and supplied, he argued, resistance groups could make a significant military contribution to ending German rule by aiding Allied forces before and during the liberation of France. He warned that if de Gaulle did not step in, the French resistance might well succumb to communist control.

Impressed by Moulin's arguments, de Gaulle sent him back to France as his official representative to the various resistance movements. Moulin's mission was to unite the groups into a single entity under de Gaulle's direction; in return, they would receive money and arms. It was an extraordinary challenge, considering the deep divisions and rivalries, both political and personal, that bedeviled the groups. Nonetheless, by the summer of 1942, Moulin had succeeded in extracting pledges of support for de Gaulle from most of the resistance organizations, including the communists, who complied only when Moulin threatened to bar them from receiving any subsidies.

De Gaulle's deepening influence with the resistance and the French public in general was strikingly demonstrated on July 14, 1942—Bastille

Day—when, in a BBC broadcast, he urged the residents of unoccupied France to turn out en masse in public demonstrations opposing the Vichy government. Hundreds of thousands of people responded, marching down the main thoroughfares of Marseille, Lyon, and other cities and towns across Vichy France, wearing the national colors, waving flags, and singing "La Marseillaise." (At the same time, de Gaulle instructed those in German-occupied France to refrain from demonstrating, to avoid giving the Germans an excuse for violent retaliation. In another impressive sign of his influence, most complied.)

De Gaulle's growing authority at home couldn't have come at a more opportune time. It provided him with a legitimacy and political base at a moment when he faced multiple challenges to his leadership, including from the president of the United States. More specifically, it allowed him to fight back against one of the Roosevelt administration's most controversial decisions of the war: the appointment of Admiral Jean Darlan, the commander of the Vichy armed forces and a notorious German collaborator, as governor of North Africa in November 1942.

THE AMERICANS' CHOICE OF Darlan came after Vichy forces in North Africa put the lie to Roosevelt's prediction that they would welcome an invasion by U.S. troops. In fact, the French mounted stiff opposition at almost every landing site. In an attempt to stop this resistance, U.S. military officials called on Darlan, who happened to be in Algiers visiting his son at the time of the attack. Next to Pierre Laval, Darlan, a former Vichy prime minister himself, was the most reviled of all Vichy officials, thanks to his complicity in the persecution of French Jews, the mass arrests of Vichy opponents, and the supplying of Rommel's troops with food and matériel.

In exchange for Darlan's engineering a cease-fire, the Allies promised to appoint him high commissioner, or governor, of North Africa. After agreeing and then reneging on the deal, Darlan, under heavy Allied pressure, finally gave in and ordered an armistice. Once in office, however, he upheld anti-Semitic laws and imprisoned de Gaulle supporters, including many who had aided the Allied operation.

His appointment, meanwhile, was greeted with a storm of protest around the globe. Unfazed by the criticism, Roosevelt told a French resistance leader visiting Washington, "For my part, I am not an idealist like [Woodrow] Wilson. I am concerned above all with efficiency. I have problems to solve. Those who help me solve them are welcome. Today, Darlan gives me Algiers and I cry 'Vive Darlan!' If Quisling gives me Oslo, I will cry 'Vive Quisling!' Let Laval give me Paris tomorrow, and I will cry 'Vive Laval!' "

In the view of many, such cynical pragmatism undermined the lofty moral position of the Allied cause. "The British in 1940 were the first to give the struggle a positive, idealistic meaning," the CBS correspondent Eric Sevareid later wrote. "The Americans, to the world's surprise, were the first to take this away." Darlan's appointment, Sevareid added, showed that the U.S. government and military "would use any means, including Fascists and Fascist institutions, to aid them in their task, regardless of how the basic issues were muddied and the future placed in jeopardy."

Members of European resistance movements, whose lives were in constant danger due in part to collaborators such as Darlan, were the most outspoken in expressing their dismay and anger. According to an SOE report, the Allies' collusion with Darlan "has produced violent reactions on all our subterranean organizations in enemy-occupied countries, particularly in France, where it has had a blasting and withering effect."

In France, the march to de Gaulle accelerated. On November 17, 1942, Jean Moulin relayed a statement from French resistance leaders calling for de Gaulle, "their uncontested leader," to be named governor of North Africa. The statement was also signed by representatives of most of France's major political parties—a signal to the Allies that the general was gathering support not only from the resistance but from traditional political forces.

De Gaulle could also count on the wholehearted backing of the European governments in exile, whose leaders feared that the Americans, having cozied up to Darlan, would cooperate with prominent collaborators in their own nations. He was supported as well by much of the British public, most members of Parliament, and the British press. Even some high-level British government officials, including several from the Foreign Office, joined the parade. In 1940, Churchill had backed the general in the face of Foreign Office opposition; now officials in that ministry were shielding him from Churchill's increasing animus, believing that de Gaulle had gained legitimacy in France and that it would not be in Britain's long-term interests to abandon him.

Yet Churchill, like nearly everyone else, couldn't help knowing that the Darlan appointment had been a huge political mistake and that something needed to be done. It wasn't long before action was taken: on Christmas Eve 1942, a twenty-year-old French military trainee burst into Darlan's headquarters in Algiers and shot him dead. There were suspicions that the U.S. and British secret services had arranged the murder, but nothing was ever proved.

To replace Darlan, the U.S. military chose a French general named Henri Giraud, who had been captured during the battle for France and, having escaped from a German prison, had allied himself with Vichy. The appointment of Giraud, who continued his predecessor's persecution of Jews and Vichy opponents, was also extremely unpopular and had little or no Allied support, except within the Roosevelt administration. "Between Giraud and de Gaulle, there is no real choice," a French resistance leader told Harold Nicolson. "Giraud is not a name at all in France. De Gaulle is more than a name, he is a legend."

It was now clear that Roosevelt was fighting a losing battle where de Gaulle was concerned. "The people of France will never accept the subordination of General de Gaulle to General Giraud," Jean Moulin declared. He called for "the rapid setting up of a provisional government in Algiers under the presidency of General de Gaulle, who will remain the sole head of the French Resistance, whatever the result of the negotiations may be." In addition to the general's support from France, thousands of Vichy French soldiers in North Africa had switched sides, joining the Free French and making de Gaulle's movement a much more potent military force.

Finally bowing, if only a little, to what most people saw as inevitable, Roosevelt acknowledged that de Gaulle could not be wholly excluded from the North Africa government. He authorized de Gaulle's association with Giraud, who invited his rival to Algiers to share leadership of the new French Committee of National Liberation.

Yet despite this temporary truce, the venomous duel between FDR and de Gaulle continued, with damaging long-term consequences that still reverberate today.

CHAPTER 14

"The Ugly Reality"

The Soviet Threat to
Poland and Czechoslovakia

BECAUSE FRANCE WAS TO BE THE PORTAL FOR THE ALLIED invasion and liberation of Europe, de Gaulle and his forces were able to defy the two most powerful Western Allies and get away with it. The French general "could afford to irritate British and American statesmen and tell them unpleasant truths to their faces," Count Edward Raczyński, the Polish ambassador to Britain, wrote after the war. "They might not like it, but they could not afford to abandon either him or France." The same, Raczyński noted, was not true of his own country, which, with the rest of eastern Europe, was treated by Roosevelt and Churchill "as something secondary, not as a vital interest of their own."

Within days after the German invasion of the Soviet Union, the British government began pressuring the Polish government in exile to make peace with the Soviets. For the Poles, the idea of the Soviet Union as an ally was a grotesque oxymoron. Less than two years earlier, Stalin, with Hitler's secret blessing, had attacked and occupied eastern Poland, annexing roughly half of all Polish territory at almost exactly the same time that the Germans annexed the other half. The Soviets' treatment of the Poles under their control was nearly as brutal as Germany's: the American diplomat George Kennan would later call it "little short of genocide."

During the twenty-one months that the Soviets dominated eastern Poland, an estimated 1.5 million Polish citizens were taken from their homes and deported in freight trains to Siberia and other Soviet regions. Thousands froze to death along the way or died of starvation and disease. Those who survived ended up in slave labor camps or were dumped onto collective farms. Most were never seen again.

Like Hitler, Stalin singled out military leaders and other members of the educated Polish elite—government officials, lawyers, landowners, priests, writers, doctors, teachers—for elimination. Indeed, officers of the NKVD (a forerunner of the KGB) met regularly with representatives of the Nazi SS to coordinate their twin repressions. In those murderous campaigns, Hitler and Stalin hoped to finish what their predecessors—the tsars, emperors, and kaisers—had begun to do in the eighteenth and nineteenth centuries: to erase Poland from the face of the earth.

Among those who disappeared in the Soviet roundup were more than 15,000 Polish army officers, including many in the army's top field command, who had been taken prisoner by the Red Army in September 1939. After the Soviet Union joined the Allies, Stalin told the Polish government in exile that he had no idea where the missing officers were. In fact, shortly after their capture, he had ordered them to be murdered. In the early spring of 1940, many of them had been taken in small groups to a clearing in the Katyn Forest, near the southwestern Russian town of Smolensk. There they were forced to kneel at the edge of huge pits and were shot in the back of the head; their bodies were pitched into the mass graves "with the precision of machines coming off a production belt." This convoy of death continued for more than five weeks.

The Poles in London would not learn of Katyn until 1943, but they were aware of the grim fate of more than a million other countrymen who had been caught up in the Soviet maw. When Churchill and Anthony Eden pressured them to sign a treaty with Stalin in July 1941 that would pledge military cooperation and restore diplomatic relations between the two countries, the Poles initially resisted.

Churchill would have none of it. Up to that point, Poland had contributed more to Britain's survival and the overall war effort than any other declared ally. But much as he valued the Poles' help, Churchill was unable to see the Soviet-Polish conflict through Polish eyes. The

Soviets were now Britain's valued allies, and he was determined to make Poland acknowledge them as such. "Whether you wish it or not, a treaty must be signed," Eden informed General Sikorski, who finally acceded.

One stipulation of the treaty actually turned out to be of great importance to the Poles—and would be of tremendous help to the Allied cause in the years to come. Under the pact, Stalin was required to release all Poles deported to his country. Although in the end he freed only a fraction of them, there were more than enough to form a new Polish army. Looking more like corpses than soldiers, tens of thousands of Poles—emaciated, toothless, many lacking fingers and toes because of frostbite—streamed out of prisons, slave labor camps, and collective farms, all of them heading for makeshift army camps on the Volga River. Their commander there was Polish general Władysław Anders, who had been wounded twice by the Soviets in 1939 and confined to Lubjanka Prison in Moscow for more than a year. In 1942, Anders moved his makeshift army, accompanied by thousands of Polish women and children, from Soviet Russia to the Middle East, where the scarecrow soldiers began to regain their health and to train in earnest. Called the Polish II Corps, they eventually numbered more than 100,000 men. By 1944, Anders's army, which would capture Monte Cassino, would be regarded, according to John Keegan, as "one of the greatest fighting formations of the war."

As important as the prisoners' release was, however, the deliberate omission from the treaty of one Polish demand was of far more consequence to the country's long-term future. The Poles wanted to include a section in which the Soviets would promise to return all of the Polish lands they had seized in 1939. Stalin, however, refused to make that pledge. In fact, from the first days of his new alliance with the West, the Soviet leader hinted that he planned not only to retain the annexed territory but eventually to gain power over the rest of Poland as well. The Soviets "neither sought nor cared about Polish friendship," Count Raczyński remarked. "Their purpose, as of old, was to gain control of Poland and subject it entirely to their will, with the view to absorbing it completely."

When the Poles expressed their anxieties about all this to Churchill, he again refused to listen. The final treaty left open the question of Poland's postwar borders.

ONE OF THE TANTALIZING "what-ifs" of World War II is what might have happened to Poland and Czechoslovakia—two vulnerable nations adjoining each other in a strategically important borderland between West and East—if they could have formed some sort of protective federation after the war. For more than a year, officials from the two countries met in London to discuss doing just that. Specifically, they examined the possibility of each nation retaining its sovereignty but cooperating on political and military matters and establishing common economic and foreign policies. As the negotiators knew, such an alliance would result in a formidable combination of manpower, arms, and fortifications, all of which would be vital in ensuring the postwar security and independence of their countries.

On the surface, Poland and Czechoslovakia seemed to have much in common. Having spent many years under foreign domination, both nations had regained their independence following World War I. And although located in eastern Europe, each leaned heavily toward the West. When it came to their national character, however, the differences between them were striking. The Czechs were regarded—and saw themselves—as a sober, sensible, middle-class people who focused on hard work and shied away from flashy heroics. "To survive is an obsession with the Czechs," *Time* magazine noted in March 1944. "It is also their greatest talent. They never had notions of grandeur. They always realized that their role is to adjust themselves to conditions not of their making—and survive."

The Poles were polar opposites. Unlike the Czechs, who had been occupied by the relatively benign Austro-Hungarian empire in the eighteenth and nineteenth centuries, most of Poland had been subjugated by the much more brutal Russians and Prussians, with only the southwestern corner held by Austro-Hungary. Even if the Poles had been inclined to get along with their occupiers—which they were not—it would have done no good. Hotheaded and rebellious by temperament, they repeatedly rose up, particularly against the Russians, and just as repeatedly were crushed. "The Poles are not troublesome as aggressors," the *New York Times* correspondent John Darnton observed, "but as victims who refuse to lie down."

The romantic, emotional Poles tended to disparage the Czechs for what they perceived as their neighbors' dullness and industriousness. "The Czechs seem to the Poles solid, heavy people, much like liver dumplings," A. J. Liebling noted in the *New Yorker* in 1942. For their part, the Czechs regarded the Poles as arrogant, foolhardy, autocratic, and suicidally reckless.

In the early twentieth century, this traditional antipathy was heightened by a virulent dispute over a highly industrialized sliver of land, called Teschen, that lay on the two countries' common border. After gaining their independence in 1918, both nations claimed Teschen, whose population was more than half Polish but that also had sizable Czech and German minorities. When the Czechs forcibly occupied a large portion of Teschen, Allied leaders at the Paris peace conference ordered them to withdraw and to divide the region fairly with the Poles. Czechoslovakia, however, got the better part of the deal, acquiring most of Teschen's land and industry.

Furious over the Czechs' high-handedness, the Poles took their revenge following the 1938 Munich Conference, when Hitler allowed them to seize Teschen from their neighbor. Whatever justice there might have been to Poland's claim to Teschen, its willingness to capitalize on Czechoslovakia's misfortune was both a moral failing and an enormous blow to Poland's reputation in the rest of the world.

Although Teschen remained a point of contention in the early 1940s, both sides in the federation negotiations in London thought a resolution might be possible. A major reason for their optimism lay in the marked differences between the wartime Polish government under General Sikorski and the prewar regime responsible for the snatching of Teschen. Sikorski, an outspoken opponent of the prewar government, and the men around him were far more liberal and democratic than their predecessors and had actively opposed their authoritarian policies. From London, Sikorski promised his countrymen that his administration would institute free elections and social reforms in Poland after the war, similar to those in prewar Czechoslovakia.

But in the end, none of that mattered. When news broke in early 1942 that Sikorski and Beneš had embarked on formal negotiations to mend their countries' relationship and consider a possible federation, Stalin made it clear he was not pleased. And when Stalin was not pleased, Beneš, who was determined to do nothing to antagonize the Kremlin, paid close attention.

Unlike the Poles, the Czechs were not immediate neighbors of Russia, had never been conquered by Russia, and lived outside the traditional Russian sphere of interest. Beneš, having by now lost all faith in the West, chose to believe that Stalin would protect Czechoslovakia's independence after the war, even though the Soviets had done nothing to aid the Czechs against German aggression in 1938 and 1939.

After he joined the Allies, Stalin did everything he could to encourage the idea that he was indeed Czechoslovakia's new best friend. The Soviet Union, for example, was the first Allied nation to recognize Beneš and his followers as the official Czech government in exile, signing the recognition agreement four hours before the British government did the same.

Stalin also made a promise to the Czechs that he refused to offer the Poles: Soviet recognition of their country's postwar independence, with no interference in its internal affairs. František Moravec, the head of Czech intelligence; Jan Masaryk; and others in the government were skeptical, but Beneš put his faith in Stalin's pledge. The Czech president "lost all realistic perspective toward Communism, blinding him to new dangers from the East," Moravec later wrote. "Throughout the war, despite the advice of many, including myself, he persisted in his accommodating attitude toward Soviets and Czech Communists in order to demonstrate his goodwill. He refused to see the ugly reality until it was too late."

But as Beneš viewed the situation, what choice did he have? He was sure that neither Britain nor the United States would do anything to help his country or the rest of eastern Europe. Wisely or not, he elected to gamble on Stalin, who did not wait long before demanding his first quid pro quo: a stepped-up sabotage campaign by the Czech resistance against the Germans, similar to the one being waged by communist insurgents in France.

LIKE THE POLES, THE CZECHS in London had left intelligence and resistance organizations behind when they had escaped from their homeland in the late 1930s. Their intelligence operation was by far the stronger of the two, designed primarily to transmit material from their prize agent, Abwehr officer Paul Thümmel, to British intelligence in London.

To Beneš's chagrin, Czechoslovakia's resistance efforts had not lived up to the country's intelligence achievements. One explanation for the underground's relative weakness, especially when compared to Poland's, was that Czechoslovakia had not been attacked and conquered; instead, it had been traded away to the enemy by its so-called allies. Many Czechs, as demoralized as Beneš by the West's desertion, saw little reason to put their lives at risk for the Allied cause.

In addition, Germany, at least in the beginning, was far more lenient toward Czechoslovakia, which had a huge armaments industry and rich agricultural lands, than toward Poland. Because the Reich badly needed the Czechs' cooperation to further its war effort, it did not treat most of the population with the savagery it showed other Slavs. Early in the occupation, "those who kept their mouths shut and heads down could go on about their lives," Madeleine Albright observed.

The SS, however, did not show the same moderation to students, intellectuals, and others who protested against German rule. After a series of peaceful demonstrations at Czech universities in September 1939, thousands of students were arrested. Some were tortured and executed, while many more were shipped off to German concentration camps. From then until the end of the war, all Czech institutions of higher learning were closed.

After the reprisals, the organized Czech resistance movement, afraid that any kind of dramatic action would touch off further retaliation, burrowed even more deeply underground. Its reticence was highly embarrassing for Beneš, who was being bombarded by urgent Soviet requests for the Czech resistance to come to the aid of the Red Army by sabotaging production of German war matériel and cutting Wehrmacht communications. Churchill and British military leaders, unable to give Stalin the second front he was demanding, were also pressuring the beleaguered Czech president to help the Soviets.

Unfortunately for Beneš, the stepped-up demands from his allies coincided with the appointment of the SS's infamous Reinhard Heydrich as governor of what was now called the Protectorate of Bohemia and Moravia. Heydrich began his rule in September 1941 with a bloody Gestapo crackdown on the already weakened Czech resistance movement. "It was futile for us to send messages home asking for an increase in resistance activity," František Moravec recalled. "We tried it. Nothing happened."

So Moravec and Beneš turned to their only remaining assets: some 150 Czech soldiers in England who were undergoing training as SOE agents. In the fall of 1941, the Czech government in exile informed resistance leaders that teams of operatives would soon parachute into Czechoslovakia to rebuild the underground and launch sabotage campaigns against enemy communications, railway traffic, and war-related industry. In an obvious attempt to counter any objections from the home front, Beneš warned that "our whole situation would definitely appear in an unfavorable light if we . . . did not at least keep up with the other [occupied countries]."

There was yet another secret mission planned. It would turn out to be one of the most daring operations of the entire war—nothing less than the assassination of the thirty-eight-year-old "butcher of Prague," Reinhard Heydrich himself.

ONE OF THE MOST powerful men in the Third Reich, the fair-haired, blue-eyed Heydrich made a vivid impression on all who encountered him. SS colleagues variously described him as "a blond god" and "a predatory animal." After meeting Heydrich for the first time, Hitler declared, "This man is extraordinarily gifted and extraordinarily dangerous." According to Heydrich's deputy, Walter Schellenberg, he "had an ice-cold intellect and was untouched by pangs of conscience. . . . Torture and killing were his daily occupations."

As the head of the Gestapo and all other SS intelligence and security

*Reinhard
Heydrich*

organizations, Heydrich was already responsible for the deaths of untold numbers of civilians in Europe and the Soviet Union. They included the victims of special SS extermination squads, known as the Einsatzgruppen, which followed German armies into Poland and Soviet lands and machine-gunned Jews, intellectuals, clergymen, political leaders, and anyone else who happened to be on their long kill list. Having been appointed in early 1941 to organize the Final Solution, Heydrich was also hard at work planning the systematic, scientific slaughter of all the Jews of Europe.

Yet it was not enough for him to be a killing machine; he also wanted major roles for himself and the SS in shaping the destiny of a Germanized Europe. To further that ambition, he was engaged in a ruthless power struggle with the German military establishment, particularly with the Abwehr, the military's intelligence operation.

The Abwehr's patrician chief, Admiral Wilhelm Canaris, despised Heydrich and the murderous thugs working for him. In September 1939, Canaris protested the Einsatzgruppen's "orgy of massacre" in Poland to Field Marshal Wilhelm Keitel, the head of the German armed forces, to no avail. Canaris was opposed "to any infringement whatsoever of the unwritten laws of humanity," noted Hermann Giskes, a high-level Abwehr officer.

Heydrich, in turn, had nothing but contempt for Canaris, complaining to Hitler and Heinrich Himmler that the Abwehr was far too weak and lenient in its dealings with the citizens of occupied Europe. To bolster his argument, he noted the slowly rising tide of resistance in France and the other captive countries controlled by the German military, where SS operatives did not have the unbridled freedom to kill that they did in the Soviet Union and Poland. Though hardly widespread, this upsurge instilled a sense of dread in top Nazi circles. "The epidemic of assassination is spreading alarmingly in French cities," Joseph Goebbels wrote in his diary in late 1941. "Our Wehrmacht commands are not energetic enough in trying to stop it."

Heydrich, who saw Czechoslovakia as his first stepping-stone to greater power, exploited the high-level angst by declaring that an "obviously large-scale resistance movement" in the protectorate not only endangered Nazi rule there but posed a major threat to the productivity of Czech industry, so essential to the German war effort. Although patently untrue, his claim convinced Hitler to fire the current *Reichsprotektor* of Bohemia and Moravia and replace him with Heydrich.

The new ruler wasted no time. Within days of his arrival in Prague, he had ordered the arrest of more than six thousand Czechs, many of them in the resistance. By the end of 1941, hundreds of those in captivity had been executed, including a former Czech prime minister, the current army chief of staff, and dozens of other top military officers. In his terror campaign, Heydrich succeeded not only in decimating the resistance movement but in severing all its radio links with London, thus cutting off the flow of intelligence to Britain from Paul Thümmel.

After making his point about the fearful consequences of rebellion, Heydrich offered incentives to those who cooperated with the German war effort. Productive workers in the armament industry, for example, were given higher wages, as well as extra rations for food, cigarettes, and clothing. In his skillful use of such carrot-and-stick techniques, Heydrich managed to stamp out virtually every sign of resistance and to boost the efficiency of Czech industry. In the spring of 1942, the pleased Hitler remarked that "the Czechs at the moment—and particularly at war factories—are working to our complete satisfaction, doing their utmost."

All this served only to further heighten Allied pressure on Beneš and led him to propose the killing of Heydrich. The assassination, he told František Moravec, would be carried out by "our trained paratroop commandos" but would be presented to the world as an achievement of the domestic resistance movement—"a spontaneous act of national desperation" that "would wipe out our stigma of passivity and help Czechoslovakia internationally."

Both Beneš and Moravec knew that the cost of Heydrich's life would be extraordinarily high. At a time when the killing of even a minor German functionary in occupied Europe invariably resulted in the executions of a dozen or more civilians, it boggled the mind to think of the potential human toll following the murder of one of Germany's most prominent officials. But when Moravec brought the subject up, Beneš replied that, regardless of the harrowing consequences, Heydrich's assassination was "necessary for the good of the country."

Only a handful of people—Beneš, Moravec, and a few other senior Czech intelligence officers—knew of the plot. No other official in the Czech government in exile was consulted, nor were the few remaining underground leaders. Beneš ordered that no written record be made, ensuring that nothing about the plan could be traced back to him.

When Moravec first approached officials in SOE's Czech section for help in training the two agents chosen to kill Heydrich, he told them only that the men had been assigned to carry out a "spectacular assassination"; there was no mention of the target. The two operatives—Jan Kubiš and Jozef Gabčík—were young sergeants in the Czech army, both of whom had fought in France in the spring of 1940 and were expert in the handling of guns and explosives. They had volunteered for the Heydrich mission, even though they knew their chances of survival were all but nil.

In late 1941, the pair underwent several weeks of rigorous training by SOE, which supplied them with equipment that included revolvers, machine guns, grenades, and suicide tablets. But the British had no role in their operational briefings and orders, all of which were handled by Moravec and his men.

A few days after Christmas, Kubiš and Gabčík were parachuted into Czechoslovakia. Throughout the winter and early spring of 1942, they lay dormant, waiting for an opportune time to carry out their mission. As they made their preparations, an additional twenty or so SOE agents were dispatched to Czechoslovakia as part of Beneš's plan to disrupt the country's weapons industry and railway networks. To a man, the new arrivals were stunned by the omnipresence of German police controls, which were far tougher than they had been led to expect. "For everyone politically active, there is a permanent Gestapo agent," one SOE operative remarked.

The sabotage campaign, perhaps not surprisingly, turned out to be a failure. Not one target was damaged or destroyed, and many if not most of the agents were caught and executed. The Germans also recovered a bounty of material parachuted in from England: arms, ammunition, incendiary devices, explosives, and five transmitters. One transmitter, however, remained in the hands of resistance members, who used it to urge London not to send any more SOE operatives. Beneš ignored the plea. The Czech president "had no intention of curtailing the program, whatever dangers there were to the survival of agents," wrote the historian Callum MacDonald. "Parachutists were expendable."

To protect their security, Kubiš and Gabčík had been told by Moravec to avoid all underground contacts and to work alone. But when they arrived in Prague, they discovered that it was impossible to follow that order. If they wanted to survive and carry out their mission, they would need the help of the resistance and the few SOE agents still at large, who hid them in a series of safe houses in Prague. It didn't take long for their protectors to find out why they were in the Czech capital. Stunned by what they considered London's recklessness, resistance leaders begged the Czech government in exile to cancel the operation.

"This assassination would not be of the least value to the Allies, and for our nation it would have unforeseeable consequences," Arnošt Heidrich, a former Czech diplomat and member of the resistance, cabled Moravec. "The ferocious repression [that would follow] would make the earlier crackdowns look like child's play. It would threaten not only hostages and political prisoners, but also thousands of other lives. The nation would be subject to unheard-of reprisals. At the same time, it would wipe out the last remainders of any organization. It would then be impossible for the resistance to be useful to the Allies."

When Moravec took Heidrich's anguished appeal to Beneš, he was ordered not to answer it. Again, there was no consultation with other top officials in the government in exile. The operation was to proceed as ordered.

ON A WARM, SUNNY MORNING in late May 1942, two young men carrying heavy briefcases stood quietly on opposite sides of a hairpin curve in a road in downtown Prague. After waiting for more than an hour, Jan Kubiš and Jozef Gabčík were growing anxious. They knew that Reinhard Heydrich traveled this route at precisely 9 A.M. each day from his country home to his office in Hradčany Castle. They knew, too, that he almost always rode without a bodyguard, confident that the cowed Czechs would never make an attempt on his life. "Why should my Czechs shoot at me?" Heydrich loftily responded when another Nazi official chided him for his recklessness. His chauffeur—a brawny six-foot, five-inch SS guard—was his only protector.

It was now close to 10:30 A.M.—and still no sign of Heydrich's black Mercedes convertible. But just as the two Czechs were losing hope, they spotted the car approaching. As expected, it slowed down to negotiate the curve. At that point, Gabčík stepped into the middle of the road, took a small machine gun from under his overcoat, aimed it straight at Heydrich, who was sitting in the back, and pulled the trigger. Nothing happened. The gun had jammed, and Gabčík had no other weapon.

As the chauffeur slammed on the brakes, Heydrich jumped to his feet, drawing a revolver from his pocket and pointing it at Gabčík. He had not seen Kubiš, who was standing behind the Mercedes and who, at that moment, pulled a bomb from his briefcase and hurled it in Heydrich's direction. The bomb exploded against the rear wheel of the car, sending it several feet into the air and showering shrapnel everywhere.

Seemingly untouched, Heydrich jumped out and squeezed off several shots at Gabčík as he sprinted away. Moments later, though, the *Reichsprotecktor* clutched his back and collapsed in the road. Hit in the spleen by shrapnel fragments, he was rushed to a hospital, where he died of sepsis eight days later.

As Arnošt Heidrich had predicted, the Reich's leaders went berserk. For the first time, a key member of the Nazi inner circle had been killed, and everyone wondered who might be next. "[The Führer] foresees the possibility of a rise in assassination attempts if we do not proceed with energetic and ruthless measures," Goebbels wrote in his diary.

Fearing he might be the next target himself, the Führer authorized a stupendous reward—1 million marks (worth more than $16 million today)—to anyone with information about the assassins' identities and whereabouts. Himmler, who burst into tears when he heard of Heydrich's death, flew to Prague to take personal charge of the manhunt. "It is our holy duty to avenge him," the SS chief exclaimed.

Some 21,000 German troops, most of them SS forces, rampaged through the Czech capital, racing from one building to the next, hammering on doors, ransacking apartments, and shooting anyone they considered suspicious or who didn't immediately obey their orders. "They're completely mad," one German detective said of the SS. An RAF pilot whose plane had been shot down and who was hiding in Prague during that period observed that the searchers "seemed almost insane."

Those killed in this orgy of violence included a number of Prague residents who had sheltered Kubiš, Gabčík, and other SOE agents. The helpers' families were also killed. Before he was executed, a teenage boy whose parents had housed the two assassins for a time was shown the severed head of his mother floating in a fish tank.

Czech Jews were targeted as well. On June 9, three days after Heydrich died, a special train left Prague carrying 1,000 Jews to Nazi death camps. Two thousand more soon followed. All those deaths, however, were not enough to satisfy Hitler, Himmler, and the SS. They needed something even more shocking—an action that would demonstrate to the occupied peoples of Europe how catastrophic the consequences of defying German rule could be. As their target, they chose a little village named Lidice, located a few miles northwest of Prague.

One of the SOE agents captured by the Germans before Heydrich's death had in his possession a letter containing the addresses of two families in Lidice. The Gestapo concluded, wrongly, that the villagers were—or had been—hiding the assassins. In the predawn darkness of June 10, 1942, hundreds of SS troops surrounded Lidice. After all its residents were routed from their homes, the men were shot on the spot and the women and children sent to concentration camps, where most of them later died. The entire village was burned to the ground, and whatever ruins remained were bulldozed. Salt was then scattered over the earth so that nothing living could take root in Lidice again.

Although the SS had not yet found Kubiš and Gabčík, their extraordinary savagery in Lidice and elsewhere, plus the huge reward offer, finally had their intended effect. On June 16, Karel Čurda, one of the few Czech SOE agents still at large, walked into Gestapo headquarters in Prague. Shaken by the extreme reprisals, angry at the seeming callousness of Czech leaders in London to the plight of the resistance, and tempted above all by the enormous bounty, Čurda revealed the identities of the assassins.

Using information provided by Čurda, the Gestapo tracked Kubiš and Gabčík to a church in downtown Prague, where they and five other parachutists from London had been hiding. For more than six hours, the Czech agents engaged in a frantic gun battle with seven hundred SS troops surrounding the church, managing to hold their pursuers off until they ran out of ammunition. Using their last bullets, Kubiš, Gabčík, and the two other agents still alive killed themselves rather than fall into the enemy's hands.

Altogether, more than five thousand Czech citizens died in the aftermath of the assault on Heydrich. The two-week bloodbath touched off a global outpouring of sympathy and admiration for the Czechs and loathing for the Nazis and their barbarism. Not surprisingly, the focus of the world's attention was the massacre at Lidice. "If future generations ask us what we were fighting for in this war, we shall tell the story of Lidice," Frank Knox, the U.S. secretary of the navy, declared. A number of towns in the United States and elsewhere were renamed Lidice in honor of the innocents who had died there.

As Beneš had hoped, Heydrich's killing and the Germans' horrific response resulted in a major propaganda triumph for the Czech cause. "I was in the U.S. at the time of Lidice, and was making no progress in our propaganda, having exhausted all the possibilities of the situation," Jan Masaryk wrote to a British friend. "Then came Lidice, and I had a new lease on life. Czechoslovakia was put on the map again." As the jubilant Moravec put it, "In the delicate matter of our contribution to the war effort, we jumped from last place to first."

On newspaper front pages throughout the world, the attack on Heydrich was acclaimed as the work of the Czech resistance movement—the most audacious act yet in its desperate campaign to free the country from German rule. According to the BBC, "the Czechs and all the other enslaved peoples must be proud in the knowledge that they have cast out fear and thus have turned the terror against the Nazis." The fact that members of the Czech resistance had done everything they could to prevent the assassination remained a closely guarded secret.

Heydrich's death also provided a rare bit of good news for the overall Allied cause, which in the spring and summer of 1942 was still suffering major defeats on nearly every front. The British, for one, showed their gratitude by formally repudiating the Munich agreement and finally treating Beneš and his government with the respect Beneš believed he deserved. "In view of the trials through which the Czech people have been passing since the death of Heydrich, we think it desirable for psychological reasons to give Beneš as much satisfaction as possible," the Foreign Office declared in an internal memo.

Yet even as the Czech president enjoyed his reclaimed prestige, his shattered country was plunged into despair and mourning. Instead of damaging the German war effort and lessening the SS's grip on Czechoslovakia and other occupied countries, Heydrich's killing, if anything, had done the opposite. "Somebody else would take his place who would be just as awful," a Prague resident observed. "Unless you could wipe out the whole of the Gestapo, it wouldn't really matter."

Amazingly, Beneš believed that Heydrich's killing would unite the Czech people and inspire many more of them to stand up against their occupiers. In fact, it destroyed what remained of a badly crippled movement. In a report in late 1942, SOE concluded that there was no longer any sign of "open resistance" in Czechoslovakia. "By his death, Heydrich fulfilled his primary ambition—the pacification of the Protectorate," the Czech historian Vojtech Mastny pointed out.

Perhaps even worse, the bloody German dragnet ended up annihilating virtually all of Moravec's intelligence networks. Paul Thümmel was captured and eventually executed, putting an end to the flow of vital military information he had provided to London over the previous four years.

As the catastrophic effects of Heydrich's killing became clearer, both the British and Beneš retreated from taking responsibility for what had happened. Churchill never mentioned it in his history of World War II, while his government insisted that it had been solely a Czech operation. Not until 1994, when selected SOE files were released, was it revealed that several top SOE officials had known beforehand who the target was to be.

Beneš, for his part, denied for the rest of his life that he had played any role in the assassination. Calling the idea "a complete fabrication," he claimed that "no order for Heydrich's murder was ever issued from London. In fact, the whole Nazi theory about the fight for freedom being conducted and ordered from London is false. All acts of resistance in the homeland [were] directed and decided by the headquarters there." Thirty years after the war's end, Moravec, in his memoirs, finally acknowledged Beneš's involvement, as well as his own.

As the conflict dragged on and the glow of Beneš's propaganda victory dimmed, he realized that the assassination and its long-term impact had become in fact a major political problem. Unable to meet Stalin's continued demands for the destruction of the Czech munitions industry and the disruption of its railway network, Beneš was determined to appease the Soviet leader in other ways. Czechoslovakia, he vowed, would never share the woeful postwar fate that he was convinced lay in wait for Poland.

CHAPTER 15

"The England Game"

SOE's Dutch Disaster

BY MID-1943, THE ALLIES HAD FINALLY HALTED GERMANY'S seemingly unstoppable momentum. A crucial turning point came in late February when the Red Army defeated the Wehrmacht at Stalingrad, ending a five-month bloodbath that had produced more than a million casualties. Three months later, the United States and Britain laid claim to *their* first major prize: the Middle East and North Africa. "In London, there was, for the first time in the war, a real lifting of spirits," Churchill wrote.

Early in the year, at a meeting in Casablanca, Churchill and Roosevelt chose Sicily as their troops' next target—an operation that would lay the groundwork for the 1943–44 Allied campaign in Italy. The two leaders also agreed to build up U.S. forces in Britain, in preparation for the long-awaited invasion of western Europe the following year.

As they began their planning for D-Day, the British chiefs of staff decided that SOE—the agency they had long scorned—would be given a role in this crucial offensive. Specifically, the chiefs wanted resistance fighters in the immediate area of the invasion and in nearby countries to support Allied assault troops by sabotaging enemy forces and facilities, particularly those involving transport and communications. SOE began working feverishly to expand its efforts in the Netherlands, Belgium, and especially in France, where the D-Day invasion was expected to take place. But it had a problem that the military brass didn't know about and that its leaders refused to acknowledge: it was in desperate trouble in all three nations.

One person well aware of the disarray was a twenty-two-year-old wunderkind named Leo Marks, who had arrived at Baker Street as SOE's new chief of codes in early 1942. Marks would spend the next eighteen months trying to alert others to the gravity of the situation, only to encounter one of the most notable British cover-ups of the war.

Leo Marks's father, Benjamin, was the owner of Marks & Co., a well-known antiquarian bookstore in west London whose regular customers included Sigmund Freud, George Bernard Shaw, and Charlie Chaplin. (The store was later immortalized in the book and movie *84 Charing Cross Road*.) When Leo was eight, his father showed him a first-edition copy of a book he had just bought containing Edgar Allan Poe's short story "The Gold Bug." Leo read it. Enthralled by Poe's tale of code breaking and buried treasure and wanting a code of his own to crack, he found one on the back page of the book he had just finished. As it happened, in every volume his father purchased, he penciled in a cipher signifying the price he had paid for it. Young Leo figured out the book's code in a matter of minutes—and found his true calling.

Thirteen years later, family connections helped him secure a place in a training class for fledgling Bletchley Park code breakers. Although undeniably brilliant, Marks was too quirky and independent even for Bletchley Park. In the early spring of 1942, he was assigned instead to SOE as head of its coding department. His primary job, he was told, was to monitor the security of agents' traffic.

Marks's first impression of SOE was its total disorganization. "No matter which country section I visited," he later wrote, "everything was in short supply except confusion." He also was struck by how poorly informed SOE officials seemed to be about the dangers facing agents in occupied Europe, especially those working as wireless operators, the agency's most important and perilous job.

Colin Gubbins, SOE's head of planning and operations, described the operators as "the most valuable link in our chain of operations. . . . Without these links, we would have been groping in the dark." Through the radio connection between London and the occupied countries, agents sent and received messages in Morse code about the status of resistance operations, upcoming sabotage targets, and plans for the dropping of operatives and weapons to resistance groups. For both the field and home office, it was an essential lifeline.

It was also highly vulnerable to detection. The early wireless sets were big and heavy. To work, they needed an outside aerial, which usually involved several dozen feet of wire, well spread out and often visible to passersby. In attics, cellars, and other hiding places throughout captive Europe, SOE operators furtively tapped out their messages on those bulky machines, trying to finish their work as quickly as possible; if they remained on the air for more than a few minutes, their signals were likely to be picked up by the Germans.

In Paris and other European cities, clerks in Gestapo headquarters worked around the clock to keep track of radio frequencies in the area. When they found signals they considered suspicious, they alerted agents cruising the city in unmarked vans containing sophisticated direction-finding equipment. The vans would then close in on the target. As Leo Marks quickly discovered, SOE officials had no idea how difficult it was to operate a wireless set when under such pressure, even for fully trained operators with considerable experience. That did not include most operators in SOE, whose training had been rudimentary at best.

Soon after joining the agency, Marks set out to lessen the danger. His first step was to get rid of the codes that the agency had been using to communicate with its people in the field. They had come from MI6, which, for the first two years of SOE's existence, had controlled its wireless circuits and provided its sets and coding. Marks was dismayed by the simplicity of the codes, which were based on classic English poems by Shakespeare and others that were "so familiar that an educated German was quite capable of recognizing them and guessing the cipher."

To replace them, he wrote poems of his own, ranging from ribald verses to tender love poems. He gave one of the latter, entitled "The Life That I Have," to a twenty-one-year-old agent named Violette Szabo, who, after being parachuted into France in 1942, was eventually captured, tortured, and killed by the Gestapo. It read:

The life that I have
Is all that I have
And the life that I have
Is yours.

The love that I have
Of the life that I have
Is yours and yours and yours.

A sleep I shall have
A rest I shall have
Yet death will be but a pause

For the peace of my years
In the long green grass
Will be yours and yours and yours.

Since then, the poem has developed a life of its own. It has been used in a movie about Szabo's life, found in poetry anthologies, reprinted on a 9/11 victims' website, and recited by Chelsea Clinton and Marc Mezvinsky at their wedding in 2010. "Every code," Marks would later say, "has a human face."

Although his poem codes did improve security to some extent, Marks was still not satisfied. He moved on to a more sophisticated method: providing agents with a number of onetime code pads imprinted on easily concealed squares of silk that could be cut off and destroyed after each use. Marks's code pad proved to be highly successful and was used by both SOE and OSS until the end of the war.

He also turned his attention and abundant energy to an even greater problem: wireless messages, called "indecipherables," that could not be read because of coding errors or mistakes in transmission. Before he arrived, wireless operators had been instructed by London to repeat any unreadable messages, thus greatly increasing their chances of detection and capture. Appalled by the order, Marks observed, "If some shit-scared wireless operator, surrounded by direction-finding cars which were after him like sniffer dogs, who lacked electric light to code by or squared paper to code on—if that agent hadn't the right to make mistakes in his coding without being ordered to do the whole job again at the risk of his life, then we hadn't the right to call ourselves a coding department." To solve the problem, he hired dozens of clerks, mostly young women, who were specially trained to tease out the meaning of indecipherables. According to Marks, the neophyte code breakers were soon "performing with the precision of relay racers and, by passing the baton of indecipherables from one eager shift to another, had succeeded in breaking 80 per cent of them within a few hours."

From his first days at SOE, the quick-witted, outspoken Marks felt considerably closer to the agents with whom he worked than to his bosses and fellow bureaucrats. Still in his early twenties, he was decades younger than most of his colleagues, and as a Jew, he felt like an outsider in SOE's old-boys-club milieu. Those running the agency, he believed, had no real sense of what their operatives faced in the field or the courage, skill, and steely nerves it took to outwit the Germans.

Marks himself took great pains to get to know each agent before he or she was dispatched to Europe. All agents, even those who were not assigned as wireless operators, had to take a basic course in Morse code so they could double as operators in an emergency. Marks wanted to make sure they understood whatever coding method they had been assigned. But he also wanted to get a sense of their personalities, which might help him and his staff decipher any future indecipherables they sent.

His deep interest in individual agents only added to his worry when things were not going well in the countries to which they were assigned. In his first year at SOE, he was most concerned about the Netherlands. And as it turned out, he had every right to be.

ALL COUNTRY SECTIONS OF SOE experienced problems in sending their first agents into the field, but the Netherlands section had a particularly difficult time. One reason was the country's geography. As the crow flies, the Netherlands is less than two hundred miles from Britain, but in many ways, it was the most isolated nation in western Europe and the most difficult for a resistance movement to operate in.

Squeezed between Germany and the stormy North Sea, the Netherlands was a virtual prison. Unlike France, it had no common borders with neutral nations like Spain or Switzerland, which were used by the British government as avenues for the smuggling of agents and others into and out of occupied territory. Its land was flat and highly cultivated, with few natural hiding places, such as forests and mountains, or large unobserved areas that could serve as improvised drop areas for weapons and agents.

Landing agents by submarine or boat was equally difficult. The Dutch coast stretched straight and bare, with no hidden coves or harbors. The beaches, meanwhile, were littered with land mines, barbed wire, huge, cross-shaped concrete blocks, and hordes of German sentries.

Holland was also the most densely populated nation in Europe; to move around unobserved was extremely difficult. Its excellent railway and road systems meant that German troops and police agents could reach anywhere, even its smallest villages, in a couple of hours. It was no accident that of all the occupied countries in Europe, Holland was second only to Poland in the high percentage of its citizens shot or deported to German concentration and death camps.

But the officials in SOE's Dutch unit, known as N Section, didn't seem to appreciate the geographical difficulties facing their agents and members of the Dutch resistance. Nor did they seem very knowledgeable about anything else to do with clandestine warfare. Richard Laming, the section's first head, had worked for MI6 in the Netherlands before the war—not exactly the most sterling qualification, considering his MI6 colleagues' abysmal behavior during the 1939 Venlo affair. Seemingly as clueless as they, he had little understanding about what was happening in the wartime Netherlands.

To fill their many gaps in knowledge about the country's geography, history, and political and social conditions, Laming and his coworkers could easily have consulted the Dutch intelligence service in London. They never did. The only role assigned to the Dutch was as a recruitment agency, identifying prospective operatives for SOE officers, who then made the final selections and supervised the agents' training. Informed only a couple of days before an agent's departure, the chief of Dutch intelligence was given no opportunity to study the agent's orders or object to anything he might disagree with.

With all its difficulties, N Section found itself months behind SOE's other country departments in infiltrating its first agents. Initially, several operatives had been scheduled to leave in the summer of 1941; at the last minute they staged a mutiny, writing to top SOE officials that they were prepared to work in the Netherlands but not under Laming's command. In their letter, the agents noted that they weren't supposed to know one another's true identity, yet N Section officials had addressed them by their real names. They also observed that the clothing they were to wear in Holland had been bought in British shops and that, although Dutch labels had been sewn in, the clothes were of a style that might as well have had "England" written all over them. The final straw, however, was that the agents had not been given a list of contacts and safe houses in the Netherlands and were told they would have to find them on their own.

Worried that the agents' insubordination might become public knowledge, SOE shipped them off to a house in a remote part of Scotland, where they were denied contact with the outside world. After a complaint was lodged by the Dutch government in exile, they were released in December 1941, but only after signing a paper promising to keep quiet about what had happened to them.

In September 1941, two new operatives finally made it to the Netherlands, with orders to recruit Dutch citizens for future resistance work. They were supposed to be picked up by boat several weeks later and taken back to England. But the prearranged rendezvous never took place, and the agents had no choice but to stay where they were. Having no wireless to communicate with London, they soon disappeared.

In November, SOE tried again, sending two more young Dutch émigrés—Hubertus Lauwers and Thijs Taconis—to The Hague. At the time of the German invasion of the Netherlands, Lauwers had been working as a journalist in the Philippines; he immediately left there to go to Britain. In London, he and Taconis, a university student who had escaped from Holland by fishing boat, were recruited by SOE. Trained as a wireless operator, Lauwers was to establish a radio link between the

Netherlands and London, while Taconis was assigned to set up new resistance networks and train their members in sabotage techniques.

After Lauwers and Taconis departed, SOE heard nothing from them for two months. Finally, in January 1942, Lauwers made contact and from then on sent regular reports about the work he and Taconis were doing. In March, however, Leo Marks discovered what appeared to be a major flaw in this seemingly successful operation. In his transmissions, Lauwers had begun omitting his security check, a signal he was supposed to use to verify that he had not been captured. Before he and other operators were sent into the field, they were repeatedly told that failure to insert the check would let London know that they were working under German control.

But when Marks brought the lack of a security check to N Section's attention, its staffers told him there was nothing to worry about. "The whole thing has been looked into," they assured him. "The agent's all right."

ACTUALLY, AT THAT MOMENT, the agent was sitting in a Gestapo prison in The Hague. For Lauwers and Taconis, nothing had gone right since their departure from Britain. In their final SOE briefing, they had discovered that N Section did not have a list of up-to-date safe houses and contacts for them. When Lauwers had asked to try out the transmitter he was to take, he was told it had already been tested and that, in any case, it was untraceable by the Germans. The false identity documents given to the agents were unmistakable forgeries: the paper they were printed on was much darker than that used for genuine documents. Admitting that the two had been poorly prepared, Richard Laming had told them that "no one will blame you if you don't go." Although rightly concerned about their prospects, they decided to proceed.

On their arrival, both men were wearing almost identical clothing that clearly was not Dutch in origin, and when they went to a café for their first meal, they tried to pay in silver coins, which they soon learned had been taken out of circulation years before. Astonishingly, they managed to escape German scrutiny and began their job of organizing what was then a comparatively weak resistance movement.

Like Norway, the Netherlands had not been invaded or occupied by a foreign power for more than a century. Its citizens had no experience with underground activities, nor were most of them inclined to engage in such actions, at least early in the occupation when German treatment of the country was relatively mild. Nonetheless, even at the start, there were scattered outbreaks of rebellion, including large gatherings in support of Queen Wilhelmina and protests by student and workers against German persecution of Dutch Jews.

As the harshness of the occupation increased, so did Dutch anger and unrest. Would-be resisters began coming together in small, disorganized groups as they had in France, using underground newspapers and leaflets to recruit new members. But, lacking discipline, resources, and a central command, many of them were soon caught and executed by the Germans or sent to concentration camps.

It was Taconis's job—and that of the SOE agents who would follow him—to bring the fragmented organizations together and meld them into a fighting force the Allies could count on to stifle the Germans in the Netherlands before and during the invasion of Europe. A daunting and highly dangerous task, it would take months, if not years, to accomplish. But Taconis and Lauwers were not able to inform London of that until ten weeks after they'd arrived. The wireless transmitter SOE officials claimed to have tested failed to work when Lauwers first tried it. After dismantling it, he discovered that the problem was a manufacturing flaw—one that took weeks to repair.

When Lauwers finally managed to contact London, he explained that fixing the transmitter, finding adequate safe houses, and assessing the state of Dutch resistance had taken up almost all of his and Taconis's time. No resistance groups were yet ready to go into action. None had yet received the weapons and explosives training they needed. And no arrangements had been made to hide and distribute the military supplies that SOE was so eager to provide.

N Section was not happy with this news. Thanks in part to the mutiny of the first batch of Dutch agents, the hapless Richard Laming had been replaced. His successor was Major Charles Blizzard, a career army officer with no experience in clandestine work, who had never met the two agents in Holland. Anxious to improve N Section's standing with SOE higher-ups and the military brass, Blizzard ignored Lauwers's pessimistic status report and ordered him and Taconis to prepare a drop zone for the parachuting of explosives and arms.

Blizzard's order was just one of several problems facing Lauwers in the winter of 1942. Another was the security of his radio transmissions to London. No one in England had informed him that he would need a long, extremely conspicuous outside aerial to send his messages. He was also concerned about the lengthy transmission times the messages required. SOE officials, who had expressed nothing but contempt for German counterintelligence efforts, had repeatedly told him that the Germans did not have the technology to trace his wireless set. After spending several months in Holland, Lauwers was not so sure.

In the early evening of March 6, 1942, Lauwer sat in an apartment in The Hague, waiting to begin his scheduled biweekly transmission to London, which was always broadcast at the same time and on one of two predetermined frequencies. Glancing over the three messages he was to send, he turned on his set and prepared to begin. Just then, the apartment's leaseholder, a Dutch lawyer who had been secretly sheltering him, flung open the front door and announced that four police vans were parked outside.

Stuffing his codes into his pocket, Lauwers hurriedly left the building and, trying to appear as calm as possible, strolled down the snow-covered street. In seconds, he was cornered by a dozen men waving pistols. "I cursed my stupidity," he later wrote. "The game was up."

ON THE SIDEWALK, a tall man with penetrating blue eyes, a sharp nose, and a thin mustache watched as Lauwers was quickly hustled into a car. Major Hermann Giskes, the forty-five-year-old head of Abwehr counterintelligence in the Netherlands, had been on Lauwers's trail for weeks. A protégé of Admiral Canaris's, Giskes had already shown an exceptional talent for tracking down Allied secret agents during his previous posting in Paris. Contrary to what SOE thought of Giskes and his colleagues, they were, in the words of the Dutch historian Louis de Jong, "extremely skilled and dangerous opponents."

The bitter rivalry between the Abwehr and the SS, so evident in Berlin and elsewhere in Europe, had been playing out in Holland, too. But in the previous few months, Giskes had managed to establish an uneasy partnership with Joseph Schreieder, his SD counterpart in The Hague. For the time being at least, the Abwehr would cooperate with the SD, the SS's counterespionage unit, in locating, capturing, and questioning agents sent from London.

In his interrogation techniques, Giskes preferred using a rapier approach—quiet, intense verbal pressure—rather than the SD and Gestapo's bludgeon, which usually involved torture. Promising Lauwers he would not be harmed, Giskes pressed the exhausted, frightened agent to agree to "play back" his wireless set—that is, to continue to send messages to London as if he were still free. At first Lauwers adamantly refused, but after many hours of interrogation, he finally gave in, confident that when he left out his security check, as he had been repeatedly told to do, SOE would realize he had been captured.

As an SOE recruit in London, Lauwers had been under the impression that he would be working for MI6—a mistaken idea that the highly secretive SOE never corrected. Like so many others who admired MI6, he thought of it as an elite intelligence service that rarely if ever made mistakes. He and his fellow Dutch agents, Lauwers later wrote, had two things in common: "a deep love of our country and a blind trust in our superiors. The long-standing reputation of the British Secret Service throughout the world and the training which the agents received brought our trust . . . to the heights of almost mystical belief."

Secure in that conviction, he resumed his transmissions to London under Giskes's watchful eye, expecting that when SOE realized his situation, it would break off all contact. In four successive messages, he left out his security check, but SOE gave no indication it understood his warnings. Increasingly desperate, he began inserting the letters "CAU" and "GHT" in his transmissions. Again no reaction. Many years later, Leo Marks would remark that "no agent in my experience tried harder than [Lauwers] to let us know he was caught. . . . Poor devil, he did his damnedest."

In fact, Lauwer's missing check *had* been noted by SOE signals operators and brought to the attention of Blizzard and his N Section subordinates. They concluded that the repeated lack of a check was insufficient evidence to prove that Lauwers was in custody. Not long afterward, they informed him that another agent would be parachuted in to join him and Taconis. When the new operative landed in late March, Giskes and his men were on hand to greet him.

Thus began *das Englandspiel* ("the England Game"), an extraordinary two-year Abwehr operation that netted more than fifty London-sent Dutch agents (several of whom had been sent by MI6 and were caught up in the SOE debacle), not to mention hundreds of tons of arms and explosives. The worst disaster in SOE history, it would virtually decapitate the Dutch resistance movement.

From March 1942 onward, a steady flood of operatives and weapons was dropped into the waiting arms of the Abwehr and Gestapo. Like Lauwers, the agents who were wireless operators agreed to "play back" their sets, believing that SOE would instantly notice their lack of a security check. And once again, the omitted checks were ignored. To a man, the newcomers' reaction to "this continuous negligence of the grossest kind" was "stupefying bewilderment," as Lauwers put it.

Initially, Giskes shared their confusion. Like Lauwers, he conflated SOE with MI6, which he long had viewed as an omnipotent agency "famous for its long experience and unexcelled skill at the conduct of underground warfare." But as *das Englandspiel* played out, he changed his mind. The agents, he wrote, were "amateurs, despite their training in England," which had been sadly negligent in preparing them for "their immensely difficult task."

In his interrogations, Giskes convinced several agents that the reason for their immediate capture was not the result of monumental incompetence on the part of SOE but because double agents in the organization's headquarters had betrayed them to the Germans. Shocked and demoralized by the lie and swayed by Giskes's promise to free them at the end of the war, a number of them acquiesced when he said they might as well tell him everything they knew. Soon, Giskes,

Schreieder, and their staffs had amassed a huge volume of information about N Section operations, including copious details about its officials, instructors, and training schools.

BY JUNE 1942, FIFTEEN AGENTS had been dispatched to Holland, all of them, in N Section's deluded view, doing a fine job of organizing Dutch resistance. Much more, however, needed to be done, and to oversee that expanded effort, the next man to be parachuted into Holland was George Jambroes, a former resistance leader who had escaped to Britain in late 1941 and who had close connections with the Dutch government in exile.

Jambroes's mission was to take command of all resistance groups in the Netherlands and form them into an underground army of saboteurs that would begin readying itself for its role in the Allied invasion of Europe. To help him in this hugely ambitious effort, Jambroes was to be sent twenty SOE agents—ten organizers and ten radio operators. He was also to work closely with the leaders of the various resistance groups, unifying them under an umbrella organization to be called the National Committee of Resistance.

On June 26, 1942, Jambroes was parachuted into Holland, with a wireless operator and several tons of weapons and explosives. They were met, of course, by a large German reception party, including Giskes and Schreieder. After an all-night interrogation of Jambroes, the Abwehr and SD were in full possession of the details of his mission. The information's significance was obvious to Giskes and Schreieder and to their bosses in Berlin: if its successes continued, *das Englandspiel* might well produce clues to the exact timing and location of the Allied invasion of Europe.

In the short term, however, the Germans faced a serious problem. After N Section had informed the SOE brass of Jambroes's "successful" arrival, a message was sent to Jambroes, ordering him to get in touch with the leaders of Ordedienst (OD), the largest and best-organized of the country's resistance organizations. The problem for the Germans, as Giskes acknowledged, was that they had no idea who Ordedienst's leaders were or how to find them. As a result, the ongoing fictional reports of success were going to be difficult to sustain. Giskes's solution was to send a cable to London in Jambroes's name, advising that the top echelons of OD had been so infiltrated by German informers that it would be suicidal to seek them out. Instead, Giskes/Jambroes proposed that Jambroes make contact with lower-level members of area OD groups, whose names would obviously be unknown to either SOE or the Dutch government in exile. N Section agreed.

In the summer and fall of 1942, SOE received a series of rosy reports from Giskes/Jambroes: the resistance groups that Jambroes had supposedly contacted were making astonishing progress in their training and now were in great need of more instructors and arms. London responded in "conveyor-belt" fashion, Giskes later wrote, dispatching twenty-seven more agents and hundreds of additional tons of equipment and supplies. By December 1942, forty-three SOE and MI6 operatives were in German custody.

For Giskes personally, the astonishing success of *das Englandspiel* was not an unalloyed triumph. He was exhausted, for one thing, having spent a succession of uncomfortable nights in marshy fields, waiting for new drops of agents and arms. During those long, cold waits, he was plagued by the fear that the British had caught on to the scheme and that bombs, not agents, would fall from the sky. But no bombs fell—only more agents and more arms for the Germans to collect.

Another downside was London's incessant demands for action on the part of the agents. Having received all those glowing reports on the development of the Dutch underground army, SOE asked to see some proof of its progress. London's calls for "open attacks on shipyards, ships, and locks became more and more pressing," Giskes remembered. He and his colleagues had no option but to use all their inventiveness to convince SOE that an active sabotage campaign against German targets was indeed under way.

Drawing on the resources of an Abwehr sabotage team from Brussels, Giskes staged a number of dummy explosions on Dutch railway lines that caused no real damage but were widely reported in the Dutch press and much discussed by Dutch railway workers. On another occasion, a barge carrying aircraft parts was blown up in broad daylight as it approached Rotterdam harbor. Rotterdam residents who witnessed the explosion erupted in cheers, unaware that the barge was a derelict and the aircraft parts came from wrecked planes and could never have been used again.

At one point, London ordered the destruction of several power stations and a key radio tower used by the Germans—targets that were clearly off limits even to faux sabotage. In the case of the power stations, Giskes/Jambroes explained to London that the three saboteurs assigned to destroy them had stumbled into a minefield before they could accomplish their mission. As for the radio tower, a story was planted in the Dutch press about a failed attempt to blow it up by "unknown criminal elements." N Section accepted the failures without complaint.

By early 1943, *das Englandspiel* had ballooned into such a huge scam that Giskes was forced to cut it back. "I was faced with the problem of keeping London . . . supplied with information about the multifarious activities of nearly 50 agents," he wrote, "and it seemed impossible that we could keep this up for long." His solution was to inform SOE that, regrettably, several of the agents had suffered fatal accidents or had been captured and killed by the Germans.

THOUGH N SECTION NEVER seemed to doubt Giskes's self-described "fairy tales," an increasing number of outsiders began expressing skepticism about the entire Dutch operation. Prince Bernhard, Queen Wilhelmina's son-in-law, informed the British of his government's worries that "all was not well" in the Netherlands. Several officers in SOE's signals department noted the lack of security checks in messages from Dutch agents. One officer was so insistent in his warnings that he was told by his superiors if he mentioned the matter again, he would be drafted into the British Army for frontline service.

Also suspicious were a number of RAF pilots who transported the Dutch agents to their drop zones. The pilots noted that they never had any problems in their flights to Holland or in making the drops. There were no German night fighters or antiaircraft guns, and the landing areas were perfectly laid out and well lighted—"too bloody perfect," in the opinion of one flier. The difficulties would begin on the return trip home. "People seemed to get a very easy run in, and then the aircraft would disappear on the way out," recalled an RAF squadron leader. "Not all of them, but it was a rough ride out." In less than a year, twelve RAF bombers were shot down by German fighters lying in wait for them on their way back from Holland—much higher losses than in similar missions to France and other countries in western Europe.

But of all the skeptics of SOE activities in Holland, none was more insistent than Leo Marks. Throughout most of 1942, he repeatedly warned Charles Blizzard and other N Section officials about the lack of their agents' security checks, only to be rebuffed. "They had a stock answer to every enquiry I made about the security of their agents: 'They're perfectly all right; we have our own ways of checking on them,'" he recalled. "I wasn't in a position to ask what they were."

Marks finally took his suspicions to Sir Charles Hambro, SOE's director. But Hambro, too, seemed unimpressed by his evidence. When the SOE chief demanded more conclusive proof, Marks turned his attention to a study of the indecipherable messages generated by Dutch operatives over the last several months. He discovered that, amazingly, not one signal from Holland had been corrupted during that period. "Why were the Dutch agents the only ones who never made mistakes in their coding?" he asked himself. "Were their working conditions so secure that they had as much time as they needed to encode their messages and didn't have to worry about Germans on the prowl?" To test his theory that the enemy was controlling Dutch radio traffic, he sent a message to Holland that ended with the common German sign-off "HH," standing for "Heil Hitler." When he received an apparently reflexive reply that also ended with "HH," Marks knew it had come from a German. (As *das Englandspiel* progressed, German radio operators had begun taking over from the captive SOE agents.)

Trying to imagine how it felt "to be in a prison cell in Holland hoping that someone in London was awake," SOE's chief of codes had finally had enough. In January 1943, he shut himself away for three days to study every message that had been exchanged between Holland and

London. When he finished, he wrote a blistering four-page report that could have been summarized, he said, in just four words: "God help these agents."

In his memo, Marks noted the receipt of several messages from Holland detailing the myriad dangers and fatal mishaps that supposedly had befallen agents in the field. Yet, he went on, "despite deaths by drowning, by exploding minefields, by dropping accidents, despite every kind of difficulty, setback and frustration, not a single Dutch agent had been so overwrought that he'd made a mistake in his coding." The question, he added, "was no longer which agents were caught but which were free."

When Marks showed his conclusions to Colonel Frederick Nicholls, the head of SOE's signaling department and Marks's immediate superior, Nicholls said that they "certainly could not be ignored" and that he would pass them on to the agency's top officials. A few weeks later, Marks was summoned to meet with Colin Gubbins, who would be named head of SOE a few months later. Gubbins told Marks that as a result of his report, an independent investigation would be launched into the Dutch agents' security. But while that was under way, Gubbins said, Marks was forbidden to discuss his suspicions with anybody but Nicholls or Gubbins himself.

Agreeing to keep quiet, Marks waited for the inquiry to begin. But as the winter of 1943 turned to spring, he saw no sign of action. Marks finally concluded that Gubbins was keeping his report under wraps because he feared its incendiary contents might mean the end of the agency. Gubbins knew that SOE's apparent achievements in Holland were considered essential for the planning of D-Day and for SOE's credibility in Whitehall. Indeed, several months earlier, Hambro had sent a telegram to OSS director William Donovan promising that "SOE will be ready by February 1943 at the latest to mount operations into France and the Low Countries, and I am confident that [this] will mark the turning point in European resistance." When he saw the telegram, Marks later wrote, "I did my best not to shout, 'Doesn't [Hambro] realize that the Low Countries' security couldn't be lower?'"

When SOE was organized in 1940, its creators mandated that if the agency ever suspected German control of any of its networks, it was to turn over the handling of those operations to MI6. Gubbins was sure that Stewart Menzies and Claude Dansey would view such information as the perfect weapon with which to kill what they considered MI6's greatest enemy. It was best, therefore, to keep such damning news secret for as long as possible.

Still, there were some changes after Marks turned in his report. Charles Blizzard was removed as head of N Section, to be replaced by Seymour Bingham, a British businessman who had grown up in the Netherlands and who had worked for MI5 in London after the war began. Dismissed in mid-1941 for heavy drinking, Bingham had then been brought to SOE by Richard Laming, who had known him in Holland. For the eleven months he headed N Section, he would turn out to be as myopic and misguided as his two predecessors.

In an even more troubling development, Marks was told that instead of canceling further drops of agents into Holland, SOE had actually decided to *increase* the number. From March to May 1943, another nine operatives were parachuted in, all of them picked up by Giskes's men as soon as they landed. By May, the Dutch prison holding the agents was so overcrowded that three of them were packed into every cell.

Before leaving London, each agent had met with Marks for the usual final briefing on codes. Those sessions were agony for Marks, who, though convinced that the operatives would soon be jumping to their doom, was forbidden by his superiors to warn them.

In July 1943, four months after Gubbins had promised an investigation, Colonel Nicholls finally informed Marks that one was under way and that he would be called as a witness. But he was also told that under no circumstances could he mention to the investigators his own report or his suspicions; those conducting the probe must be allowed to make up their own minds about the true state of affairs in the Netherlands. When Nicholls asked Marks if he had any questions, he replied, "Only one, sir. Is this to be a genuine inquiry or an in-house cover-up?"

To Marks, the answer was obvious.

AS MARKS LATER POINTED OUT, SOE had become a real threat not only to its own agents in Holland but to many members of the Dutch resistance, whose "lives were needlessly thrown away." Using information gleaned from the captured agents by Abwehr interrogators, Dutch collaborators working with the Germans were able to infiltrate many resistance groups, leading to large-scale arrests and executions.

In March 1943, a collaborator posing as an SOE agent made contact with Koos Vorrink, a former Dutch prime minister who had become one of the country's top resistance leaders. The collaborator told Vorrink that their country's government in exile wanted to know the names of the members of a large underground group of prominent Dutch politicians that Vorrink had created. The ex–prime minister complied with the request; by the following afternoon, he and more than 150 other key resistance figures were in prison.

As for the fifty-four London-sent agents who ended up in German custody, the SS did not live up to the promise it had made to Giskes to spare their lives. In September 1944, as invading Allied forces approached Holland, most of the agents were transported to the infamous Mauthausen concentration camp in Austria. Shortly after their arrival, they were machine-gunned in a granite quarry inside the camp. Of the original fifty-four, only four, Lauwers among them, survived.

The Germans' *Englandspiel* triumph also eviscerated the Dutch resistance movement itself. Scattered remnants continued to operate, but without weapons and effective direction, there was little they could accomplish. "The attempt of the Allied secret services to gain a foothold in Holland had been delayed by two years," Giskes noted in his postwar memoirs. "The establishment of armed sabotage and terror organizations, which might have . . . crippled our defenses at the critical moment of invasion, had been prevented."

After the war, rumors swirled in Holland that the *Englandspiel* disaster was, in fact, the work of a traitor in SOE's N Section or perhaps a deliberate sacrifice of Dutch agents by the British to mislead the Germans about a possible Allied invasion of the Netherlands. Despite all the astonishing wartime blunders made by SOE and MI6, the myth of the infallibility of the British secret services continued to hold sway in Holland, as it did in much of the rest of the world. For many, it was simply impossible to fathom how the devastation caused by *das Englandspiel* could have been the result of stupidity and ineptness.

The British response to an investigation of *das Englandspiel* by the Dutch government four years after the war's end muddied the story even further. When a parliamentary commission from Holland arrived in London in October 1949 to begin its inquiry, MI6, which had been made the custodian of SOE wartime records, claimed that most of the N Section archives no longer existed. Indeed, those reports—along with many other SOE records that, according to one agency operative, were "time bombs waiting to explode"—had been destroyed in a mysterious fire in early 1946. Among the missing papers were lengthy postwar reports by Leo Marks about *das Englandspiel* and N Section's overall dismal performance in the first three years of the war. "I'd worked too long for SOE to believe that the fire was accidental," Marks said years later. His was a belief shared by many former colleagues.

Stymied in its quest for written evidence about what had really happened, the Dutch investigating commission was also allowed little formal access to SOE and MI6 officials who'd been involved with secret operations in Holland. Richard Laming, N Section's hapless first chief, was the only former official who provided official testimony, which largely consisted of explaining the intricacies of Britain's Official Secrets Act. Off the record, several ex-SOE officials, including Colin Gubbins, agreed to meet with the commission; all denied that SOE had had any part in Germany's highly successful roundup of its agents in Holland.

The Dutch commission declared in its final report that "errors of judgment" had been responsible for *das Englandspiel*—"errors" that some in the British government had done their best to keep secret during and after the war. The commission added that no evidence existed for charges of "treachery on either the British or Netherlands side." As the British historian M.R.D. Foot put it, "The truth is more mundane: the agents were victims of sound police work on the German side, assisted by . . . incompetence in London."

In 1943, as it happened, that same sorry scenario was playing out in France, with far more serious implications.

CHAPTER 16

"Be More Careful Next Time"

SOE's Debacle in France

IN EARLY 1943, LEO MARKS WAS PREOCCUPIED WITH THE DISAStrous situation in Holland and could think of little else. But on a cold, gray afternoon in February, he forced himself to preside over a final coding session with a problem agent who was due to go to France the following month.

Few people at SOE had anything good to say about twenty-six-year-old Francis Cammaerts. His instructors acknowledged that he was highly intelligent but complained that he had no flair for subversion or leadership—major problems for someone assigned to organize and head a sabotage network in France. "Rather lacking in dash," read the final report on Cammaerts. "He lacks drive and has a somewhat negative personality." According to his coding instructor, he was "a plodder who does his best to follow instructions but seems unable to grasp the basic principles."

Having read Cammaerts's lackluster file, Marks resolved to spend as little time with him as possible. But soon after entering the briefing room, he changed his mind. The fair-haired, extraordinarily tall Cammaerts "wasn't plodding at all," Marks concluded. In fact, he was much like Marks himself—passionate, skeptical, and fiercely independent—traits unappreciated by the SOE brass, as Marks knew from personal experience.

Cammaerts's reputation as a plodder, Marks observed, stemmed from his refusal to accept anything at face value. He tested "the logic of it all . . . taking nothing and no one for granted, least of all his various instructors," Marks added. SOE's codes chief left his session with Cammaerts "feeling very sorry for anyone who made the mistake of writing this man off."

Francis Cammaerts

Marks's evaluation turned out to be exactly right. Francis Cammaerts would become one of the most successful SOE agents operating in western Europe during the war. The son of a noted Belgian poet and a British actress, he had been born and raised in England. Like many others his age, he had been appalled by the senseless slaughter of the Great War. He had become a pacifist while at Cambridge and was teaching history at a small, innovative grammar school in southeast London when World War II broke out. After registering as a conscientious objector, he was assigned to work as a laborer at a farming commune in eastern England.

In 1942, Cammaerts's younger brother, Pieter, was killed in a RAF bombing mission over Germany. Shattered by his brother's death, Francis decided that although he still believed in the general idea of pacifism, he must get involved in this war. Through a friend, he was put in touch with SOE.

From the beginning, he, like Marks, was bemused by the slapdash, careless nature of the place. Although the need for secrecy was endlessly preached, files were left open on desks, and agents who were supposed to be kept apart for security reasons were constantly bumping into one another. Convinced that he and his colleagues were not being given adequate training to match wits with the enemy, Cammaerts later remarked, "I never believed that we amateurs could play clever buggers with the German security services; they were the pros."

Even more dismaying to him was the chaos of SOE's French operations at a time when D-Day planners, in a top secret decision, had chosen the Normandy coast as the epicenter of the invasion of Europe. France's resistance movement was thus slated to play an important part in the initial stages of liberating the Continent, hindering German troop movements by sabotaging power supplies, bridges, railways, roads, arms dumps, and communications.

The invasion's planners envisioned the resistance effort as a unified, coordinated campaign. But where France was concerned, SOE was as divided as it could possibly be, thanks largely to the contentious relationship between the British government and de Gaulle's Free French. Whereas other countries had one SOE section in London to direct their resistance activities, France had six, only two of which—F and RF sections—would, in the end, play significant roles. F Section had been created in late 1940 after the Foreign Office, at Churchill's behest, had instructed SOE to operate in France without Free French participation, in an effort to appease anti-German, anti–de Gaulle elements in the Vichy government. Most of its agents were French-speaking British subjects, with a sprinkling of other nationalities, including Americans and Canadians. De Gaulle, who was not supposed to know about F Section's existence, quickly found out about it and protested that the creation of British-run resistance networks in France infringed the sovereignty of his country.

In addition to further poisoning the British–Free French relationship, de Gaulle's exclusion from SOE activities was anathema in French noncommunist resistance circles, which were already beginning to rally around the general as their leader. To mollify them and de Gaulle, SOE created RF [Republique Française] Section, a British-staffed department that worked with de Gaulle's intelligence and sabotage operations. Its job was to supply money, transportation, weapons, and communications equipment to agents recruited and trained by the Free French. But this did not make de Gaulle happy either. Intensely resentful of his dependence on the British, he hated the fact that SOE officials could veto his agents' orders and had control over Free French communications with France.

Adding to his anger were efforts by F Section and MI6 to recruit prospective agents from the growing tide of young Frenchmen coming to Britain to join the general's forces. During the war, all new arrivals from Europe and elsewhere had to undergo questioning by British security services at an interrogation center in southwest London. As a result, those services could sweep up the most promising potential agents for themselves before de Gaulle's men had a chance to contact them.

With all these feuds and rivalries, it's hardly surprising that there was fierce internecine warfare between the Free French and SOE, as well as within the agency's French sections. At one point, according to Leo Marks, RF Section officers and their Free French counterparts were not allowed to speak to each other and had to meet in safe houses as if they were in France. Infuriated by all the bureaucratic jealousy and intrigue, one RF agent exclaimed to Marks, "I can tell you what [this war in SOE] is not about. It's not about killing Germans or helping agents to survive or shortening the war, though that's what they all pretend it is."

ORGANIZED EARLIER THAN ITS Free French counterpart, F Section still got off to a slow start. Its first agents were not sent to France until the spring of 1941, and their numbers were not significant until the summer of 1942. Even then neither they nor RF's agents had been able to accomplish much by early 1943. After the war, Field Marshal Gerd von Rundstedt, Germany's supreme army commander in the west, noted that "during the year 1942, the underground movement in France was still confined to bearable limits. Murders and attacks on members of the Wehrmacht, as well as sabotage, were common, and trains were frequently derailed. A real danger for the German troops and real obstruction of troop movements did not, however, exist."

Like N Section in Holland, F Section was pressured hard by the SOE brass to accelerate its efforts. It resembled the Dutch operation, too, in the weakness of its London-based staff members, particularly those at the top. Its head, Maurice Buckmaster, was an Eton-educated businessman with no training or experience in clandestine warfare. A senior manager of Ford Motor Company in France before the war, he was recruited by SOE because of his knowledge of French industry. Tall and genial, the forty-one-year-old Buckmaster "brought the optimism of a sales director into Baker Street," said a colleague. No one seemed to know why or how he had been selected to lead one of the agency's most important sections. Colin Gubbins later acknowledged that Buckmaster had gotten the job because "there was nobody else."

According to Philippe de Vomécourt, one of the relatively few native French agents in F Section, Buckmaster was a remote department head, unable to comprehend the "fears and excitements of those who worked in the field." He also left much to be desired as a leader. "He was not firm enough in his orders to us—leaving dangerous doubts in our minds about who was responsible and who was subordinate in the field," noted de Vomécourt, an aristocrat who built up a key sabotage network in south-central France. "And he was not determined enough in fighting his superiors on our behalf. He was too readily submissive to higher ranks."

De Vomécourt, who was described by an SOE official as "one of the most exceptional men ever to go into the field," also faulted Buckmaster for the poor quality of some of the agents he dispatched: "those whose French accents were so bad that they were a danger to themselves and to those with whom they worked." Under severe pressure to send over an ever-increasing number of operatives, Buckmaster "lowered his standards to recruit men and women who really did not possess all the necessary qualities and qualifications for the work."

Adding to the problem, in de Vomécourt's view, was Buckmaster's and F Section's dangerous lack of understanding of the conditions in occupied France: "the all-important details of existence, a knowledge of which might mean the difference between life and death."

WHEN FRANCIS CAMMAERTS LANDED in France on March 22, 1943, he felt like Alice on the other side of the looking glass. Compared to this strange new world, the bureaucratic warfare in London seemed like child's play.

France was now a much more dangerous place than in the first years of the war. The discipline and politeness of its German occupiers had long since vanished. With the Allied triumphs in North Africa and at Stalingrad, the Reich, once seemingly omnipotent, was now under threat and on the defensive, which made life even more perilous and menacing for those under its thumb.

Eight months before Cammaerts arrived, the SS in France had wrested control of policing and security from the German military, and it had immediately set out to destroy the increasingly audacious French resistance movement. "If there had been any bridle upon the terror before 1943, it was swept away now," recalled de Vomécourt.

In its first year after taking control, the SS and French police arrested some 16,000 resistance members, many of whom were tortured and put to death. Also taking part in the purge was a brutal new French paramilitary force called the Milice, whose members, according to Vomécourt, were "almost to a man, thugs on the make."

When Cammaerts arrived, one of his first assignments was to check out the security and effectiveness of a large F Section network called Carte, whose territory covered much of southeastern France, including Provence and the French Riviera. From the moment he landed on French soil, an hour's car drive away from Paris, he was horrified by Carte's egregious indifference to safety. Met by a reception committee of five young Frenchmen late at night, he was bundled into a car and driven to Paris, which was under a strict nighttime curfew. "There appeared to be very little security consciousness in [the Frenchmen's] chatter," he recalled. "My whole reception couldn't have felt more insecure [and] dangerous."

Once in Paris, Cammaerts was taken to the apartment of a French-man who was a key member of Carte. Following several hours of dis-cussion, he left for a safe house after making plans to meet the Frenchman again the following morning. But when he returned, he found that his contact had been arrested by the Gestapo. Leaving Paris immediately, Cammaerts, posing as a French citizen, traveled south-east by bus to the tiny village of Saint-Jorioz, some sixty miles from Lyon.

In this beautiful setting, surrounded by the snow-covered peaks of the French Alps, several F Section officers, acting as organizers for Carte, had set up an ad hoc headquarters. When Cammaerts arrived in Saint-Jorioz, his concerns increased when "five or six young men, looking not a jot like the villagers, got off when I did and made straight for a nearby villa" where Carte members congregated. The house "was public," he recalled, "far from silent, full of mysterious comings and goings, unsafe."

As Cammaerts knew, the avoidance of any kind of central gathering place was a cardinal rule of clandestine work. He quickly concluded that Carte was a shambles, lacking any sense of security or organiza-tional rigor. "That kind of stupidity, to my mind, was unpardonable," he later said. "Lots of people lost their lives because of it." Concluding that the network was beyond repair, he decided to move on and set up a group of his own elsewhere. Weeks later, the Germans descended on Saint-Jorioz, arrested Carte's SOE liaison officers, and rolled up the network.

Cammaerts, meanwhile, traveled to the verdant Rhône Valley, noted for its Roman ruins and hillside vineyards, where he began to create what was to be code-named the Jockey network. Traveling by bicycle, on foot, and only infrequently by car or bus, the tall, gangling young agent covered hundreds of miles and reached out to scores of farmers, tradesmen, mechanics, grocers, teachers, and other residents of southeast France, many of whom joined him in the fight. Within a few months, he had organized about fifty small groups to be trained in sabotage techniques; he also set up drop zones for weapons, explosives, and other supplies from Britain.

Cammaerts, it turned out, had an instinctive talent for covert operations. Unlike many of his SOE colleagues, he was extremely security-conscious. Adopting methods long used by communist organizers and resistance fighters, he established cells of no more than fifteen men; each cell was to have minimal contact with the others. To join a cell, a new member had to be recommended by someone already in it.

Members of the network were given other strict rules. The use of telephones was forbidden, as was carrying written messages or reports on one's person. Agents were not to sleep in the same place more than two or three nights in a row, and they were not to gather in the same spot. They were to avoid spending nights in hotels and boardinghouses as much as possible, because anyone staying in them had to fill out a detailed registration form, which the Germans could easily acquire. If one member of a cell was arrested, its other members must immediately change locations.

Above all, no one was to divulge his or her real identity—a rule that Cammaerts himself religiously followed. "From the moment of landing in France, for my own good, but especially for the safety of anyone I came in contact with, no one would ever know my real name," he recalled. His code name was "Roger."

In the course of organizing his network, Cammaerts formed an extraordinary bond not only with the members of his groups but with the hundreds of ordinary French citizens—men, women, and children—who sheltered, clothed, and fed him during his eighteen months in the country. As he traveled throughout the region, he was passed from one family to another, putting his life in their hands while risking theirs.

There were hundreds of thousands of these caregivers all over France—people who never carried a gun or threw a grenade but whose willingness to provide safe houses for those who did made them invaluable members of the resistance. "Without any of the vainglorious conceit that characterized some of the self-styled 'chefs,' and with no thought of reward or glory, they had placed their homes at our disposal . . . thereby endangering their lives far more than any member of an armed maquis band," noted Xan Fielding, an SOE officer who worked closely with Cammaerts.

Such people, Cammaerts said later, were the heart of the resistance. Most of them won no medals or honors after the war, nor were books written about them, unlike the top resistance leaders and various SOE agents, including Cammaerts himself. "What has continuously irritated me has been the talk about the resistance as if it was created by a few heroes and heroines," he remarked. "And they've tried to make me a hero, whereas the most important thing was the heroism of the people we were living with. . . . They were sacrificing everything—children, partners, elderly relations, their land."

As Cammaerts and others repeatedly asserted, there could have been no resistance movement without these unheralded French supporters. "The resistance in France was born of the French themselves," said Philippe de Vomécourt. "The French people would have been less effective had it not been for the arms and materials, the instructors and organizers, sent from London. But there would still have been a resistance. The agents from London, on the other hand, would have been helpless without the never-failing help and courage of the ordinary people of France. There were many, many more in the resistance than have ever been counted."

Cammaerts's insistence on tight security stemmed in no small part from the heavy responsibility he felt for putting so many people's lives in danger. While unusual in his caution, he was not alone; several other SOE organizers were just as vigilant. Not surprisingly, their operations turned out to be the most effective when the Allied invasion finally took place.

Among them was twenty-nine-year-old Pearl Witherington, who headed the 1,500-man Wrestler network in central France—the only female SOE agent to take on such a leading role. "I was very much awake to what was going on around me because you never knew, wherever you were, if anybody was watching or listening," Witherington observed after the war. "You had to be careful of everything." Even the slightest gesture or movement, such as instinctively looking right before crossing a street, could bring disaster to an English agent who momentarily forgot the French rules of the road.

For her, as for Cammaerts, wariness and suspicion were constant companions.

OBSESSIVE ATTENTION TO SAFETY, however, was the exception, not only in SOE but also among the Free French and the French underground as a whole (with the prominent exception of communist cells). There was, for example, widespread disregard for the need to have as little contact as possible with other groups or individuals in the movement. The bustling city of Lyon, in southeast France, for example, was known as the capital of French resistance because so many of its leaders, including Jean Moulin, tended to congregate there. "You couldn't go ten meters without running into an underground comrade whom you had to pretend not to know," one leader noted. Clandestine meetings in Lyon were held in only a few familiar places, making detection considerably easier for the French police and Gestapo.

Indiscretion was endemic, too, among the resistance rank and file, most of whom were young men and women who had joined this clandestine world with little idea of the tactics required for simple survival. "At the start, it must be confessed, we thought of the whole business as a game," Philippe de Vomécourt said of himself and his movement counterparts. "A serious, deadly one, but a game, nevertheless, [filled with] amusement, excitement, and adventure. The Germans, however, never thought of it as a game." In organizing these inexperienced resisters, SOE agents were supposed to instill a sense of discipline and security. Many, however, lacked both the training and temperament to do so. Included in that group was a thirty-two-year-old barrister named Francis Suttill, who parachuted into France in October 1942. The son of a British father and French mother, Suttill had been given the task of forming a new network based near Paris and covering a wide swath of central France. Called Prosper, which was Suttill's code name, the network was to replace Carte as F Section's leading operation.

For several months, Suttill traveled throughout the region, inviting local resistance leaders to join his circuit and form one huge organization. By June 1943, Prosper was by far the largest SOE network in the country, boasting more than sixty local groups in an area that included Chartres, Orléans, and Compiègne.

Prosper, however, turned out to be too vast and unwieldy for its own good. Unlike Cammaerts, Suttill did not insist on a policy of strict compartmentalization among the groups within his network. Instead of remaining isolated from one another, members of the various sub-circuits often gathered together in the same safe houses, restaurants, and cafés in Paris and its environs—the most Gestapo-ridden area of the country. As Cammaerts later put it, "the Prosper folks [were] living a fantasy life."

That fantasy life, such as it was, ended in June 1943, the worst month of the war for the French underground. On June 21, Jean Moulin, the movement's most important and powerful figure, was captured by the Gestapo, along with six key resistance leaders with whom he was meeting at a safe house in a Lyon suburb. Less than two weeks later, Francis Suttill and two close SOE associates—his courier and radio operator—were arrested in Paris. Within days, the Gestapo had slashed through the Prosper network like a giant scythe, arresting hundreds of local resistance members and seizing dozens of arms caches throughout central and northern France. They also acquired two wireless sets, one of them belonging to Gilbert Norman, Suttill's radio operator.

A local operative in Paris immediately informed London of the mass arrests. A few days later, Maurice Buckmaster received a message that said, "The entire Prosper organization is destroyed. No element of it should be touched." Yet on June 27, three days after Suttill's capture, F Section received a message from Gilbert Norman's wireless set declaring that he was still free.

The message, however, came with a major flaw, which Leo Marks identified as soon as he read it: Norman had omitted his security check. Marks immediately informed Buckmaster of the missing check, stressing that Norman was probably transmitting under Gestapo control, but Buckmaster refused to believe it. Without telling Marks, the F Section chief ordered the following reply to be sent to Norman: "You have forgotten your double security check. Be more careful next time." He then instructed his staff to continue sending messages to Norman.

Thus began France's version of *das Englandspiel,* a ten-month operation run by Josef Kieffer, the SS chief of counterintelligence in Paris. A highly competitive man, Kieffer was determined to equal or surpass the extraordinary success of the deception campaign run by Hermann Giskes, Kieffer's Abwehr counterpart in the Netherlands. And SOE unwittingly did its best to help Kieffer achieve his ambition.

Amazingly, no one in F Section knew anything about the doubts and suspicions long swirling around N Section and its agents. For all of SOE's general lack of security in London and the field, its headquarters was highly secure in at least one regard: the agency's country sections worked in almost complete isolation, sharing neither information about their operations nor warnings about possible blown networks with one another.

Although Leo Marks knew more about the Netherlands disaster than anyone else in SOE, he had been sworn to secrecy by Colin Gubbins and couldn't warn Buckmaster about the looming threat to his section. Nonetheless, Marks and staffers from the signals department continued to insist that something was very wrong with Gilbert Norman and the messages he was exchanging with London.

More than a month after Norman supposedly returned to the air, Buckmaster finally accepted the fact that the wireless operator was in German custody. But he insisted that the section continue sending messages to him, in the hope that doing so would save Norman's life. Buckmaster, who cared deeply about the fate of his agents, "never wanted to believe anyone was captured," a colleague said. To him, "all his geese were swans."

But there was another reason for behaving as if Norman were still at liberty. The RAF had been lobbying hard to stop all SOE supply flights to France, using the collapse of the Prosper network as its rationale. Halting the flights would have been calamitous for SOE's campaign to prove itself as an important instrument of war. In the view of Colin Gubbins and other top agency officials, it was vital to act as if the Prosper disaster were a thing of the past and to flood France with even more agents and weapons.

"Strategically France is by far the most important country in the Western Theatre of War," Gubbins noted in a memo. "I think therefore that SOE should regard this theatre as one in which the suffering of heavy casualties is inevitable. But it will yield the highest possible dividends. I would therefore increase to the maximum . . . SOE aid to the French field from now on and maintain it until D-Day."

As a result, many more agents were sent to central and northern France—and ended up in German custody within a few days of their arrival. They included two French-speaking Canadians dispatched to the Ardennes Forest to form an important new network called Archdeacon. Once they landed, however, the agents—an organizer and radio operator—disappeared from view for more than a month. "No one has the slightest knowledge of the Ardennes group, which must be considered lost," an F Section staffer wrote.

But in early August 1943, Archdeacon's wireless operator finally sent a message saying that the network was up and running. Although he, like Norman, left out his security check, F Section chided him for doing so and plunged ahead. From that point on, SOE agents and arms descended on that German-controlled circuit, as well as on other new networks that were blown even before they were created.

A collateral victim of the Archdeacon sting was Noor Inayat Khan, the ethereal, absentminded WAAF officer who had been judged temperamentally unsuitable for clandestine work by her SOE instructors. With an increasingly acute shortage of radio operators in France trumping any thought of caution, Buckmaster ordered that Khan's training be cut short and that she be sent to France as soon as possible. When the commander of the training school protested that she was "not overburdened with brains," Buckmaster replied, "We don't want them overburdened with brains," thereby establishing his own deficiency in that department.

Khan arrived in Paris on June 16, 1943, less than two weeks before the collapse of Prosper. She escaped the German dragnet and, as one of the few SOE agents still free, became virtually overnight one of F Section's foremost operatives and the Gestapo's most wanted British agent in Paris. Even though her sense of security had not noticeably improved, she managed to remain free throughout the summer and early fall, transmitting regularly to London. Then, in early October, F Section instructed her to link up with Archdeacon. When she did, she was promptly arrested by Gestapo agents, who acquired not only her transmitter but a notebook on her bedside table in which she had recorded every message she had received and sent since arriving in France.

Like Khan, almost all the hundred-plus SOE operatives who ended up in Gestapo custody in France were eventually sent to German concentration camps, where most suffered horrific deaths. Among them were fourteen women, including Khan and twenty-four-year-old Andrée Borrel, Francis Suttill's courier. Khan, who was held in chains for months after her capture, was shot in the back of the neck at Dachau. Borrel, one of four SOE female agents injected with poison at Natzweiler concentration camp in western Germany, was still alive when pushed into a cremation oven, according to eyewitnesses. Before she succumbed, she reportedly scratched one of the guards and shouted, *"Vive la France!"*

Ever since the war, the work and tragic fates of these agents have been recounted in countless magazine articles, books, and films. Left in the shadows were the thousands of French citizens who also lost their lives, thanks to their cooperation with the British-trained operatives, whether as couriers, operators of safe houses, saboteurs, or in other capacities. In the Prosper debacle alone, it's estimated that more than a thousand French men and women were arrested, the majority of whom were killed by the Germans.

Brian Stonehouse, one of the few F Section operatives to survive the hell of a concentration camp, blamed the huge casualty list in France on the lethal carelessness of London. Yet the SOE disaster there was not solely the result of British ineptness or of German counterintelligence skills, although both were considerable. Even before the Prosper roundup, suspicions were circulating of a traitor in SOE itself. Before his arrest, Francis Suttill, who was hanged at Sachsenhausen, told a friend, "The Germans seem to have known all our movements for some time now."

The target of Suttill's distrust was a dark, curly-haired Frenchman who controlled all SOE air traffic in and out of the Paris region—an operative who happened to be a particular favorite of Maurice Buckmaster's.

EVERYONE AGREED THAT HENRI DÉRICOURT was astonishingly good at what he did. In the Paris area, most agents were not parachuted in; instead, they arrived in small, light RAF planes that could take off and land on very small fields. It was Déricourt's responsibility to find those landing areas and to organize the safe arrivals of incoming agents and the departures of operatives returning to London. A delicate and perilous job, it also was an amazing feat of logistics—controlling the landings and takeoffs of RAF aircraft within twenty-five miles of German-occupied Paris. So much could go wrong. Yet with Déricourt in charge, nothing ever did.

Even the ever-wary Francis Cammaerts was impressed by the thirty-three-year-old Déricourt, who greeted him in March 1943, when Cammaerts first arrived from London. "Déricourt's operation was smooth, quiet, unfussy, and absolutely on time," he recalled. "[It] took place without interference. The security was perfect—which was not surprising when you knew the facts."

Before 1939, Déricourt had had a checkered flying career, working at various times as a barnstormer, mail pilot, and test pilot. After the war broke out, he flew for a small Vichy airline while moonlighting as a black marketeer and a part-time intelligence agent for the U.S. embassy in Vichy. When he expressed an interest in working full-time for the Allies, U.S. embassy staffers helped smuggle him to London.

Officials in MI5, Britain's domestic counterintelligence and security agency, who questioned Déricourt when he arrived were suspicious of him from the start. For one thing, he freely admitted that before the war he had been acquainted with a high-ranking German intelligence officer in Paris. In fact, he frequently socialized with the officer, General Karl Bömelburg, the SS chief of counterespionage for all of France, even after the war began. MI5 was unaware of how close Déricourt's relationship with Bömelburg actually was, but it knew enough about him to deny him clearance for secret service work.

Nonetheless, he was immediately taken on by Maurice Buckmaster, who badly needed an experienced air transport officer in northern France now that preparations for D-Day were heating up. Sent back to

France in January 1943, Déricourt was an immediate success, handling with unruffled ease the arrivals and departures of agents from Prosper and other networks, as well as the dropping of more than two hundred containers of arms and explosives.

Most of the agents who encountered Déricourt liked and trusted him, so much so that many gave him top secret reports to be carried back to London by plane, including descriptions of upcoming sabotage operations, lists of operatives' addresses, and photographs of potential targets. Others, however, were skeptical, noting Déricourt's persistent questioning of virtually every operative he happened to meet. Shortly before Tony Brooks, the twenty-year-old head of a major network in southeastern France, departed for a brief stint in London, Déricourt asked him, "Aren't you the organizer of Pimento?" When Brooks insisted that he was just a courier, Déricourt kept up a barrage of questions: "Where are you staying? When did you arrive? When do you expect to return from London?" Brooks refused to say anything more.

"You're not supposed to be interrogated by people who are just supposed to receive you," said Harry Despaigne, another agent who was badgered by Déricourt to tell him where and what his mission was in France. "I wouldn't talk to him, and I left Paris the same day."

A few months after Déricourt arrived in France, doubts about him blossomed into denunciations. Henri Frager, the head of a network in northwestern France, told SOE he believed that Déricourt had been copying the contents of agents' reports to London and passing them on to the Germans. At the same time, MI5 sent F Section a report from the Free French security service saying that "since the armistice, Déricourt had started to frequent German circles in France."

Nicholas Bodington, Buckmaster's deputy in F Section, swiftly came to Déricourt's defense. As it happened, Bodington, a former Reuters correspondent in Paris, had been friends with Déricourt before the war; according to some accounts, he had known General Bömelburg as well. Bodington insisted to Buckmaster that the allegations against Déricourt were untrue, calling them "typical French backbiting."

It didn't take much to convince the F Section head that Déricourt had been wrongly accused. It was unthinkable for him even to consider the possibility that this man, who had done such a brilliant job in ferrying his agents to and from France, could be a traitor. After the war, Buckmaster, who could never bring himself to believe that Déricourt could have been a German operative, claimed that the allegations had been motivated by simple jealousy: "[Déricourt's] efficiency . . . was staggering, and it was his very success that raised the ugly idea that he was controlled."

In fact, every accusation made against Déricourt was true. Within three days of his arrival in Paris in January 1943, he had made contact with Bömelburg, who arranged for him to stay at the Bristol, a luxury hotel in Paris where prominent German officials were quartered. Not long afterward, Déricourt officially became an agent for the SS general, identified in German records as BOE/48. Bömelburg in turn introduced him to his immediate subordinate, Josef Kieffer, with whom Déricourt also worked.

From then on, Déricourt provided the SS with detailed information about every SOE agent who traveled into and out of the Paris region, along with dates and locations of landing operations. The Germans agreed to Déricourt's proviso that they not intervene in any of his operations, although Gestapo agents were stationed nearby and had incoming agents under surveillance practically from the moment they landed.

As Frager had charged, Déricourt also passed on to the Gestapo the reports and other documents that agents gave him to send on to London. The material, which supplied German counterintelligence with invaluable information about upcoming SOE operations, was photographed and returned to him within twenty-four hours, so it could be dispatched to SOE headquarters with no one the wiser.

Thanks to Buckmaster's fateful decision not to investigate Déricourt, SOE operatives kept landing under his auspices until February 1944, unaware that Gestapo agents were watching them as they stepped off the plane. Almost all of them were arrested on the day of their arrival or very soon thereafter.

DURING AND AFTER THE WAR, rumors circulated that Déricourt wasn't the only one guilty of betrayal. Allegations were also leveled at none other than Claude Dansey, the deputy chief of MI6. One of Dansey's accusers was Colin Gubbins, the new head of SOE, who apparently believed that his archrival had had a role in Prosper's destruction. According to William Stephenson, who headed Britain's wartime intelligence operations in the United States, Gubbins "formed the impression that Dansey was involved" in the collapse of the network. Gubbins told Stephenson that "Dansey had betrayed to the enemy a number of his key agents in France," Stephenson recalled. He added, "Because Gubbins was not a man who lied or exaggerated, I believed this, for I myself had formed the impression that Dansey was an evil man who would stop at nothing to get someone out of his way."

Echoing that view, Gubbins's deputy, Harry Sporborg, told an interviewer after the war, "Make no mistake about it. MI6 would never have hesitated to use us or our agencies to advance their schemes, even if that meant the sacrifice of some of our people."

There's no question that Dansey and Stewart Menzies were still trying to do all in their power to shut down SOE operations in France and the rest of western Europe, believing that SOE was compromising the safety of their agency's intelligence efforts in those countries. If Gubbins's organization in France "could be suppressed, our intelligence would benefit enormously," Menzies was publicly quoted as saying in January 1943. Up to that point, however, he had failed in his efforts to convince Churchill to put an end to SOE.

There's also no question that Claude Dansey was icily indifferent to the fate of agents in the field. The writer Somerset Maugham, who served as an MI6 agent in World War I and wrote *Ashenden,* a collection of interconnected short stories based on his experiences, reportedly used Dansey as a model for one of the book's main characters: Colonel R, a British intelligence officer who recruits the protagonist. "A lot of nonsense is talked about the value of human life," Colonel R tells his protégé in one of the stories. "You might just as well say that the counters you use at poker have an intrinsic value, their value is what you like to make it; for a general . . . men are merely counters and he's a fool if he allows himself for sentimental reasons to look upon them as human beings."

When news of the Prosper disaster broke, Dansey rushed into the office of Patrick Reilly, a young diplomat working temporarily as Menzies's assistant. "With delight all over his face," Dansey asked Reilly if he had heard the news. When Reilly said no, Dansey acted "as if this was the most important moment of his career." "SOE's in the shit," he exclaimed. "They've bought it in France. The Germans are mopping them up all over the place." Reilly, who after the war served as British ambassador to France and the Soviet Union, recalled feeling "quite sick inside. I realized that Dansey was the most evil and the most wicked man I had met in public service, and nothing since has made me change my mind."

In an effort to capitalize on Prosper's destruction, Menzies swiftly submitted a damning report on SOE activities in France to the British chiefs of staff and the Joint Intelligence Committee. In it, the MI6 chief declared that "resistance groups [there] were to a considerable extent compromised and that the enemy must be in possession of much detail regarding those groups. . . . [They] cannot be counted on as a serious factor."

The Joint Intelligence Committee agreed, recommending to Churchill that MI6 assume control of SOE and that RAF supply flights to France be stopped. To Menzies's and Dansey's dismay, the prime minister vetoed both ideas. The French resistance movement, he insisted, was vital to the war effort. Though enemy reprisals had indeed been terrible, he added, it must be remembered that "the blood of the martyrs was the seed of the Church."

DID CLAUDE DANSEY INDEED play a role in the ruin of Prosper? Such speculation continued to circulate after the war and was the subject of a 1988 book entitled *All the King's Men*. In it, Robert Marshall, a former BBC producer, claimed that Dansey had foisted Déricourt on Buckmaster and that Déricourt was in fact working as a triple agent, reporting back to MI6 on German counterintelligence operations in France. The problem with that theory is that there is no hard evidence to back it up.

But it's also true that this was not the only time Dansey's name had been coupled with a British agent who turned out to be a traitor. In August 1941, at his behest, MI6 had sent a wireless operator named Bradley Davis to join Marie-Madeleine Fourcade's fast-growing Alliance network in France, one of the agency's most important intelligence operations. Fourcade and her colleagues were stunned when they first saw Davis. He was, she recalled, "the most ridiculous, most grotesque parody of a 'typical' Frenchman, attired . . . in a short jacket and waistcoat, striped trousers, a spotted cravat, a stiff shirt with cutaway collar beneath a little goatee beard, pince-nez, and as a crowning glory, a bowler hat. Was this British Intelligence? The boys burst into roars of laughter."

To Fourcade, however, Davis, the first Englishman to work in the field with Alliance, was no laughing matter. From the beginning, he acted suspiciously, asking too many questions and showing too much interest in everyone who came to see her. When he finally left to work as an Alliance radio operator in Normandy, she breathed a sigh of relief.

Not long afterward, news came of widespread arrests of Alliance operatives in Normandy and Paris. One of Fourcade's agents in the French capital sent word to her that Davis, whose code name was Blanquet, had betrayed them to the Gestapo. When she relayed the information to Kenneth Cohen, her MI6 liaison officer in London, he dismissed the suspicions and argued that Blanquet was "operating admirably in Normandy, sending us first-class information." She asked Cohen if he was certain that Blanquet was transmitting from Normandy, adding that several of her own agents had seen him in Paris.

MI6 did not pursue the matter, however, and the arrests of Alliance operatives continued. But in the end, Fourcade and her lieutenants amassed enough evidence to convince MI6 higher-ups that Blanquet had indeed been responsible for the capture, torture, and deaths of dozens of their colleagues. "It's incredible, incredible!" Léon Faye, Fourcade's deputy, exclaimed. Fourcade agreed, later writing, "We had good grounds for amazement. The first man sent to us by British Intelligence was working for the Nazis." In an urgent cable, Cohen instructed Fourcade and her men to find and execute Blanquet. They did so. During an interrogation before he was killed, Blanquet admitted to being a British fascist who had infiltrated MI6 so he could work with the Nazis in France.

Anthony Read and David Fisher, who wrote a largely sympathetic biography of Claude Dansey, called the Bradley Davis/Blanquet affair "one of Dansey's most serious mistakes"—a blunder that "brought almost total disaster to the Alliance network." Other historians have suggested the possibility of treachery in the higher reaches of MI6. Again, there is no evidence to support such a claim. The truth is that MI6, for all its complaining about the amateurism of SOE, was, more often than not, as inept, careless, and downright stupid as its hated rival in its selection of agents and conduct of intelligence operations.

As for Fourcade, her MI6 liaison officer was relieved to note that Davis's betrayal "did not dim her loyalty to the British." That may have been true. But it certainly disillusioned her, as did a several-month stint in London beginning in July 1943. With the Gestapo hot on her trail in France, Dansey and his cohorts were anxious to bring her in for a rest and debriefing. She, in turn, wanted assurances from them of more money and other aid for her crippled organization.

During her stay in London, another, more massive wave of arrests swept through Alliance, this one netting several members of her inner circle, including her close deputy and friend Léon Faye. Fourcade was devastated. "Since September 16, my beloved Eagle [Faye's code name] had fallen, and with him more than 150 members of my network, including the veterans," she wrote. "Every time I crossed out the name of a friend, I experienced the feeling of having wielded the executioner's axe. I was dying of grief."

Fourcade and all the friends she lost had repeatedly risked everything over the previous three years to provide the British with the detailed intelligence they demanded—from the condition of the Luftwaffe and location of German army units to the number and movement of ships and submarines in French ports. But "while the Gestapo was hunting us down," she noted, Dansey and the other British officials she met had been busying themselves with another kind of war, waged against their fellow bureaucrats in Whitehall and occasionally against the Free French. Like other newly arrived French citizens who had personally defied the Germans, Fourcade was disheartened by London officialdom's detachment from the tragic realities of life across the Channel.

"Everyone I speak to seems to be fighting the war at a distance, a mental as well as a spatial distance," Emmanuel d'Astier de La Vigerie, the founder of the Libération-Sud resistance movement, noted of his own stay in London. "The Resistance and its adversaries were reduced to files, locked away in huge metal cabinets."

In the future, neither Astier nor Fourcade would waver in their commitment to working with the British; both their operations would be of significant help to the success of the Allied invasion of France. Yet neither ever got over their disappointment with the London-based officials with whom they associated. "Never in my life had I felt so hurt," Fourcade later wrote. "To use our information as part of the continuing struggle for power seemed to me to be quite inhuman. . . . Cannon fodder for British Intelligence, that's all we were."

CHAPTER 17

"Heroism Beyond Anything I Can Tell You"

Rescuing Allied Airmen

ON A BLAZINGLY HOT DAY IN AUGUST 1941, THE BRITISH vice consul in the Spanish city of Bilbao was roused from his afternoon nap. He must return to the consulate at once, he was told. A group of Allied escapees had just arrived from France.

At the consulate, he found four men and a petite, dark-haired young woman dressed in blue fisherman's trousers. The woman, who spoke French and identified herself as Andrée de Jongh, acted as their spokesman and leader. The men, she explained, were a Scottish soldier left behind at Dunkirk and three Belgian officers who wanted to join their country's forces in Britain.

Many other British servicemen were still hiding in Belgium, the twenty-four-year-old de Jongh added. Some, like the Scot, were survivors of the fight for France and the Low Countries; others were British pilots and crewmen shot down on recent bombing missions. In the past few months, she had set up an escape line through Belgium and France, manned mostly by friends of hers, to help these men get back to England. If the British government would give her money to help pay for the line—for mountain guides, safe houses, food, and rail fares—she could bring out many more. But, she made clear, the line would have to remain under her control.

Although the men who accompanied her verified de Jongh's story, the vice consul was somewhat skeptical. How could this slip of a girl organize such a complex operation on her own? More specifically, how had she managed to get these men across the dizzying heights and fast-flowing rivers of the Pyrenees—a dangerous and grueling trek for the most experienced outdoorsman?

Nonetheless, he finally came to believe her. In a cable passing on her request to London, he described de Jongh, who was known as Dédée, as "a girl of radiant integrity, as well as something of a beauty and physically hard as nails." His message was sent to MI9, a small, clandestine agency (and subsection of MI6) created in late 1940 to act as a liaison with British prisoners of war and to help rescue Allied servicemen trapped behind enemy lines. MI6's Claude Dansey, convinced that de Jongh was a German plant intent on sending enemy agents into Britain, initially rejected her proposal. But Lieutenant Colonel James Langley, the twenty-five-year-old head of MI9's escape section, supported her, and in the end the British government agreed to finance her operation—a shoestring venture that would soon evolve into the most important Allied escape network in western Europe.

James Langley

Airey Neave

JAMES LANGLEY AND HIS MI9 colleague Captain Airey Neave had a vested interest in the success of de Jongh's escape line and others like it. Early in the war, both men had been caught behind German lines and had themselves been rescued by fledgling escape networks. The Cambridge-educated Langley, an officer in the Coldstream Guards, had been wounded during the 1940 fighting in France—an injury so severe that his left arm had had to be amputated. Nonetheless, he had managed to escape from a German hospital in Lille, helped by a French nurse who had found a safe house for him. He had later acquired fake identity papers and had been passed from one sanctuary to another until he finally reached neutral Spain.

For his part, Neave, who had joined Langley at MI9 in 1942, had been captured after being hit by machine-gun fire at Dunkirk and had later been sent to prisoner-of-war camps in Germany and Poland. In April 1941, he had escaped from the Polish camp, only to be arrested soon afterward and interrogated by the Gestapo. He had then been transferred to the "bad boys'" punishment camp at Colditz, a forbidding medieval castle near the German city of Leipzig used to house persistent Allied escapers.

Surrounded on three sides by sheer rock cliffs, Colditz was said to be impregnable. But Neave and his fellow prisoners were determined to prove otherwise—an attitude that underscored the Germans' dubious logic in depositing the most expert and seasoned escapers in one place. After two escape attempts, the twenty-five-year-old Neave finally succeeded in early January 1942, when he and a Dutch army lieutenant, disguised as German officers, broke out of the heavily guarded fortress during a snowstorm and reached Switzerland four days later.

Getting Neave, the first Briton to escape from Colditz, back to England was a high-priority operation, but it still took him four months to make the perilous 1,500-mile journey. In the course of that trip, he was guided by members of the Pat O'Leary line, the first major escape group organized after the fall of France. Based in Marseille, it bore the code name of one of its founders, a Belgian doctor whose real name was Albert Guérisse.

IN ALL, SOME 7,000 BRITISH, American, and other Allied servicemen, most of them aircrew, were spirited out of occupied Europe during the war. Escape networks existed in all the captive nations, but they were most active in Belgium, France, and the Netherlands. Those were the countries over which hundreds of England-based planes flew on their daily bombing runs to Germany from 1941 until the war's end. The losses of aircraft were heavy, up to 15 percent on some days, but many crew members of downed bombers were able to parachute out before their planes crashed, leaving them stranded in enemy territory. At a time when trained bomber crews were in desperately short supply, it was vital for the Allied war effort to retrieve as many airmen as possible and bring them back to England to continue the fight.

By 1943, an American or British airman who was shot down over northwest Europe and was lucky enough to evade immediate capture had a good chance of finding a safe house, where he would be hidden until sent to a collection point in a nearby large city, such as Brussels, Paris, or The Hague. There he would be supplied with false identity papers and suitable clothes, given strict instructions on how to behave, then escorted, usually by train, to the Spanish frontier, where he and other evaders would be shepherded by guides over the Pyrenees. Hermann Göring, who noted that "it takes less time to build a plane than to form a crew," was as keenly aware of the value of those men as were their commanders in England. The Luftwaffe chief ordered the escape lines to be crushed, using any means necessary.

Even though participation in the escape networks was arguably the most dangerous form of resistance work in occupied Europe, thousands of people, from the very young to the very old, took part. The most perilous job of all was handled mostly by young women, many of them still in their teens, who escorted the servicemen hundreds of miles across enemy territory to Spain. Unlike resistance fighters, who were in hiding much of the time, these women did their work out in the open, riding on trains and other public forms of transportation with foreigners whose appearance and actions were often all too distinctive. "Nothing could have expressed more powerfully the spirit of resistance to Hitler," Airey Neave later wrote of the Frenchwomen who had accompanied him through France. According to Neave's biographer, his "admiration of the girls who carried out this dangerous task grew and unquestionably reshaped his attitude to women." Having shed much of the chauvinism endemic among Englishmen of his generation and class, Neave would emerge almost four decades later as a top aide to Margaret Thatcher and the main organizer of the movement that led to her selection as Conservative Party leader and election as Britain's first female prime minister.

At MI9, Neave and Langley provided funds and radio communications to those involved in running the escape lines. They also furnished liaison officers to escort evaders out of Spain and back to England. Yet throughout the war, the two young men were never given the resources to adequately supply and protect the thousands of operatives who worked under their auspices. Considered a sideshow by Dansey and other MI6 higher-ups, their tiny agency was always short of funds, so much so that it occasionally had to rely on financial gifts from a sympathetic British businessman to keep it in operation.

Of the many frustrations that Langley and Neave experienced, none was as painful as their inability to provide as much help as they thought necessary for Dédée de Jongh's burgeoning network.

DE JONGH'S MISSION TO save Allied soldiers had begun in the chaotic days after the 1940 German invasion of Belgium. When her father, a schoolmaster in Brussels, told her that King Leopold had surrendered to the Germans after only eighteen days of fighting, he began to weep. "I'd never seen my father cry before—never," she later said. "I was in despair and, at the same time, enraged. I said to my father, 'You are wrong to cry. You'll see what we do to them.'"

With the passion that earned her the nickname "the little cyclone," she immediately set out to do what she could to thwart the enemy. A commercial artist by profession, she had been working as a nurse for Allied troops wounded in the fighting in Belgium. After recruiting a group of friends and acquaintances to harbor escapees, she began to

*Andrée
de Jongh*

smuggle injured British soldiers out of the German-controlled hospitals in and around Brussels and into safe houses she had arranged.

The crew she enlisted to form the Comet line, as it came to be known, was, in one historian's words, "a cross-section of all that was young, lively, and freedom-loving in Brussels." Its members came from a wide variety of backgrounds and occupations, ranging from students to aristocrats to garage mechanics. What they had in common, besides their youth (few were over twenty-five), was their hatred of the Germans and their affection for de Jongh. As was true of other escape lines, the majority of Comet line workers—couriers, guides, and operators of safe houses—were women.

While Langley and Neave admired the group's joie de vivre and courage, as exemplified by its leader, they were frustrated by Comet's refusal to allow MI9 to have any say in its operation. "The last decision always rested with the men and women in the field, as from the outset Dédée had made it clear that she would brook no interference from outside," Langley wrote. "The line was Belgian, would be run by Belgians and any help would be gratefully received; but payment of money was simply reimbursement of expenses and in no way gave us the right to issue orders."

Again and again, de Jongh turned down MI9's offer to send a wireless operator to Brussels so she could communicate directly with London rather than relying solely on periodic meetings with British liaison officers in Spain. With a radio connection, she would have been able to alert MI9 immediately in case of trouble, and the agency could let her know about possible traitors and German agents posing as helpers or Allied airmen.

De Jongh's repeated refusal to accept a radio and operator was "always a very sore point" that "nearly drove me frantic," Langley said. In retrospect, however, her unwillingness to forgo such assistance might not have been such a bad thing, considering the disasters that plagued the wireless operators sent by London to the Netherlands and France.

Though Langley never stopped worrying about de Jongh, he had to admit that this young amateur knew what she was doing. Within months, she had put together a far-flung line that started in Brussels and snaked its way south through France to the Pyrenees. It worked like an assembly line: downed airmen were taken to the closest safe house, which was usually in a village or on a farm, then were transported to collection points in Brussels and Paris. From there small groups of evaders were escorted hundreds of miles by Comet line workers to a chain of safe houses in the foothills of the Pyrenees. From that point, Spanish guides—many of them German-hating Basques— escorted them over the mountains to Spain.

De Jongh was involved in every aspect of the operation. She organized the safe houses, worked with photographers and forgers to produce false Belgian and French identity papers, and personally guided dozens of airmen to the Spanish frontier and over the Pyrenees. Affectionately referring to her British and American charges as "parcels," she was known as the "postmistress"; her real identity was kept a closely guarded secret.

Concealing these young men was an extremely difficult task. As one British intelligence officer observed, "It is not an easy matter to hide . . . a foreigner in your midst, especially when it happens to be a red-haired Scotsman or a gum-chewing American from the Middle West." Finding clothes to fit Americans, who often were a head taller than Britons or other Europeans, was a particular problem. So was finding enough food, at a time when food was increasingly scarce for everyone. Coming from a country that had scant experience with the privations and dangers of war, many American evaders had a hard time understanding why they were given so little to eat while on the run; Britons, whose nation had been strictly rationed since 1940, were less demanding. British airmen were also generally calmer and more disciplined than their American counterparts, who were more independent and outgoing—and sometimes less willing to accept de Jongh's demand that every order they were given had to be obeyed without question.

Before setting out on their perilous journeys, Comet line workers did their best to teach their charges how to blend in while in public. In one memorable case, a bowlegged Texan was instructed in the fine art of walking like a European. Since most of the evaders spoke neither French nor Flemish, they were ordered to keep quiet at all times in public settings. U.S. airmen were told to keep their hands out of their pockets—a particularly American trait—and coached on the way Europeans use silverware when eating, keeping their fork in their left hand rather than shifting it back and forth during a meal. Both Britons and Americans were instructed never to smoke cigarettes produced in their own countries—the smoke had an odor significantly different from that of European cigarettes—and never to be seen with a bar of chocolate, which was impossible to find in occupied nations during the war. As the evaders were repeatedly told, the slightest mistake could draw the attention of one of the many French, Belgian, and German security officials who patrolled buses, trains, and other forms of public transportation.

During their time together, it was not uncommon for Comet line operatives and the servicemen in their care to develop close if fleeting relationships. "I loved them like they were my brothers, my children even," de Jongh said. "We would have done anything for them, even giving up our lives." For their part, the airmen were intensely moved by the selflessness and courage of those who rescued them. "I fell in love with them totally, absolutely," said Bob Frost, a British sergeant who, at age nineteen, was escorted by de Jongh to Spain in the autumn of 1942. Many years after the war, Frost declared, "I have nothing but the utmost respect for the people who worked in the Comet line. They knew the price if they were caught. It was heroism beyond anything I can tell you."

When Airey Neave questioned returning airmen about their escape experiences, he recalled, their eyes filled with tears when talking about de Jongh and her associates. "They were afraid for her," Neave said. "So were all those who knew the terrible risks she ran, including Jimmy and me."

For Comet, as for other escape lines, the threat of discovery was ever present, arrests were uncommonly frequent, and casualties were heavy. If captured, Allied servicemen were sent to German prisoner-of-war camps, where Geneva Convention rules applied. When escape line members were caught, they faced torture, the horrors of a Nazi concentration camp, and/or execution. James Langley once estimated that, for each serviceman spirited back to England, at least one "Belgian, Dutch, or French helper gave his or her life."

Well aware of the Gestapo's determination to smash Comet, Neave and Langley urged de Jongh and her father, who worked closely with her, to come to England before it was too late. But, as always, she "kept to her own rules," refusing even to consider the thought of leaving. "To the last," Neave recalled, "she made her own decisions."

THE LARGER AN ESCAPE line grew, the more vulnerable it became. Checking out the backgrounds of the hundreds of people who worked on the lines or who used them to get back to England was a virtual impossibility. Not surprisingly, then, the Gestapo and Luftwaffe secret police had considerable success in infiltrating the lines with agents impersonating Allied fliers. Equipped with documents and dog tags recovered from killed or captured airmen, most of the agents were German nationals who had lived in Britain or the United States for long periods and spoke fluent English.

The Germans were also aided by French, Belgian, and English collaborators who joined the lines as workers. The Pat O'Leary line, run by one of the leading escape organizations in France, was betrayed to the Gestapo by Harold Cole, a British Army sergeant who earlier had been captured by the Germans and persuaded to turn traitor. Because of Cole, Albert Guérisse and the operation's other founders were arrested and sent to concentration camps, and the line was destroyed.

As for the Comet line, its workers, too, had suffered repeated arrests since its creation, although its founder led somewhat of a charmed existence, managing to elude the Nazis for more than eighteen months. Then in early January 1943, de Jongh set off with three American pilots on her eighteenth trip to Spain. A heavy snowstorm caught them just before they crossed the Pyrenees, and they were forced to seek shelter for the night at a safe house in the foothills. The next morning, the Gestapo, acting on a tip from a local farmworker, burst into the house and arrested all the occupants.

Allied evaders were repeatedly warned that, if captured, they were not to reveal the identities and locations of those who had hidden or otherwise helped them. Their only obligation, they were told, was to give their name, rank, and serial number, just as if they had been captured in combat. But the extreme stress of Gestapo interrogation proved too much for some, including one of the pilots caught with de Jongh, who disclosed the names of his guides and the owners of the safe houses who had sheltered him.

While the three pilots were sent to a POW camp, Dédée de Jongh disappeared without a trace—one of the tens of thousands of political prisoners in occupied Europe who, under a decree from Hitler, vanished into the "night and fog" of Nazi concentration camps.

A few weeks later, the Gestapo again struck hard at Comet, this time arresting nearly one hundred of its operatives in Brussels and elsewhere in Belgium. Among the few who managed to escape was a twenty-eight-year-old woman named Peggy van Lier, a close friend and associate of de Jongh's. Although interrogated by Gestapo agents, she persuaded them that she was not involved with Comet. Fearing she would be picked up again, friends spirited her to Gibraltar, where she was sent by plane to London. There, Jimmy Langley was waiting to greet her. He was instantly smitten by the red-haired, blue-eyed Belgian, falling in love with her, he later told friends, the moment she stepped off the plane. The two were married the following year, had five children, and lived contentedly together in Suffolk for the rest of their lives.

Lier was lucky to get out when she did. Just a few months after she arrived in London, Comet was hit again, this time because of a French collaborator working for the Gestapo in Paris. De Jongh's father, who was the head of Comet in the French capital, was rounded up, along with her sister and dozens of other associates. Most of them, including de Jongh's father and sister, were executed.

"It seemed incredible that Comet could survive, but survive it did," Langley noted. Despite the calamitous loss of its founder and so many other key members, the network was able to rebuild itself, enlisting the help of new couriers and guides, setting up more safe houses, and maintaining the flow of Allied servicemen back to England.

Although it would suffer additional losses, the line would continue its rescue operations until Belgium was liberated in the fall of 1944. In the three years of Comet's existence, its workers sent back more than 800 servicemen, most of whom soon returned to active duty. Of that number, 118 were escorted to Spain by de Jongh herself.

ALTHOUGH THEY WERE THE largest and best-known escape lines, the Comet and Pat O'Leary organizations had no monopoly on such work in northwestern Europe. Dozens of other, smaller groups operated there, and most of them, like Comet, were organized and managed by women. The founder of France's Marie-Claire line, however, was in a class by herself.

Airey Neave called Mary Lindell, Comtesse de Milleville, "one of the most colorful agents in the history of MI9." That description hardly does her justice. An elegant, imperious, outspoken woman with a passion for adventure, Lindell gave fits to the Gestapo and Claude Dansey alike. *"Pour une femme, Marie-Claire est un grand homme"* ("For a woman, Marie-Claire is a great man"), proclaimed a Catholic priest who helped her in her work.

Neave and Langley first learned of Lindell on July 27, 1942, when

Mary Lindell with a former French resistance fighter after the war.

the British consulate in Barcelona informed them that an English-woman had just arrived in Spain from France and wanted MI9 to send her back as the organizer of a new escape line. A week later, she arrived at Langley's flat in London for an interview with him and Neave. Clad in a French Red Cross uniform emblazoned with several rows of French and British decorations, the fortyish Lindell "looked much younger than her age," recalled Neave, who greeted her at the door. "Despite her gruff manner, she was still beautiful." As he showed her into the flat, he realized immediately that "she was used to getting her own way. . . . Her tone was peremptory and English in every inflection. I might as well have been the butler answering the door."

Lindell, who came from a wealthy family in Surrey, would have readily agreed with Neave's description of her as an amalgam of "fear-lessness, independence, and not a little arrogance." In fact, that's how she described herself. "My godfather died and left me quite a nice little lot of money, so I was really quite independent," she later said. "That, I suppose, is why I'm arrogant and independent now, because from fifteen onwards I never knew what money was. It just was there."

After the outbreak of World War I, the nineteen-year-old Lindell traveled to France and volunteered for the French Red Cross. For the next four years, she tended Allied wounded in forward dressing stations and field hospitals, often accompanying medics to the front lines, just a few feet from the German trenches, to treat injured soldiers. Known as the "bébé anglais," she was awarded the Croix de Guerre for gallantry under fire.

After the war, she married a French aristocrat whom she had nursed when he was wounded, and, as the Comtesse de Milleville, had three children and became a major figure in Parisian society. When France fell in 1940, Lindell also became a one-woman escape line. She hid British officers in her flat, then drove them across the Vichy frontier and on to Marseille to deliver them to what later became the Pat O'Leary line. She obtained the necessary permits and gasoline coupons for her trips from German officials by claiming that she needed them for her Red Cross work. Later, she would note that the German military "simply loved titled people, especially in uniform, with lots of medals and ribbons."

The Gestapo in Paris, however, were not quite so dazzled. Suspicious about the constant flow of people into and out of her flat, they arrested her in January 1941. At the end of her trial, Lindell received an extremely lenient sentence of nine months of solitary confinement in a Paris prison. "I found this completely ridiculous," she recalled, "so I remarked to the [German] court, 'Just sufficient time for me to have a baby with Adolf.'" Her lawyer and translator nearly fainted. When the judge demanded to know what she had said, her lawyer jumped up before the translator could respond and exclaimed, "The accused said that in the circumstances, she considers the sentence fair."

After serving her sentence, Lindell was released. Although determined to continue her escape-line activities, she knew she could not do so in Paris. She decided to escape to England to seek training and other assistance to enable her to work in another part of France.

Neither Neave nor Langley was enthusiastic about the prospect. Both were adamantly opposed to the idea of sending anyone back to occupied Europe who was known to the enemy; doing so, Langley said, would "endanger her own life and those of others." He and Neave also thought of Lindell as a loose cannon and found her difficult. What particularly annoyed them, according to Langley, was that she preferred "to use a battle-axe rather than the more usual feminine charm when dealing with difficult males." After their first meeting with Lindell, Langley turned to Neave and declared, "I've nothing to say at the moment except that I want a very large whisky and soda."

Determined to get her way, Lindell refused to listen to MI9's objections, saying she would continue to raise hell until she was allowed to return to France. In desperation, Neave and Langley turned for help to Claude Dansey, who agreed to tell her she could not go back. For once in his life, however, Dansey encountered someone as hard-nosed and tough as he. When he pointed out that he was only trying to save her life, Lindell retorted that the only thing he was interested in saving was his own reputation. Emerging from his meeting with Lindell, Dansey told Langley, "Spare no effort to get her to France as soon as it is humanly possible."

After two months of training, she was sent back in October 1942 to create a new escape network in southwestern France, near Limoges. When he escorted her to the airfield, Airey Neave discovered that others had a far different opinion of Lindell than he and his colleagues in London did. As she was introduced to the young RAF pilot who was to fly her across the Channel, he took both of her hands in his and said, "I just wanted to say thank you for going over there. . . . All the boys have tremendous admiration for what you are doing."

On her return, Lindell discovered that while she had been in London, the Germans had sentenced her to death. Changing her identity to the Comtesse de Moncy, she rented several rooms at a small hotel in the market town of Ruffec and made them the headquarters and collection point for her new line, which she called Marie-Claire. Over the next year, Allied airmen were brought to Ruffec by Lindell's associates, then escorted over the Pyrenees to Spain.

The Marie-Claire network, however, was best known for its rescue of the survivors of Operation Frankton, one of the most daring British commando raids of the war. In December 1942, a small group of Royal Marine commandos, using canoes, paddled up a river leading to Bordeaux and placed limpet mines on five German ships anchored there, then exploded them, causing heavy damage. Though most of the commandos drowned or were captured, two managed to get away. Following instructions they received in London, they traveled nearly a hundred miles to Ruffec, where they made contact with the Marie-Claire line.

Lindell, meanwhile, had been seriously hurt in a bicycle accident a few days earlier and was hospitalized with five broken ribs and a head injury. But when one of her agents told her about the commandos, she insisted on slipping out of the hospital and overseeing their rescue herself.

Before she sent them south, along with a couple of RAF airmen, she told them, "We've got only one rule for Englishmen in our care—*no girls*. From past experience, we know that once they meet a pretty girl, everything goes to hell. So we shall take care to keep them away from you." She and her organization were true to her word—and got them safely into Spain.

Lindell ran her line for more than a year before being betrayed to the Gestapo in November 1943. She was arrested at a railway station on one of her many trips south. Put on a train bound for Paris, she managed to divert the attention of her guards long enough to leap off the train, but as she did so, she was shot in the head and cheek. Astonishingly, she survived and, after several surgeries, was sent in September 1944 to Ravensbrück, an infamous concentration camp north of Berlin. The only camp designed specifically for women, it was built on an estate owned by Heinrich Himmler. Of the 132,000 women and children sent there during the war, more than 70 percent died—of starvation, torture, beating, hanging, shooting, horrific medical experiments, and, beginning in November 1944, in a newly installed gas chamber.

From the day she arrived at Ravensbrück, Lindell refused to follow orders. Not surprisingly, she became known as *"die arrogante Engländerin"* ("the arrogant Englishwoman") to camp officials, who seemed as cowed by her haughty aggressiveness as her MI9 superiors were.

Thanks to her nursing experience, Lindell was assigned to the camp hospital, where she was credited with saving the lives of several prisoners by stealing medicines meant for German personnel and administering them to the inmates instead. Near the end of the war, when the Swedish Red Cross arranged for the evacuation of Scandinavian prisoners from Ravensbrück, Lindell browbeat the camp commandant to include in the evacuation all the British women in the camp, several of whom had been involved in the French resistance. Among them was twenty-three-year-old Yvonne Baseden, an SOE agent dangerously ill with tuberculosis, whose life Lindell saved by insisting she leave the hospital and get onto an evacuation bus.

Yet another of the gray, emaciated, wraithlike inmates liberated from Ravensbrück at the war's end was Dédée de Jongh. Even though she had freely admitted to creating the Comet line after her capture in 1943, she managed to avoid execution because the Germans could not bring themselves to believe that such a pretty, delicate-looking young woman could have devised such an intricate operation.

BY THE END OF the war, it had become commonplace for many people in France and the Low Countries to aid downed Allied airmen—so commonplace that, according to MI9, some twelve thousand citizens of those nations had done it. Other estimates put the number far higher.

Allied fighter and bomber crews, in turn, were briefed by MI9 on how to evade capture and seek out rescuers if shot down. They were given a compass, silk maps of the countries they were flying over, and other emergency supplies to sustain them while looking for help. If downed, they were told, their most immediate objective was to shed their parachute and get clear of their landing spot as soon as possible.

As the war progressed, that goal became increasingly easy. Whenever the thunder of planes was heard in Belgium, Holland, or northern France, crowds of people would gather outside to watch the enormous armadas pass overhead. If parachutes were seen, onlookers would hop on bicycles or run in the parachutes' direction in the hope of getting there before the Germans.

In February 1943, an American pilot shot down on his way back from a raid on the French port of Saint-Nazaire was spirited away by several women who raced up to him as he landed. In August of that year, two American sergeants downed near Brussels were each greeted by crowds of several dozen Belgians and passed on to escape lines.

That same month, another U.S. sergeant, who was shot down near the French town of Toulouse, was swarmed by some thirty Frenchmen as he drifted to the ground. They immediately removed his parachute and uniform and gave him civilian clothes to put on. "In four minutes," he said, "I looked like any one of the Frenchmen who had rescued me, and my flying suit and equipment had disappeared."

In Belgium, a woman who had spent more than three years in a German concentration camp for aiding the escape of downed Allied fliers observed more than fifty years later that she never regretted the heavy price that she and other members of the Belgian resistance paid for helping the Allied cause. "The airmen who come for reunions feel they can't thank us enough," she said. "We say if it wasn't for the English, we might be German now."

CHAPTER 18

A Giant Jigsaw Puzzle

European Spies Prepare for D-Day

W HEN ABWEHR OFFICERS IN NORTHERN FRANCE CAP-
tured several members of the French resistance in late 1943, they were
astonished by what they found in their captives' belongings: detailed
plans of the port of Saint-Nazaire on the Brittany coast. When an
Abwehr colleague at the port saw the documents, he threw up his
hands. They were precise drawings of giant submarine bunkers and
locking installations then being built at Saint-Nazaire by the Germans.
Allied spies, the officer complained, were forever sneaking in and gathering information about the construction. No matter what he and
other security officials did, they "were quite unable to prevent a recur-
rence of such incidents."

As D-Day approached, Saint-Nazaire was hardly the only German
military facility in France to suffer such intrusions. Everywhere on the
country's coasts, it seemed, intelligence agents from occupied Europe
were penetrating top secret enemy bases and stealing fortification blueprints and other material that the Allied planners of D-Day had asked
them to gather. In effect, the agents were collecting the pieces of a
giant jigsaw puzzle that, when put together, would give the Allies a
minutely detailed picture of the German defenses they would face during the invasion. "As fast as the intelligence came in," one British historian wrote, "it was followed swiftly by greater demands. The more the
information poured in, the more the demands grew."

Abwehr officer Hermann Giskes, who had been reassigned to France after his *Englandspiel* triumph in the Netherlands, was a firsthand witness to the European spies' success. "We had no illusions about the difficulty of stopping this illegal activity," he later remarked. "These hydra grew new heads quicker than we could cut them off. . . . It was obvious that we were only intercepting a fraction of the material which was getting to the enemy." He called the situation "catastrophic." In October 1943, *Life* magazine noted that "practically all the Allied war plans for the invasion of Europe are based on information about the conquered territory supplied by underground intelligence systems. Most of the information would never have reached London if the exiled governments were not there."

Although agents from throughout occupied Europe helped in the collection of this intelligence mosaic, most of it came from French and Polish operatives. Marie-Madeleine Fourcade's Alliance network was particularly active, despite its recent crippling loss of agents. One Alliance operative—a French naval engineer working at Lorient, another German submarine base on the Brittany coast—provided a flood of information about the complex, including the number and movements of the U-boats based there. Another Alliance agent was hired as a workman to paint the offices of the Todt Organization, the Reich's engineering and construction agency, in Caen, a city close to the English Channel that would become a pivotal battleground during the upcoming campaign in France. In the course of his work, the painter managed to spirit out plans for the area's German fortifications.

Perhaps the most spectacular intelligence exploit of all was the product of still another Alliance operative—a painter and art teacher in Caen who bicycled down the coastline of Normandy and made sketches and drawings along the way. The fruit of his efforts was a fifty-five-foot-long map showing the position of every German gun emplacement, fortification, and beach obstacle along the coast, together with details of German army units and their movements. Smuggled to London in March 1944, the map proved an invaluable resource for Allied military commanders directing the invasion.

The Poles provided additional key intelligence, notwithstanding the loss of Interallié, their most important intelligence network in

France. Roman Garby-Czerniawski, the Polish air force officer who founded Interallié in 1940, was betrayed to the Abwehr by his French girlfriend in November 1941, and the operation was rolled up. It was soon replaced by a new Polish-run organization called F-2, which by 1944 had nearly three thousand operatives, most of them French, working in ports, railway stations, armaments plants, and even German war production offices. Like the Alliance workers, F-2 agents provided a cornucopia of information about the German order of battle, coastal fortifications, and defense lines, as well as train, ship, and submarine positions and movements. Many of the agents were forced laborers whom the Germans had brought into places like Saint-Nazaire and Lorient to do construction and other menial jobs.

Other invaluable sources of intelligence were Polish slave laborers working in armaments factories, shipyards, and major industrial plants inside the Reich itself. On numerous occasions, the material they provided led to the Allied bombing of important strategic targets. "We can state categorically that the Polish Intelligence Service is extremely active," a German military report gloomily noted in July 1943. "It is already operating in a vast number of protected German factories through the workers employed there. This creates great threats to the production of military matériel. Such threats are multiplied, since Polish Intelligence, supported by the fanaticism of the Polish resistance, works with much skill and is difficult to contain."

DESPITE HIS ARREST BY the Abwehr in 1941, Roman Garby-Czerniawski made his own crucial contribution to the success of D-Day and the subsequent Allied offensive in France. One of the colorful band of double agents used by the British in their D-Day deception campaign, he managed to help persuade Hitler and the German high command that the Allies would launch their invasion of Europe from the Pas de Calais area in northern France, more than 200 miles away in Normandy.

Garby-Czerniawski began his career as a double agent (in his case, a triple agent) soon after his arrest, when he offered to collaborate with the Germans, specifically by going to Britain and spying for them. De-

Roman Garby-Czerniawski

claring that he would "do for Germany what he had been doing in France against them," he promised to gather information on British aircraft and tank production, troop deployments, and, above all, Allied plans for the invasion of the Continent.

Agreeing, the Abwehr smuggled Garby-Czerniawski into Britain. Once there, he claimed he had escaped from German custody and was treated as a returning hero by the Polish government in exile. A few weeks later, however, he revealed the truth to the head of Polish intelligence, who, with General Władysław Sikorski's blessing, handed him over to MI5. It, in turn, passed him on to its so-called Double Cross Committee, whose mission was to devise a plan to mislead the Germans about the location of the upcoming D-Day landings. Although suspicious at first of Garby-Czerniawski's true intentions, the committee finally concluded that "his loyalty is entirely to his own country," deeming him "a loyal and patriotic Pole." They gave him the code name "Brutus."

Of all the former German operatives in Britain whom the committee enlisted to pass on disinformation to the Reich, Garby-Czerniawski, as a former head of an intelligence network himself, was by far the most skilled and persuasive. "From their knowledge of him, the Germans will expect him to achieve the impossible or bust," reported his MI5 handler. "Brutus is a professional spy and an artist at producing the most detailed and illustrated reports." Judging from the cable traffic between him and Germany, it was clear that his supposed masters had complete trust in his information and regarded him as one of their top double agents. According to MI5, his reports were "studied not only by operational [intelligence] sections, but by most prominent persons in Berlin, including Hitler and Goering."

In his first major deception, Garby-Czerniawski convinced Hitler and his men that the Allies were seriously considering an invasion of Norway, thereby ensuring that the 350,000 German troops stationed there would not be deployed to northern France. But his biggest triumph lay in a series of detailed reports to Berlin of a massive U.S. army, under the command of General George Patton, that was supposed to be training in England for an attack on the Pas de Calais. The army, of course, was fictitious, but the Germans unquestioningly accepted its existence. Even when the Allies did actually land at Normandy on June 6, 1944, Hitler continued to put his faith in Brutus, believing that the main assault would be at the Pas de Calais, thus keeping tens of thousands of German troops away from the real front.

ENSURING THE SUCCESS OF D-Day was only one of the major accomplishments of the army of European intelligence agents who worked for the Allies. Another was stopping Hitler from succeeding in his bold, last-ditch efforts to destroy London and prevent the D-Day armada from crossing the English Channel.

Although the Führer was never able to develop an atomic bomb, he came up with a plan to mount an avalanche of devastating attacks on England with two new terror weapons devised by his scientists: long-range rockets and pilotless jet aircraft armed with bombs. Since 1936, the German military had been conducting tests on what came to be known as the V2 rockets and V1 "buzz bombs" at Peenemünde, the world's largest missile experimentation center, on Germany's Baltic coast.

In mid-1943, Hitler gave top priority to the mass production of both weapons, pouring huge amounts of money into it and assigning thousands of slave laborers to the task. Calling the V2s and V1s "the new weapons that will change the face of the war," he told his top military officials that by the end of 1943, London would be leveled, Britain forced to capitulate, and any planned invasion of the Continent rendered impossible. The attacks would begin on October 20, 1943, he declared. The V2 rocket would be the first to launch.

Unfortunately for the German leader, his apocalyptic vision remained just that—a dream. V1s and V2s were indeed fired at London, creating considerable damage and loss of life. But they never came close to leveling the city, let alone stopping D-Day, thanks to reports from European agents that allowed the British to disrupt the weapons' development and production. Thanks to those spies, "no one could say we had been caught by surprise," Churchill wrote in his memoirs.

The prime minister first received word in April 1943 of the Germans' alarmingly rapid progress in developing the V1 and V2. His informant was Dr. Reginald Jones, a young physicist from Oxford who served as assistant director of scientific intelligence at the Air Ministry and unofficially as Churchill's chief adviser on scientific warfare. Jones, in turn, received his intelligence from a wide array of European operatives, some of them forced laborers at Peenemünde. Although non-Germans were not allowed into the research labs or near the launch sites, the workers were close enough to observe weapons tests and the layout of the complex, including workshops, airfields, and factories.

The first detailed information about Peenemünde came from two citizens of Luxembourg conscripted as construction workers there. Each smuggled out reports about the development of the V2, including a map of the locations where it was assembled and from which it was fired. More intelligence was supplied by Polish agents, one of whom worked for a crew that installed telephone lines at a Peenemünde research facility. He confirmed that test flights of the V1 were also being carried out. When the British requested a detailed map of the entire complex and surrounding areas, Polish intelligence replied that it was "a bit of a tall order." A few weeks later, the map arrived in London.

Of the many reports Reginald Jones received about Peenemünde, one was particularly notable. Obviously written not by a laborer but by someone close to high-ranking German officials, it contained a host of details about the V2 rocket: the identity of the military officers overseeing its trials, the sound it made ("as deafening as a Flying Fortress"), the weapon's deficiencies, the location and description of its launching pads ("so sited that they can methodically destroy most of Britain's large cities during the winter"). When Jones asked about the source of what he called "this extraordinary report," all he learned at the time was that it came from *"une jeune fille la plus remarquable de sa génération"* ("the most remarkable girl of her generation").

Jeannie Rousseau

As Jones discovered later, the document's author was a lovely twenty-four-year-old blonde with a photographic memory named Jeannie Rousseau, whose father was a high-ranking municipal official in Paris. A graduate of the University of Paris and an intelligence operative for the Alliance network, she was, the *Washington Post* later wrote, "one of the most effective—if unheralded—spies of World War II."

Fluent in German, Rousseau, whose code name was "Amniarix," worked as an interpreter in Paris for a syndicate of French industrialists who often met with German military officials to discuss thorny commercial issues like the Reich's commandeering of French business inventories. In the course of her dealings with various German officers, she overheard scraps of conversation about secret weapons tests somewhere in eastern Germany.

The Germans soon started inviting the pretty Parisian to their evening social gatherings, where they ate, drank, and talked freely about their work, which included the secret weapons. Playing the role of a coquettish, dim-witted blonde, Rousseau "teased them, taunted them, looked at them wide-eyed, insisted they must be mad when they spoke of the astounding new weapon that flew over vast distances, much faster than any airplane." Over and over, she exclaimed, "What you are telling me cannot be true!" Finally, one of the officers had had enough of her playful skepticism. "I'll show you," he said, pulling from his briefcase drawings of the rockets and documents detailing, among other things, how to enter the Peenemünde test site, the passes that were needed, and even the color of each pass.

Rousseau wrote down everything she learned that night, as she did after subsequent get-togethers with her talkative German friends. Within a few weeks, she had acquired a voluminous amount of information about both the V1 and V2, all of which Alliance dispatched to London.

Having been alerted by Rousseau and many others, the British confirmed the existence of the secret weapons through a series of reconnaissance flights over Peenemünde. On the night of August 17, 1943, more than five hundred RAF bombers pounded the complex, heavily damaging its research center and production facilities and destroying all blueprints of the V2s. Although Wernher von Braun, the head of the weapons' research and production teams, survived, more than a hundred scientists, engineers, and other staff members were killed.

As Churchill later noted, the raid "had a far-reaching influence on events." The production and testing of both weapons were pushed back several months, long enough to prevent an attack from interfering with the Normandy landings. Fearing more bombing raids, the Germans moved the V2 tests from Peenemünde to an area near Blizna, a small village in southern Poland, that they believed was beyond the range of Allied bombers.

That may have been true, but, by moving to Poland, they were now in the lair of the most skilled, extensive spy organization in all of occupied Europe. Just a couple of weeks after the first V2 trials at Blizna, London received detailed reports about them from Polish intelligence agents. The Poles also set up a special team whose assignment was to beat German patrols to the scene of their crashed rockets, where its members would scoop up and analyze weapon fragments, pieces of radio and other guidance equipment, spilled fuel, and anything else that might be helpful to Allied understanding of the missiles.

In the early summer of 1944, a V2 fell on a riverbank near Blizna but did not explode. Before the Germans could retrieve it, the Poles hid it, then took it apart and spirited the parts away. The head of the team, an engineer named Jerzy Chmielewski, later somehow transported the dismantled missile to an improvised landing field two hundred miles to the southeast. The British dispatched a plane from Italy to pick up the parts, along with fragments from other crashes, and carry them back to London.

Initially, the V2 was to be used almost simultaneously with the V1, which could have had calamitous consequences for England. But thanks to the raid on Peenemünde and continued difficulties with the V2's production and testing, the Germans repeatedly had to postpone its use. Instead, as Churchill and his men discovered from reading Jeannie Rousseau's reports, the V1 was to be deployed first.

In the fall of 1943, a flood of information poured in about the construction of what appeared to be launching sites in a number of locations near the northern French coast. Shaped like ski jumps, they all seemed to be pointing directly at London. One French agent, who worked as a draftsman at one of the sites, copied all its blueprints and sent them to the British capital.

Beginning in December 1943, the U.S. Eighth Air Force, operating from bases in Britain, launched massive bombing raids to knock out the V1 sites wherever and whenever they appeared. The Germans finally gave up on their construction, switching to prefabricated mobile launchers. It was from those platforms that V1 bombs were finally fired at England, beginning on June 13, 1944, eight months after Hitler's planned launching date and one week after the Allies successfully landed on the beaches of Normandy.

"Were the Germans able to perfect these new weapons six months earlier, it was likely that our invasion of Europe would have encountered enormous difficulties and, in certain circumstances, would not have been possible," General Dwight D. Eisenhower, supreme commander of the invasion forces, later wrote. "I am certain that after six months of such activity, an attack on Europe would have been a washout."

For nearly three months, thousands of the pilotless missiles—called "buzz bombs" because of the noise they made—showered down on London and its outskirts, killing 5,500 residents, injuring some 16,000, and destroying about 23,000 houses. Most people considered the new onslaught to be far worse than the Blitz. In his memoirs, Churchill recalled the unbearable strain that the V1s exacted on his war-weary compatriots: "The man going home in the evening never knew what he would find; his wife, alone all day or with the children, could not be certain of his safe return." The V1, noted Evelyn Waugh, was "as impersonal as a plague, as though the city were infected with enormous, venomous insects."

But though the losses were heavy and the fear and worry excruciating, the damage caused by the V1s was considerably less than it might have been. The British could not prevent them from being launched, but in the fifteen months that they had known about the weapon's existence, they had been able to plan countermeasures to greatly lessen its impact. Adaptations were made to the Allies' fastest fighter planes to allow them to overtake the missiles and either fire at them at close range or tip them over with a wing. The pilots who engaged in this extremely dangerous "aerial shooting gallery," as one flier called the midair interceptions, grew to be quite adept at it. Of the more than 8,500 V1s fired at London, fewer than 30 percent overall reached their targets. By August, less than one bomb in seven—about 15 percent—got through to the London metropolitan area, thanks in large part to the fighters and also to the improved performance of antiaircraft guns located on the English coast. Early in September 1944, the V1 campaign came to an abrupt end when Allied troops fighting in France overran the areas containing the buzz bombs' launching sites.

Londoners, however, enjoyed only a few days of relief. On September 8, from sites in still occupied Holland, the Germans unleashed the V2 rocket, a forerunner of modern missiles, which tormented the British capital until just a few months before the end of the war. To most people, the V2s—which traveled faster than sound and approached their targets in total silence—were even more terrifying than their predecessors. More than five hundred of them exploded in and around London, rocking the city like an earthquake and killing nearly three thousand people.

Again, though, the death toll and scale of damage were far less than they would have been had Germany been left unhindered. Without the delays caused by the Peenemünde raid, the rockets would have been fired months earlier and from shorter ranges. After the Allies overran northern France in midsummer, the Germans were forced to launch the V2s from improvised platforms in Holland, nearly twice as far from London and with much less accuracy. "Although we could do little against the rocket once it was launched," Churchill observed, "we postponed and substantially reduced the weight of the onslaught."

Roman Garby-Czerniawski helped add to that effort. When asked by the Germans about the accuracy of the missiles, he told them, falsely, that most of the rockets were falling several kilometers short of London. German scientists then changed the V2s' trajectory, causing many of them to overshoot the capital and explode in less populated areas.

AFTER THE WAR, Churchill paid tribute to the "excellence" and "gallantry" of the countless European intelligence agents who had risked everything to ensure the success of D-Day and help save London. Yet most of them had engaged in their perilous work without ever learning if it had had any effect at all. Many years later, Jeannie Rousseau would describe the "lonesomeness, the chilling fear, the unending waiting, the frustration of not knowing whether the dangerously obtained information would be passed on—or passed on in time."

A young Belgian intelligence agent made the same point at the end of a report about German radio communications that ended up on Reginald Jones's desk. "We have been working so long in the dark that any reaction from London about our work would be welcome to such obscure workers as ourselves," the operative wrote. "We hope this will not be resented since, whatever may happen, you can rely on our entire devotion and on the sacrifice of our lives." Shortly thereafter, the Belgian was captured by the Gestapo and later executed.

Dozens of other agents suffered similar fates. The French artist who drew the fifty-five-foot map of the Normandy coast, for one, never knew about the triumphant outcome of his work. With fifteen other resistance members, he was arrested and on June 7, 1944—the day after the Normandy landings began—was shot. Jerzy Chmielewski, the Polish engineer in charge of dismantling the V2 rocket downed near Blizna, was caught by the Gestapo and executed in Warsaw in August 1944.

In a poignant irony, several of the operatives who had reported to the British about the V1s and V2s at Peenemünde were killed when the RAF bombed the complex. "A substantial proportion of our bombs fell to the south of the establishment itself," Jones recalled, "and particularly on the camp which housed foreign laborers, including those who had risked so much to get the information through to us."

Another Allied bombing raid—on a German factory making electronic components for the V2's guidance and control systems—resulted in the deaths of hundreds of inmates at Buchenwald. The factory, where many of Buchenwald's prisoners were forced to work, was adjacent to the concentration camp.

The man who made that raid possible was Pierre Julitte, once a staff officer for Charles de Gaulle in London and now himself a Buchenwald inmate. Tired of Free French intrigues, Julitte had returned to occupied France as an intelligence agent in 1942 and been captured a year later. After being sent to Buchenwald and assigned to work in the factory, he quickly realized what he was assembling: parts for a guidance system for "a self-propelled projectile, navigating in space and remote-controlled by radio," which turned out to be the V2. Julitte smuggled out a report to de Gaulle's London headquarters in which he described the components and urged that the factory be bombed, knowing that he and his coworkers would probably die if the Allies did as he suggested.

The raid, which was conducted on August 24, 1944, destroyed the factory and killed some five hundred workers. But Julitte was not among them. Although he had had no advance warning of the raid, he managed to get out of the factory as the bombing began and was only slightly injured.

Jeannie Rousseau, meanwhile, continued her reports to London, which now included intelligence she collected while making occasional business trips to Germany with members of the industrialist syndicate for which she worked. By the spring of 1944, she had become so important to the Allied scientific intelligence effort that British officials decided to bring her to London for an extensive debriefing. She was to be picked up by a boat off the coast of Brittany, but the operation went awry and she was captured by the Gestapo.

Rousseau spent the last months of the war in three German concentration camps, among them Torgau, whose inmates worked in a factory making armaments, including parts for the V2. When she arrived at Torgau, the twenty-five-year-old Frenchwoman refused to set foot in the factory and convinced a number of other newly arrived inmates to do the same. "We will go and pick your potatoes but we won't make your bombs," she told the camp's commandant. She was confined to a punishment cell for several weeks, where she received daily beatings.

Rousseau's war ended at Ravensbrück. Weighing only seventy pounds and close to death, she was rescued by the same Swedish Red Cross team that evacuated Mary Lindell and dozens of other Ravensbrück inmates. She was taken to Sweden, where she slowly recovered her health. In 1946, she returned to France and married Henri de Clarens, a French aristocrat turned resistance fighter who was himself a survivor of Auschwitz.

After more than thirty years of staying out of the public eye and trying not "to stir up old memories," Rousseau, now the Vicomtesse de Clarens, agreed to a meeting in 1976 with Reginald Jones, who informed her in detail of the extraordinary contributions she and other intelligence agents from occupied Europe had made to the eventual Allied victory. Her encounter with Jones, whom she fondly referred to as "dear Reg," was "a great personal experience but also shed a light on the past," she later wrote in a foreword to Jones's wartime memoir. "From what he tells us, our efforts were worth it."

CHAPTER 19

"A Formidable Secret Army"

The Resurrection of SOE

ON APRIL FOOLS' DAY 1944, THE ABWEHR'S HERMANN Giskes sent a mocking cable to SOE announcing the official end of his *Englandspiel* campaign. "YOU ARE TRYING TO DO BUSINESS IN THE NETHERLANDS WITHOUT OUR ASSISTANCE STOP," the cable read. "WE THINK THIS RATHER UNFAIR IN VIEW OF OUR LONG AND SUCCESSFUL COOPERATION AS YOUR SOLE AGENT STOP . . . SO LONG STOP."

Actually, as Giskes well knew, *das Englandspiel* had been dead since the fall of 1943, when two Dutch agents he had captured managed to escape from prison and flee to Switzerland. There, they had reported to the Dutch embassy that the Germans had taken control of SOE's entire operation in the Netherlands. That news was quickly transmitted to the Dutch government in exile in London, which passed it on to Whitehall. By then the RAF had already suspended its flights to the Netherlands because of the abnormally high casualties its crews had suffered there.

At the prompting of MI6's Stewart Menzies and Claude Dansey, the British government's Joint Intelligence Committee launched another official inquiry into SOE's activities, again with the goal of shutting it down and transferring its functions to MI6. SOE officials "would gladly have murdered me," Victor Cavendish-Bentinck, the committee's head, recalled. "I arranged with [a colleague] that if I suddenly died, he was to carry out an autopsy."* Once again, however, Winston Churchill rejected the committee's recommendation and sided with SOE, which argued that the fiasco in the Netherlands had been an exception and that its activities in other countries had not been penetrated—a statement that was a long way from the truth.

Despite Giskes's heavy-handed attempt at humor, there was nothing funny about a deception that had exacted such a terrible cost: more than fifty SOE and MI6 agents arrested and killed, as well as fifty RAF crewmen and several hundred members of the Dutch resistance who lost their lives. Yet even as Giskes crowed about *das Englandspiel's* successes, he knew it had yielded surprisingly little intelligence of value to the Germans. Indeed, it had failed to achieve its "supreme objective," according to Giskes: uncovering details of the forthcoming Allied invasion of Europe.

Even more unsettling for the Reich, SOE, chastened by the disaster in Holland, had made a startlingly successful comeback. From late 1943 on, it began cooperating much more closely with the European governments in exile. Finally coming to grips with the seriousness of the threats facing its agents in the field, it also adapted its operations to conform to actual conditions on the Continent. Wireless operators, for example, were no longer forced to transmit at the same time and on the same frequencies every week. Each transmission was limited to five minutes or less, which also helped decrease the risk of being detected. Even more important, the operators were able to jettison their big, bulky sets for new, smaller portable ones that ran on batteries rather than local electricity grids that could be monitored by the Germans.

Meanwhile, resistance movements in France and the Low Countries were also showing new signs of life. "Despite severe setbacks, they had made preparations for the invasion which could not be controlled, much less destroyed," Giskes observed. "They had learnt how to mobilize forces . . . which could be set in motion at the moment of landing to form a formidable secret army behind the German front, an army which would appear everywhere but which could not be pinned down."

---

*   He wasn't joking.

In Holland, the resistance made an energetic fresh start with the help of a transformed N Section. Understandably enraged by *das Englandspiel*, the Dutch government in exile demanded considerably more input into clandestine operations in the Netherlands—demands that the humbled and embarrassed SOE was in no position to resist. N Section agreed to inform the Dutch before every new operation. Dutch officials were allowed to monitor SOE training courses for Dutch agents and to provide their own briefings. The Dutch also were shown all wireless traffic between N Section and the field, including SOE orders to agents, and were asked for their comments and observations.

The new collaboration produced immediate and striking results. From early 1944 on, N Section dispatched more than fifty agents to Holland—none of whom was detected by the Abwehr or Gestapo—as well as large quantities of arms and explosives. Extreme care was taken with security checks, identity papers, clothing, coding, safe houses, and other details about which SOE previously had been so careless. "Drops of agents continued blind," Giskes said. "Transmitters were being replaced without being captured. It was evident that London was now profiting by the experience which it had bought so dearly."

The relationship between MI6 and Dutch intelligence officials, which at times had been extremely hostile, also showed considerable improvement. Like SOE, MI6 officers began treating their Dutch counterparts more like partners than like useless subordinates. These partnerships, as Giskes admitted, resulted in "a highly efficient espionage organization, possessing good courier lines and numerous radio links between Holland and England. We never succeeded in destroying or even seriously weakening this operation." Its value was reflected in the greatly increased quantity and quality of intelligence reports from Holland. From 1940 until late 1943, only sixty-eight had been sent; from then until the end of the war, there were more than ten thousand.

After years of frustration and betrayal, the Dutch resistance movement was enjoying a period of explosive growth, stimulated primarily by Germany's compulsory labor program. Just as in France and elsewhere in occupied Europe, tens of thousands of young Dutch citizens went into hiding rather than comply with orders to work as forced laborers in German factories and agriculture.

Throughout the country, German security forces conducted what the Dutch called *razzias*—door-to-door searches to root out the young evaders, who came to be known as "underdivers." During the *razzias*, the Germans kicked down doors and shot through floors and into walls and wardrobes. When they found underdivers, they often killed them and whoever was sheltering them on the spot.

In response to the *razzias*, a group known as the National Organization for Assistance to People in Hiding sprang up across the country. Known as LO, it soon became the largest resistance organization in Holland. At first, its main goal was to protect underdivers and find hiding places for them. But it soon broadened its focus, working to shelter members of other groups hunted by the Nazis, including Jews, downed Allied airmen, and resistance fighters.

LO's work included the moving of those under its care from cities to more sparsely settled, safer agricultural areas, as well as collecting money, food, and clothing for its charges. It created its own teams of forgers, responsible for creating false identity papers and food rationing coupons. It also organized small assault bands to raid official food distribution centers. Some of the groups went on to engage in actual sabotage, destroying railroad tracks and munition dumps and cutting telegraph lines.

Despite heavy casualties, LO expanded rapidly, as did several other Dutch resistance organizations. By 1944, virtually every town of any size had at least one such group. In a report to the German high command in Holland, local officials across the country said that, in their opinion, the Dutch population would actively support Allied troops in the event of an invasion. "To a greater extent than ever before, hatred of the Germans had taken hold of the Dutch people, and Dutch and Germans faced each other as belligerents committed to fight to the finish," the historian Werner Warmbrunn noted.

Among the mushrooming number of resistance members was a thirteen-year-old aspiring ballerina named Audrey Kathleen Ruston, later to be known to the world as Audrey Hepburn, who lived with her mother and brothers in the resort town of Arnhem. The future movie star was intimately acquainted with the underdiver phenomenon. Her older brother was an underdiver himself, having escaped the *razzia*s to join the underground. Her younger brother was not so fortunate: caught by the Nazis, he was sent to work in a munitions factory in Berlin. Audrey herself worked as an occasional courier for the resistance.

UNTIL LATE 1943, MEMBERS of the Belgian underground found themselves mired in troubles strikingly similar to those afflicting their Dutch counterparts. Like N Section, SOE's Belgian section had been headed by amateurs who knew little or nothing about clandestine warfare, among them Hardy Amies, who later became Queen Elizabeth II's favorite fashion designer.

As in the Netherlands, the Abwehr in Belgium was able to play back the wireless sets of several SOE operators it had captured, allowing the Germans to mount a successful counterespionage campaign that snagged a dozen or so additional London-sent agents, as well as large caches of weapons and explosives. The Abwehr also penetrated several intelligence networks controlled by MI6 and rounded up resistance members affiliated with those operations.

The German infiltration of SOE and MI6 operations in Belgium was discovered in 1942, but neither recovered until 1943, when the two British agencies instituted the same kind of drastic reforms in Belgium that had worked in Holland. At the same time, the Belgian resistance movement exploded in size, again spurred by Germany's plan for forced labor conscription. By early 1944, many thousands of labor evaders had joined the resistance.

"In parts of Belgium, conditions verged on civil war," Hermann Giskes reported. "The number of ambushes, attacks and incidents involving use of explosives by the Belgian underground increased slowly but steadily. . . . Any [German] who was known to be engaged in intelligence work against the enemy or working against its agents could expect to receive a salvo of machine gun bullets round every corner. Bloody frays in which Abwehr officers and other [German security]

officials were shot to bits increased in number and forced us to take additional security measures."

IN FRANCE, THE SITUATION was murkier. While SOE-run resistance
groups thrived in the south, life was harder in the north, thanks to the
collapse of Prosper and other networks there. Unlike SOE's Dutch and
Belgian departments, no changes were made in the leadership of F Section, and Maurice Buckmaster, despite many warnings, continued un-
knowingly to dispatch new operatives to German-run networks until
May 1944, less than a month before D-Day.

In February 1944, new allegations of treachery were made against
Henri Déricourt, and Buckmaster was forced to bring him back to
London for interrogation. When questioned about his dealings with
the Germans in Paris, Déricourt, according to one witness, replied,
"Well, of course, I have to cooperate with the Germans and give them
some black market oranges from Spain and be friendly toward them, so
that I can get on with my work for you."

Inexplicably, the investigators cleared Déricourt of any wrong-
doing, although they did bar him from any further participation in
SOE operations. As one observer put it, Déricourt "had powerful protectors in London," among whom were Buckmaster and his deputy,
Nicholas Bodington. Free to go, he transferred to the Free French, for
whom he flew reconnaissance missions until the end of the war. "There
was no shadow of a doubt that he was a traitor," Francis Cammaerts
later observed, "and he never paid for it properly."

The German deception campaign in France had run its course by
D-Day, and Hitler himself had the idea of sending a message to F Section boasting of its success. "We thank you for the large deliveries of
arms and ammunitions which you have been kind enough to send us,"
said the cable, signed by the Gestapo. "We also appreciate the many tips
you have given us regarding your plans and intentions which we have
carefully noted." Unfortunately, it added, "certain of the agents have
had to be shot."

Although a few SOE operatives survived and continued to operate in the north, they and the resistance fighters with whom they worked would play a relatively small part in the D-Day landings and the ensuing offensive. In the south, where resistance had infected much of the population, it was a far different story.

Francis Cammaerts discovered that for himself when he returned to France in February 1944 after spending several weeks in England. The RAF plane in which he was flying was shot down over the Drôme Valley, in the southeast, near the French Alps, and Cammaerts parachuted out, landing near a farmhouse. A few months earlier, he might have been wary about approaching an unknown house for fear of being turned in to the Germans. Now, however, "I knew that . . . nine out of ten people would welcome me with open arms, one out of ten might be frightened and send me away, and one in a thousand would ring up the police." He walked up to the farmhouse and knocked on the door. The farmer who opened it exclaimed, "Oh, you're an airman!" Ushering Cammaerts in, he shouted to his wife, "Go and get the wine out! We'll make him an omelet!"

Like giant spiderwebs, several SOE sabotage networks, each made up of hundreds of resistance fighters, crisscrossed the region. In the southeast, there were Cammaerts's Jockey circuit and Tony Brooks's Pimento group. To the west, George Starr, a former mining engineer, ran Wheelwright, and, in the area around Bordeaux, Roger Landes, an ex–land surveyor, was in charge of Actor. In early 1944, several new circuits were created in the south by veteran SOE agents transferred to France from other posts.

For all of them, the D-Day assignment would be the same: prevent enemy reinforcements from reaching the landing areas by destroying all forms of German transportation and communication. But, extensive as the circuits had become, they faced a major difficulty in preparing for their mission: none of them had been provided with enough arms and explosives. In late 1943 and early 1944, more than 150 reception committees throughout France stood by each month for munitions drops, but fewer than ten operations were actually carried out. Spurred on by MI6, RAF officials had insisted that they could not spare planes from bombing raids.

As D-Day loomed, the morale of the resistance fighters and their

SOE organizers plummeted. How could they possibly have an impact on the outcome if they had no means of doing so? In January 1944, two men—a fiery French resistance leader and a rebellious SOE officer—separately decided they'd had enough of British bureaucratic stone-walling. Each of them went straight to the one man who could transform the situation in an instant: Winston Churchill himself.

"Without Churchill, there was nothing to be done," noted the Frenchman—Emmanuel d'Astier de La Vigerie, the forty-four-year-old head of the Libération-Sud movement. "English secret services wanted to treat the French uprising as if it did not exist. Popular struggle meant nothing to them."

Astier was a charismatic, romantic figure, just the type to appeal to Churchill, who later described him to FDR as "a man of the Scarlet Pimpernel type." Astier felt the same about Churchill. "Like de Gaulle," he later wrote, "Churchill was a hero out of the Iliad, the lone and jealous governor of the British war effort."

In his meeting with the prime minister, Astier spun stories of the thousands of maquis in his movement who were desperate to fight Germans but had no arms or ammunition. Churchill was enthralled. "Brave and desperate men could cause the most acute embarrassment to the enemy," he declared, "and it [is] right that we should do all in our power to foster and stimulate so valuable an aid to Allied strategy."

Now that Astier had Churchill on the hook, it was up to SOE agent Forest Yeo-Thomas to reel him in. Like the Frenchman, Yeo-Thomas, who worked in SOE's RF Section, had much in common with Churchill: he was volatile, impetuous, physically courageous, obstinate, often unruly—and, most important, a passionate lover of France.

Born into an English family with extensive business interests in France, Yeo-Thomas, who was known as Tommy, had spent much of his life there. Before the war, he had worked as general manager of Molyneux, one of Paris's most noted fashion houses. Following a brief wartime stint in the RAF, he had joined SOE.

"After Churchill, the man Tommy most admired was de Gaulle," said Leo Marks, a close friend of Yeo-Thomas's. "To his superiors' astonishment, he was able to criticize the Free French to their faces without causing a national temper tantrum and was the only Englishman actually welcomed" at de Gaulle's headquarters. According to Marks, Yeo-Thomas refused to obey SOE regulations forbidding officers of different country sections from exchanging information. "Tommy was always prepared to compare notes on the Gestapo . . . with anyone in SOE of whatever nationality."

In early January 1944, Yeo-Thomas vented his fury over the resistance fighters' lack of arms in a series of meetings with officials from various British ministries. "Our present puny efforts are as likely to succeed as a man trying to fill a swimming pool with a fountain pen filler," he stormed to Air Ministry higher-ups. With them, as with everyone else he met in Whitehall, he got nowhere.

On February 1, two days after Astier's meeting with Churchill, Yeo-Thomas managed to wangle an appointment with him as well. The prime minister gave him five minutes to make his case, then listened to him for more than an hour. Again Churchill was treated to riveting tales of men and women who risked torture and death to "carry messages through the crowded, police-ridden streets of Paris and to wait for agents to land in the windy wilderness of central France." In his presentation, Yeo-Thomas emphasized the resisters' appalling lack of arms and the desperate need for hundreds more RAF supply flights to make up the shortfall.

When Yeo-Thomas finished, Churchill remarked with a slight smile, "You have chosen an unorthodox way of doing things, and you have short-circuited official channels. It might mean trouble for you, but I shall see that no such thing happens." He ordered the Air Ministry to supply SOE with at least a hundred aircraft capable of carrying out 250 missions over France each month. Almost overnight, the arming of the French resistance had become a top British priority.

Throughout the next four months, RAF bombers regularly swooped over drop zones in France, parachuting more than three thousand tons of arms and supplies to resistance fighters waiting below. The planes also brought in dozens more SOE radio operators and organizers to help mold the maquis into an effective fighting force.

At the same time, Churchill put a definitive end to MI6's ceaseless efforts to obstruct SOE. He was responding to yet another complaint by Stewart Menzies that the campaign to arm the French resistance was diverting aircraft from his own intelligence operations. In addition, Claude Dansey had submitted a memo suggesting that SOE had greatly exaggerated the strength of the underground in France and that fewer than two thousand resisters were likely to take up arms against their occupiers. Impatient with all the bureacratic games being played, Churchill issued a statement of unequivocal support for SOE. Menzies and Dansey finally got the message: there was to be no more talk of abolishing SOE or handing it over to MI6.

With the infusion of arms and ammunition, the French resistance groups stepped up their sabotage efforts. Francis Cammaerts's and Tony Brooks's networks blew up trains, railway tracks, turntables, and locomotive sheds. So did the groups run by Roger Landes and George Starr. SOE circuits also destroyed hydroelectric plants, cut phone and power lines, and ambushed German military units. "From January 1944, the state of affairs in southern France became so dangerous that all commanders reported a general revolt," Field Marshal Gerd von Rundstedt recalled. "The lives of German troops there were seriously menaced and became a doubtful proposition." He added, "It was impossible to dispatch single members of the Wehrmacht, ambulances, couriers, or supply columns without armed protection."

The Germans, who now viewed the maquis as a genuine military threat, moved strong reinforcements into southern France, including Gestapo squads and several Waffen-SS regiments, which ranked among the Reich's most fearsome military units. "By now, France was a furnace," Philippe de Vomécourt observed. "Everyone knew that D-Day was imminent. The Germans were killing indiscriminately in an attempt to wipe out the Resistance."

In a report to London in March 1944, Cammaerts wrote, "These are very difficult days. The Germans are attacking everybody, even those who are only slightly suspect. A reign of terror [exists], with farms burnt, shootings, and hangings. In centers where resistance is strong, there is a state of siege." Cammaerts himself had a bounty of 3 million francs on his head.

The vicious German reprisals added to a major new difficulty facing Cammaerts and other SOE organizers: how to keep their increasingly restless fighters under control. After waiting months for the invasion to take place, the maquis began to question whether it would ever happen. Those in charge worried that their underlings might decide to take matters into their own hands.

At the same time, concerns were mounting in London about the effectiveness of the resistance groups once the Allied assault began. The most skeptical were the Allied military commanders, who never had put much faith in the usefulness of these unconventionally trained partisans.

"No one man could know, either in London or in France, just how many men there were, how strong they were, how well they would perform as individual units," Philippe de Vomécourt observed. "It was not, and never could be, disposed like a trained and well-organized regular army. But what we did know, and what we were to prove, was that however untidy and irregular the underground forces might be, they were more single-minded in their courage and resolution than any regular army could ever be."

CHAPTER 20

"The Poor Little
English Donkey"

Stalin and Roosevelt
Flex Their Muscles

ON FEBRUARY 22, 1944, THREE WEEKS AFTER HIS DECISION to arm French resistance fighters, Winston Churchill appeared before the House of Commons to make a major announcement. Among the many spectators who crowded into the visitors' gallery to hear him was Jan Nowak, a twenty-nine-year-old member of the Polish resistance. Nowak, who had organized the underground's radio network in Poland, had been dispatched by Home Army commanders to London several months before, carrying with him reports documenting German atrocities in Poland, including detailed evidence about the extermination of Jews in Nazi death camps.

Like many if not most of his compatriots, Nowak idolized Churchill. Poles, he later observed, "lived by faith in the Allies, in Churchill and Roosevelt." To a devastated people struggling to survive the savagery of German occupation, the United States and Britain were the embodiment of "the ideals of justice, truth, and freedom." Yet once Nowak heard what Churchill had to say that day, his trust in the Allies and what they represented gave way to bitter disillusion.

Six weeks earlier, Soviet forces, having pushed the Germans out of Russia and now surging west toward Germany, had crossed the border of prewar eastern Poland, reentering the Polish lands they had seized in September 1939. In his speech to the House, Churchill had announced that Britain supported Stalin's intention to keep that territory once the war was over.

Nowak's first emotion was shock, followed by "a rage I could hardly contain." He understood that the British could do nothing to stop the Soviets from overrunning Poland, but Churchill's statement was "a public offer to agree in advance to Russian annexation of almost half of our country even before they [completely] occupied it."

The prime minister's declaration was the culmination of two years of behind-the-scenes maneuvering over the issue. Ever since early 1942, Stalin had urged Churchill to agree to his claims for eastern Poland. At first Churchill had rejected the idea, but under the strain of repeated British military defeats and the fear that Stalin might seek a separate peace with Hitler if his demands were not met, he finally gave in.

To Churchill, it was clear that the Russian leader's acquisition of Polish territory would involve "the forcible transfer of large populations against their will into the Communist sphere." By agreeing to this, he acknowledged in 1942, Britain would turn its back on fundamental principles of freedom that were the supposed bedrock of the Allied cause.

In private, he agonized over the decision. He was genuinely concerned about Poland's fate but would take no action on its behalf. The reality was, as it always had been, that Britain's political and military interests were inextricably tied to the future of western Europe, along with Greece and the Balkans. It had no such interests in eastern European countries, such as Poland.

Above all, the British were not prepared to risk their alliance with the Soviet Union for the sake of the Poles. Both Churchill and Roosevelt were willing to give Stalin whatever he wanted to ensure that the Soviets continued to bear the brunt of the fight against Germany on the eastern front.* The two Western leaders "found it convenient, perhaps essential, to allow Stalin's citizens to bear the scale of human sacrifice which was necessary to destroy the Nazi armies, but which their own nations' sensibilities rendered them unwilling to accept," the historian Max Hastings noted.

---

*   Of all the wartime casualties suffered by the Big Three, the Soviets accounted for 95 percent.

In fact, Churchill and Roosevelt had already secretly conceded Poland's eastern territories to Stalin at the Big Three Conference at Tehran in November 1943, three months before Churchill's public announcement of the fact. Early in their partnership, FDR had made it clear to the prime minister that he would do nothing, beyond defeating Germany, to help Poland. For the president, there were no U.S.-Polish treaties to honor and agonize over, no debt to Polish pilots or troops for helping his country survive.

It mattered little to Roosevelt that many of the Poles' contributions—such as the breaking of Enigma and the intelligence collected by Polish spies throughout Europe and much of the rest of the world—were of great importance to a future Allied victory. Nor did it matter to him—and certainly not to Stalin—that Polish sabotage had played a vital role in relieving pressure on Soviet forces at a time when such help was most desperately needed. Germany's main supply and communications lines to Russia went through Poland, and the Poles, by blowing up hundreds of bridges and destroying or derailing more than seven thousand trains over the course of the war, had been responsible for massive delays and disruptions of German rail transports. In response, Germany operated trains through Poland only under increasingly heavy guard or diverted them away from Poland altogether, which caused even more delays.

In March 1943, Roosevelt told British foreign secretary Anthony Eden that it was up to the Americans, Soviets, and British to decide Poland's borders; he, for one, had no intention of "go[ing] to the peace conference and bargain[ing] with Poland or the other small states." Poland was to be organized "in a way that will maintain the peace of the world." At Tehran, Roosevelt told Stalin he was in complete agreement with him on moving Poland's frontiers to the west.

It was at Tehran, Churchill remarked, that he first realized what a small country Britain itself was and how little say it would have in world affairs from then on. "Here I sat, with the great Russian bear on one side of me with paws outstretched, and, on the other side, the great American buffalo," he told an acquaintance. "Between the two sat the poor little English donkey." Making the same point later to de Gaulle, Churchill remarked, "I am the leader of an unbeaten nation. Yet every morning when I wake, my first thought is how I can please President Roosevelt and my second is how I can conciliate Marshal Stalin."

Boxed in by his larger and more powerful allies, Churchill tried to do the impossible: bargain with Stalin and reconcile the differences between the Soviets and the Poles. In exchange for eastern Poland, he urged the Russian leader to pledge his support for a free if truncated Poland and an independent Polish government. "As long as I live," Churchill told Polish prime minister Władysław Sikorski, "I shall not depart from the principles, which I have always respected, of individual freedom and the right of large and small states to independence." Although Churchill managed to convince himself that Stalin would abide by such a quid pro quo, the Poles knew better. The events of April 1943, in their view, provided definitive proof of that.

On April 13, German radio had made a stunning announcement: the discovery of the bodies of more than 4,000 Polish officers packed in mass graves in western Russia's Katyn Forest, an area still occupied by the Germans at that point in the war. According to the Nazis, who provided a wealth of evidence, the murderers had been Russian. News of the massacre struck Poles like a hammer blow. For twenty months, the Polish government had been searching for those officers and 10,000 others, all of whom had vanished after the Soviet invasion of Poland in 1939. The Poles had repeatedly pressed the Soviets for information about the missing men; again and again, Stalin and his men had denied any knowledge of their whereabouts.

On April 15, Sikorski formally asked the International Red Cross to conduct an independent investigation. Dismayed by Sikorski's action, Churchill, who had little doubt as to the Soviet guilt, nonetheless pressured the Poles not to make an issue of the atrocity and to withdraw the request. Conciliation, Churchill said, was "the only line of safety for the Poles and indeed for us."

Sikorski, who had been criticized by some in his government for making too many concessions to the Soviets, refused Churchill's demand. "Force is on Russia's side, justice on ours," he declared. "I do not advise the British people to cast their lot with brute force and to stampede justice before the eyes of all nations." On April 26, 1943, the Soviets, using the appeal to the Red Cross as a pretext for taking a step they had been planning for months, formally severed diplomatic relations with Poland. When the British government attempted to repair the break, Stalin replied that it could be mended only by dismantling the current Polish government in exile.

General Władysław Sikorski

For the Poles, the Katyn revelations were followed three months later by another shattering blow: the death of Sikorski in a plane crash off Gibraltar. The RAF aircraft carrying the Polish leader and his party dived into the sea less than a minute after taking off from an airstrip on the British-controlled redoubt in Spain. Immediately after the crash, a young Polish officer who witnessed it sobbed as he repeated again and again, "This is the end of Poland." There was a certain prescience to his words.

Over the previous three years, Sikorski, who served as both prime minister and commander in chief of the Polish armed forces, had emerged as one of the most influential and highly respected leaders of occupied Europe. "He was unmistakably the 'doyen' of the exiled governments," noted the British historian William Mackenzie. "His status was not so very far below that" of Roosevelt, Churchill, and Stalin. William Strang, a top official in the British Foreign Office, had told

Count Edward Raczyński, Poland's ambassador in London, that he and his colleagues "regarded Sikorski as a very great man—in fact, the greatest of all the European statesmen whom war has driven into exile."

In his diary, Raczyński noted that Sikorski's death had had "a tremendous impact on Allied, neutral and enemy circles. It has shown us that he had a more assured international position than his Polish compatriots were wont to realize." Echoing that view, Harold Nicolson wrote that the general "was the only man who could control the fierce resentment of the Poles against Russia, and force them to bury their internecine strife. He is one of those rare people whom one can describe as irreplaceable."

Many Poles in London, including Raczyński, suspected sabotage, although an investigation by the British government concluded that the crash had been an accident. Some proponents of the sabotage theory believed that the Soviets had engineered the crash, noting that the Soviet spy Kim Philby was then MI6 chief for the Iberian Peninsula.* Whatever the truth, there was no question that Stalin viewed Sikorski as a major obstacle to his getting what he wanted. Before Katyn, the general had pushed hard for cooperation between the Poles and Soviets, but, as Raczyński noted, "the Soviets didn't want to negotiate and compromise. They wanted to deal with a Polish leader who would give them justification for taking over Poland and imposing their own regime."

Whether he was involved in Sikorski's death or not, Stalin certainly profited from it. After Sikorski died, the Polish government in exile splintered into warring camps, with no strong leader to unite the various factions or exert the kind of influence that the general had had with the British government and the other Allies.

Churchill, meanwhile, refused to bow to the increasingly obvious reality that Stalin meant to control all of Poland, not just the eastern territories. Two weeks before his February 22 announcement, the prime minister told his War Cabinet that, "with the Russians advancing into Poland, it was in our interest that Poland should be strong and

---

* In 1969, the British government ordered a review of the wartime inquiry's findings. The staffer responsible for the investigation reported that "security at Gibraltar was casual, and a number of opportunities for sabotage arose while the aircraft was there." He added, "The possibility of [Sikorski's] murder by persons unknown cannot be excluded."

well supported. Were she weak and overrun by advancing Soviet armies, the result might hold great dangers in the future for the English-speaking peoples." He ordered the RAF to triple the number of aircraft flying arms and supplies to Poland—to a grand total of twelve per month.

His order made no sense except to offer hope—false, as it turned out—to the Poles. Twelve bombers could hardly deliver the enormous quantity of weapons and ammunition needed by the Polish Home Army to ward off its enemy, whether that enemy was Germany or, at some point, the Soviet Union. As Jan Nowak came to understand during his stay in London, Poland had no place in British and U.S. strategic planning. The Western Allies would never send any of their forces to liberate Poland, as they were now preparing to do for France and the Low Countries.

Haunted by guilt, Churchill could not bear to acknowledge any of that. Neither could General Colin Gubbins and the other higher-ups at SOE, who had encouraged the Polish Home Army from the beginning in its plan to mount a full-scale national uprising against the Germans in conjunction with the Allied offensive in western Europe. According to the Poles' blueprint, the uprising would require Polish bomber and fighter squadrons, now flying with the RAF, to ferry to their homeland thousands of Polish paratroopers, also based in Britain.

Gubbins had known since 1941 that the plan was an impossibility, but his own close attachment to the Poles and, more important, the Allies' reliance on Polish sabotage and intelligence had kept him from telling them the truth. Gubbins's biographer puts it plainly: "Neither SOE nor the Polish GHQ [in London] had it in them to tell the Home Army that, owing to the impossibility of providing adequate air support, their plans were plain rubbish."

Making the situation worse, Gubbins ordered Peter Wilkinson, a staffer in SOE's Polish section, to engage, as Wilkinson later observed, in "make-believe joint planning with the Polish general staff, working out the logistic requirements of a full-scale airborne invasion of Poland which both they and I knew could not possibly take place. For my part it was a thankless task and I felt deeply frustrated and depressed by the futility of the whole exercise."

Even the British chiefs of staff, whose resistance to aiding a Polish uprising had been unwavering, were much less adamant in their public pronouncements on the issue than they were in private. Unwilling to antagonize the Poles and jeopardize the flow of valuable military intelligence they were providing, the chiefs talked vaguely of "the desirability of preparing the Secret Army in Poland for action coordinated with the military operations of the Allies." The Poles, the chiefs added, "should be supplied with the largest possible quantity of equipment," subject only to "the availability of suitable aircraft." There was a catch-22, of course: no suitable aircraft were ever available in the numbers needed.

In late 1943, a War Office staffer declared that "the time is fast approaching when we must tell the Poles firmly and without ambiguity what the fate of their main plan for the support of the Secret Army is to be. As I understand it . . . the Combined Chiefs have already turned down the plan as it stands. If this is so, then the quicker the Poles are told, the better." But no such warning was given, and Home Army leaders in Poland, unaware of the machinations in London, continued to believe that their Western allies would come to their aid.

In London, meanwhile, Czech president Edvard Beneš paid close attention to the events in Poland. Stalin's aggressive behavior and the Western Allies' acquiescence reinforced the Czech leader's determination to cut a deal with the Soviets as quickly as possible in regard to the postwar fate of his own country.

Having already concluded that he could not rely on Britain or the United States for support, Beneš traveled twice to Moscow in 1943 to negotiate with Stalin and Czech communist leaders who had taken refuge in the Soviet capital for the war's duration. The communists were, in effect, a Moscow-based shadow cabinet to the London-based government in exile.

Beneš's second trip there, in late December, resulted in the signing of a treaty with the Soviets that called for a strong communist presence in any postwar Czech government. If Beneš had not gone to the Soviet capital and agreed to the treaty, Jan Masaryk later told a friend, "the Czechs would have found themselves in the same situation as the Poles."

IN THE MONTHS LEADING up to D-Day, Britain's relationship with the Free French seemed as fraught and tangled as its dealing with the Poles. But de Gaulle and his men clearly had an advantage over the Polish exiles. Though de Gaulle himself was still shunned by both FDR and Churchill, British policy makers were focusing more and more on the postwar world, in which good relations between Britain and France were considered vital.

In mid-1943, Roosevelt had lobbied hard for Churchill to withdraw all British support from de Gaulle, claiming that the French general "has been and is now injuring our war effort and . . . is a very dangerous threat to us." As before, Churchill found himself in an exceedingly difficult position. If he did what Roosevelt wanted, he would face stiff resistance from the British people and many officials in his own government. While the United States, from the relative safety of the North American continent, could easily afford to write off France as a postwar player, Britain thought it essential that its closest European neighbor be as strong as possible after the war to help balance a possibly renascent Germany and increasingly powerful Soviet Union. On the other hand, while the war continued, Churchill needed the United States far more than he did France.

Swayed by Roosevelt's anti–de Gaulle arguments, Churchill, describing the general as "this vain and even malignant man," urged his Cabinet to consider "whether we should not now eliminate de Gaulle as a political force." The Cabinet, strongly influenced by Anthony Eden, rejected the idea, declaring that "we would not only make him a national martyr but would find ourselves accused . . . of interfering improperly in French international affairs with a view to treating France as an Anglo-American protectorate."

Preparations for the invasion of France, dependent as they were on intelligence from the French underground, might also be placed in jeopardy if the man regarded by the resistance as its leader were shunned by the Americans and British.

Eden and the Foreign Office worked to persuade Churchill to recognize the French Committee of National Liberation—the Algiers-based organization that de Gaulle coheaded with General Henri Giraud—as the main governing body of North Africa and other liberated French colonies, as well as the sole voice of free France. The European governments in exile had all granted recognition to the committee, as had Canada, Australia, and South Africa; the Soviet Union was poised to do the same. Most members of Parliament and much of the British press had also advocated recognition. So had General Dwight D. Eisenhower, the commander of Allied forces in North Africa and the future supreme commander of the Allied Expeditionary Force in Europe.

Although Roosevelt continued to resist the pressure, Churchill finally succumbed, telling the president he might have to break with him over the issue: "I am reaching the point where it may be necessary for me to take this step so far as Great Britain and the Anglo-French interests . . . are concerned." With that, facing virtual unity among the other Allies, FDR agreed in late August 1943 to a severely limited U.S. recognition of the French committee. (On the same day, the British government issued its own, less restrained statement of recognition.)

At the same time, the president refused to halt his efforts to get rid of de Gaulle—to the general's intense anger and resentment—and to boost the standing of Giraud, inviting him to the United States and receiving him with full honors at the White House. Roosevelt's campaign had no effect. In November 1943, Giraud was forced out as the committee's cochairman, and de Gaulle took full control.

As 1944 dawned, the prickly French general was clearly a figure to be reckoned with. In addition to commanding the French resistance, he was now regarded by millions of ordinary French men and women as their leader. His Free French forces now totaled more than 400,000, many of whom had fought in North Africa and Italy. Like the Poles, the French forces had distinguished themselves in the Italian campaign: in the spring of 1944, they would smash the southwestern flank of the Germans' Gustav Line in central Italy and help open the door to Rome. In addition, seven new French divisions had begun training in Britain, preparing for the fight to reclaim France. De Gaulle also headed a greatly enlarged empire of French possessions that had deserted Vichy and come over to the general's camp. They included French North Africa and the West African country of Senegal, with its vital naval base at Dakar—the prize that de Gaulle and Churchill had unsuccessfully sought three years before.

Roosevelt, who cared about none of this, was determined to exclude de Gaulle from any role in France's liberation and governance. He told Churchill that the general and his forces must be kept in the dark about the planning for D-Day, including the landings' actual date. He also barred de Gaulle and his committee from participating in the administration of their country once it was freed. According to FDR, U.S. military forces should govern France until it could hold postwar elections, a decision that Eisenhower and other U.S. military leaders strongly opposed. "An open clash with de Gaulle would hurt us immeasurably," Eisenhower wrote, "and would result in bitter recrimination and loss of life."

Finally Churchill confronted FDR, telling him in May 1944 that de Gaulle could not be left entirely out of Operation Overlord; he must be invited from Algiers to London, brought up to date on the operation, and be included in discussions about the future administration of France. "Otherwise," he explained to Roosevelt, "it may become a very great insult to France." After FDR reluctantly gave his approval, de Gaulle arrived back in England—less than forty-eight hours before D-Day was launched.

His encounter there with Churchill did not go well. "The prime minister, moved by a sense of history," greeted the French general "with his arms outstretched," Anthony Eden noted. "Unfortunately de Gaulle did not respond easily to such a mood." That was, to put it mildly, an understatement.

Bitterly resentful of being shut out of the invasion of his own country, de Gaulle exploded with rage when Churchill told him that Eisenhower would deliver a broadcast message to the French people on D-Day and asked him to do the same. Eisenhower's proclamation, which had already been printed, called on the French nation to follow the orders of the Allied invasion force; it contained no mention of de Gaulle or his men. As de Gaulle saw it, his country, rather than being liberated, was about to be occupied. He refused to follow Eisenhower's broadcast with one of his own, and his talk with Churchill turned into a nasty verbal brawl. At its end, according to de Gaulle, Churchill shouted at him, "We are going to liberate Europe, but it is because the Americans are with us to do it. For, let me tell you! Any time we have to choose between Europe and the open seas we shall always be for the open seas. Any time I have to choose between you and Roosevelt, I will always choose Roosevelt."

Eden and the other British officials who were present were appalled at Churchill's outburst. "I did not like this pronouncement," Eden later recalled, "nor did [Labor Minister Ernest] Bevin, who said so in a booming aside. The meeting was a failure." After de Gaulle had gone, Churchill, shaking with rage, declared that the general was guilty of "treason at the height of battle" and ordered him to be sent back to Algiers, "in chains if necessary."

De Gaulle, for his part, would never forgive or forget Churchill's tirade that day—just another manifestation, he thought, of the prime minister's and FDR's shabby treatment of him and his country throughout the war. The long-term result of those thorny relationships would be serious, lasting damage between de Gaulle and the English-speaking powers. According to Roosevelt biographer Jean Edward Smith, writing in 2008, "FDR's pique against de Gaulle poisoned the well of Franco-American relations, the legacy of which continues to this day." The same was true for the British-French relationship.

At the time, however, Eden and French officials did yeoman's work in calming the two men down. Thanks to those efforts, de Gaulle, putting aside his anger at least temporarily, agreed to deliver a broadcast to the French people, and Churchill's order to expel him from the country was rescinded.

That evening, several British officials, feeling considerable apprehension, escorted de Gaulle to Bush House for the recording of his broadcast. He had refused to give them a written version beforehand, and, fearing the worst, the officials grouped themselves around the glass walls of the recording studio "in deathly silence." As the general began speaking, they looked at one another in amazement.

"Without a trace of nervousness, he delivered a superb broadcast," recalled Robert Bruce Lockhart, the head of the British government's propaganda operations. "He began with a reference to England, which, when all seemed lost, had stood alone against the greatest military machine the world had ever seen." It was only fitting, de Gaulle said, that "this old bastion of freedom" should be the springboard for the "liberation of France and all Europe." De Gaulle's paean to England "carried the conviction of sincerity in every word," Lockhart wrote. The propaganda chief's eyes welled with tears, and, self-conscious about his lack of control, he looked at his colleagues and found that they were teary-eyed, too.

There was only one problem with the speech: de Gaulle had referred to himself and his committee as the government of France. With a written transcript in hand, Lockhart rushed to the Foreign Office to show the speech to Eden and point out the difficulty it would cause. After reading it through, the foreign secretary remarked, "I'll have trouble with the prime minister about this, but we'll let it go." Lockhart noticed that Eden was smiling.

CHAPTER 21

Settling the Score

Europe's Liberation Begins

As DAWN BROKE ON JUNE 6, 1944, THE MIGHTIEST ARMADA in history knifed through the windswept waves of the English Channel toward France. In the thousands of warships—and the serried ranks of bombers and fighters overhead—one could see the full power and grandeur of the Western Alliance.

The invasion forces bearing down on the beaches of Normandy were, for the most part, British, American, and Canadian. But the countries of occupied Europe played a significant role on that historic day as well. The invaders carried with them detailed maps of the German fortifications on the coast to which they were heading—maps based on intelligence supplied by European agents. The ships ferrying and protecting the D-Day troops included Norwegian, Polish, Belgian, and French vessels, while Dutch, Belgian, Czech, Polish, and French pilots and aircrew flew overhead.

As impressive as the June 6 spectacle was, it was only the first wave of what was to come. Over the next three months, nearly 2 million Allied soldiers and airmen—more than 200,000 of them from occupied Europe—would take part in the effort to break out of Normandy and fight their way across France. The French 2nd Armored Division would be the first major force from the exiled allies to land in continental Europe. A Polish armored division would follow soon afterward, as would smaller Belgian, Czech, and Dutch units.

The European troops looked forward to the coming fight on the Continent with a passion unmatched by their American, British, and Canadian counterparts. For them, the chance to help liberate Europe would make up for the humiliation of their countries' defeats and demonstrate their loyalty to the Allied cause. Above all, they yearned to liberate their nations and exact retribution against their occupiers. It was time to start settling the score.

AS THE ALLIED FLOTILLA approached the beaches of Normandy on that cloudy June morning, the BBC's European Service broke into its scheduled programs to announce the D-Day landings. It was only fitting that the people of Europe should get their first news of the invasion from the broadcasters at Bush House. From June 1940 on, the BBC had helped them shed their despair and begin to believe in the possibility of liberation. Five years later, that possibility was finally on the verge of reality.

BBC announcers read the message—in French, English, Dutch, Flemish, Norwegian, and Danish—from General Eisenhower, in which he declared that the landings in Normandy were "but the opening phase of the campaign in Western Europe." The general's broadcast was followed by recorded messages from de Gaulle, King Haakon, Queen Wilhelmina, Belgian prime minister Hubert Pierlot, and Grand Duchess Charlotte of Luxembourg.

The night before, the BBC had performed another of its key wartime roles—as a conduit of information between London and the European resistance movements. Since the summer of 1941, the European Service, in addition to providing news and commentary, had broadcast specific coded messages to resistance members and SOE operatives in the field. The idea had originated with Georges Bégué, the first SOE agent parachuted into France. Worried that frequent use of his wireless set would give him away to the Germans, Bégué suggested to London that some of its instructions to him be sent in the form of short, prearranged phrases or sentences broadcast by the BBC, whose meaning only he and F Section would know.

The BBC messages did not replace Morse code transmissions, but

they became an important additional method of communication between London and those in the field. In France, Jean Moulin arranged
for the widespread distribution of radio sets to resistance groups around
the country, who were then instructed to listen to the BBC for certain
private communications. As Georges Bidault, a resistance leader and
future French prime minister, later put it, "In the undergrowth of the
moors, in friendly streets of shadowy towns, word arrived from across
the Channel and spread in miraculous fashion; and so a web was woven,
invisible to the enemy."

The concept of personal messages rapidly spread to every country
section in SOE, as well as to MI6. After the BBC broadcast its nightly
news programs to the occupied nations, it would send out a series of
brief, cryptic personal reports. Most of them sounded nonsensical to
those who did not know their meaning: "Dandelions do not like the
sardine," "Father Christmas is dressed in pink," "Louis has to see the
pastor," "The milk is boiling over," "Jan, you have to cut your mustache."

From this stream of apparent gibberish, an agent would pick out the
one sentence that meant something to him or her, and no one else
would be able to decipher it. The message could indicate a number of
different actions or situations: an impending parachute drop; the start
of an operation; the dispatch of arms, supplies, or agents; the signal
that an agent or courier had arrived safely in London or the field; the
warning of someone's arrest. Proving to be both efficient and fool-
proof, the personal messages became an integral part of agents' communications and substantially reduced wireless operators' airtime—and
thus their chance of being detected.

They also fulfilled a function that Bégué hadn't foreseen: they en-
abled operatives in various countries to say to people whose help they
needed but who doubted their identities as British agents, "You make
up a short message—it doesn't matter what—and I'll arrange for it to
be broadcast a week from now on the BBC." In the words of Ben Cow-
burn, an SOE organizer in France, "That was the first manifestation of
power: you'd been able to give an order to tell this formidable British
broadcasting company what to say. . . . And you were somebody from
then on."

Most Europeans had tremendous trust in and affection for the BBC, which proved to be another great aid for SOE officers. In France, "active resisters were a very small minority but the majority of the French people listened to the BBC," said Harry Rée, another SOE organizer there. "So on the whole, you could be pretty certain that anyone you didn't know, if you asked for help in a difficult situation and said you were English, would help. They might be frightened and not help you for very long but they would certainly not give you up."

On an average night, the personal messages from the BBC's French service would take no more than five minutes to read. On the eve of D-Day, they lasted more than half an hour. One by one, they tumbled through the air: "The dice are on the table," "He has a falsetto voice," "It is hot in Suez," "Napoleon's hat is in the ring," "John loves Mary," "The arrow will not pass," "The giraffe has a long neck." To members of the French resistance, each was a summons to battle.

Throughout the country, thousands of resistance fighters left their homes and businesses, collected arms and explosives from hiding places, and embarked on the prearranged sabotage assignments that the coded messages had ordered. In Normandy and elsewhere, employees of the government-run telephone and telegraph company cut telephone lines, forcing German forces to use radio transmissions as their sole method of communication—transmissions that could be easily intercepted and decrypted by Bletchley Park. In the early days of the invasion, more than seventeen thousand messages were intercepted daily, including detailed information on troop and supply movements. Once they had pinpointed exact locations of enemy units, the Allies called in air strikes.

Overall, the postinvasion sabotage efforts of the resistance were far more successful than anyone had thought possible. From June 6 on, the Germans could no longer rely on control of their own rear areas or lines of communication. On the first night alone, SOE and other sabotage teams carried out 950 of 1,050 planned disruptions of railway traffic throughout the country. All the main routes leading to the Normandy beaches were cut.

In the north, SOE's Farmer network severed the tangle of railway lines near the industrial town of Lille, rendering them useless until the end of the month. SOE courier Pearl Witherington, who had taken over part of the Stationer network after its organizer was captured by the Gestapo and renamed it Wrestler, was in charge of three thousand saboteurs who cut railway lines throughout the Indre region, in west-central France.

The dozens of resistance groups in the south, meanwhile, brought railway traffic in their areas to a virtual halt, preventing the several German divisions stationed there from moving quickly to reinforce the German defenses in Normandy. Some, like Francis Cammaerts's Jockey and Tony Brooks's Pimento networks, consisted of thousands of men. Others were much smaller. A good many groups were communist-run, while some were organized and manned by local resisters with no outside affiliation at all.

Whatever their origin, French saboteurs played havoc with German rail and other traffic, blowing up railway lines, barricading roads, derailing trains, immobilizing locomotives, and destroying fuel dumps and bridges. According to one historian, "the entire French railway system was so shot through with subversion that the Germans practically had to abandon its use." In addition to troop reinforcements, crucial supplies like ammunition, fuel, and food were greatly delayed in reaching Normandy. As a result of these shortfalls, the Allies were given the time they needed to consolidate their beachheads in the crucial first hours and days of the assault.

The stop-and-go journey of one German armored division from the south of France to the beachheads serves as a prime example of the effectiveness of the saboteurs and the ferocity of the Germans' reprisals. The 19,000 troops of the fearsome 2nd SS Panzer Division "Das Reich" were regarded, according to Max Hastings, as "among the most formidable fighting soldiers of World War II." Equipped with the latest heavy tanks, the unit had been sent in early 1944 to Toulouse, in the southwest of France, to rest, train, and refit after months of hard duty on the Soviet front.

During the division's stay there, its tanks were stored under heavy guard in the nearby town of Montauban, but the railway flatcars that had transported the tanks were left unguarded on railway sidings several miles away. Taking advantage of the opportunity, local saboteurs, among them a pair of teenage sisters, siphoned off all the axle oil from the flatcars and replaced it with ground carborundum, a fine abrasive powder made of stone.

When Das Reich's commanders received orders on June 7 to proceed immediately to the Normandy front, they sent for the flatcars, all of which broke down on their way to Montauban. As a result, the tanks were forced to travel by road, which took far longer and severely damaged their treads. At least 60 percent of the tanks were unserviceable by the time the division reached Normandy. Along the way, the Das Reich troops were incessantly harried by guerrilla fighters. As Eisenhower later wrote, "They surrounded the Germans with a terrible atmosphere of danger and hatred which ate into the confidence of leaders and the courage of soldiers."

In normal times, it would have taken the division no more than three days to reach Normandy. In the chaos of June 1944, the journey lasted seventeen days. What's more, Das Reich reached the battlefield "in a state of extreme disorganization and exhaustion," having been bombed by Allied planes as it approached the beachhead. The division, which had suffered heavy losses by then, did not actually begin to fight until July 10, far too late for it to have any impact.

But its snail-like sluggishness in making its way to the front was not just the result of sabotage. It also slowed down because its commanders had received orders from Berlin to kill as many maquis as possible as it headed north. Das Reich's troops, the order said, "must immediately pass to the counter-offensive, to strike with the utmost power and rigor, without hesitation."

Hitler's rage at the maquis's fierce resistance took priority over Germany's need to summon as many reinforcements as possible, as quickly as possible, to Normandy. The Führer's "obsession with retaining every foot of his empire once again betrayed him," Max Hastings noted. "In the first vital days after the Allied landings, the German struggle to hold France against Frenchmen employed forces—above all, the 2nd Panzer Division—that could have made a vital contribution on the battlefield." On June 16, Field Marshal Gerd von Rundstedt urged Berlin to abandon all of France south of the Loire River and order the sixteen divisions located there to the Normandy front. That, replied Berlin, was "politically impossible." Instead, Das Reich and the other divisions focused first on liquidating resistance groups. "Not even the most optimistic Allied planner before the invasion," Hastings wrote, "had anticipated that the German high command would be so foolish as to commit major fighting formations against *maquisards*."

The Germans' brutality, already extreme, escalated further. "They burned, pillaged, and killed," Philippe de Vomécourt noted in writing about the German terror campaign in eastern France. "They slaughtered the innocent and the guilty (those belonging to the Resistance) alike. They shot a man working on a hedge. They murdered seven woodcutters going home after a morning's work on the forest."

Das Reich, for its part, took out its rage on an entire village. On June 10, SS troops from the division marched into Oradour-sur-Glane, set in the woods and fields near Limoges, about 150 miles north of Toulouse. On that beautiful Saturday afternoon, Oradour, which had been an oasis of peace throughout the war, was bustling with people out for a leisurely lunch or doing their weekly shopping. Their normal Saturday routine was suddenly interrupted when the town crier beat his drum to summon all the village's inhabitants to its small central square.

Once all were assembled, the women and children were separated from the men and herded into a church, which was set ablaze. As the flames shot high, SS troops opened fire on the screaming villagers trapped inside, while soldiers formed a cordon outside the church to make sure no one got out alive. The men of Oradour, meanwhile, were shoved into nearby garages and barns, where they were mowed down by machine-gun fire. More than 600 people died that day, including 190 children and babies.

The wholesale killings in Oradour were followed a month later by another spasm of German savagery, this time aimed at more than three thousand resistance fighters who had established themselves on a mountainous plateau called Vercors, near the city of Grenoble in the French Alps. On July 3, the maquis at Vercors had declared the plateau a free republic, with its own laws, currency, and flag—"a foolish but very understandable act, given their passionate need to erase the shame of 1940," noted Francis Cammaerts, who by then had been appointed head of all Allied sabotage missions in southeastern France. The Vercors maquis, convinced that the Allies would send them supplies and arms and would soon reinforce them with regular troops, planned to make a stand against the Germans on the plateau. But they had neither the training nor the artillery or other heavy weapons needed to act as a conventional fighting force. Their job was to harass the enemy and stay on the move, not to pin him down. In a pitched battle with the Germans, there was no way they could win.

Cammaerts, foreseeing imminent disaster, frantically urged London to dispatch men and heavy weapons to Vercors, predicting a bloodbath if they were not sent. Both SOE and de Gaulle's Algiers headquarters dropped several hundred containers of rifles and other light weapons, none of which could hold off a concerted German attack.

On July 20, dozens of gliders appeared in the sky over Vercors. The maquis were overjoyed, thinking the Allies had finally come to save them. Then they saw that the gliders were German. Thousands of crack SS troops had been sent to crush the rebellion, and crush it they did. Over the next three days, more than 650 resistance fighters were killed. The Germans also raped, tortured, and murdered more than 250 inhabitants of a nearby village.

Like Oradour and Lidice in Czechoslovakia, Vercors would take its place in history as an unforgettable symbol of Nazi barbarity.

IF ONLY THE MEN of Vercors had waited until August to launch their republic, they and their dreams of freedom might have survived. On August 15, ten Allied divisions, made up of U.S. and French forces, landed on the beaches of southern France in a campaign called Operation Dragoon. Within seven days, the troops had stormed up the Rhône Valley and reached Grenoble, 180 miles to the north. As the military historian Rick Atkinson observed, the enemy in southern France "never had a chance." One reason was the aid given by local resistance forces.

Thanks to intelligence provided by underground groups, the invaders knew beforehand "the underwater obstacles, we knew *everything* about that beach and where every German was, and we clobbered them!" said Colonel William Quinn, the chief intelligence officer of the U.S. Seventh Army. The French, Quinn added, "told us everything we wanted to know."

CBS's Eric Sevareid, who accompanied the Dragoon forces, said that "the Allies had never before had such precise information about the German defenses and the location, numbers, and condition of their troops. When we landed, all our officers carried maps indicating not only the location of every farmhouse but the name of the farmer living there, over an area of hundreds of square miles. Within two days, all our invading forces, including airborne men in the back country, were linked in one solid front."

Having received coded alerts from the BBC about the landings, resistance fighters in several coastal towns and cities rose up against their occupiers as the Allies invaded. In Saint-Raphaël, "the shopkeepers were upon the Germans' backs" as the first troops hit the beaches, Sevareid noted. By the time troops reached nearby Saint-Tropez, they found that local Frenchmen had already captured or killed more than a hundred Germans and that the German garrison there had been surrounded. When Allied forces arrived in Marseille, much of that city, too, was in the hands of its inhabitants. According to Field Marshal Henry Maitland Wilson, the supreme Allied commander in the Mediterranean, "the resistance reduced the fighting efficiency of the Wehrmacht in southern France to 40 percent at the moment of the landings."

As German troops retreated up the Rhône Valley, members of Francis Cammaerts's Jockey network were "yapping at their heels like angry terriers closing in on a fox" and helping clear the way for the French and American infantry following closely behind. When an American tank unit reached the town of Gap, about seventy miles from Grenoble, it expected to have to fight its way in. But it found the Germans already gone, and, instead of a battle, it took part in a victory parade.

"What the resistance achieved in the Alps of France is quite straightforward," Cammaerts declared after the war. "The troops that landed on August 15 got through to Grenoble in seven days . . . because there was no fighting. The Alps had already been taken over by the resis-

French resistance fighters guarding captured German soldiers.

tance. That to my mind is an enormous achievement, which saved tens of thousands of lives."

As it happened, Cammaerts almost didn't live to see that day. Two days before Operation Dragoon began, he and two other SOE officers were arrested at a German roadblock. They were taken to the nearest Gestapo headquarters and interrogated. Although his interrogators had no idea that they had caught the notorious "Roger," they decided that Cammaerts and his colleagues were indeed spies and ordered them shot.

When Christine Granville, Cammaerts's SOE courier, found out about the arrests, she headed immediately to the jail where the three men were being held. Granville, a twenty-six-year-old native of Poland whose real name was Krystyna Skarbeka, was well-known in SOE circles for her beauty, charm, and extraordinary audacity. Confronting the official in charge, she told him that the U.S. troops' arrival was imminent and that he would be tracked down and killed if Cammaerts and the others were shot. He agreed to release them but demanded a ransom of 2 million French francs in return. Two days later, after a courier from Algiers had brought her the money, she handed it over, and the three SOE officers were freed, just two hours before their scheduled execution.

PRIOR TO D-DAY, the Allied military commanders had been highly skeptical about the effectiveness of the French underground once the landings occurred. Some generals thought that the resistance fighters would be more of a hindrance than a help, while others expected that any support they provided would last only a few days at best. "It is probable," according to one report from Eisenhower's headquarters, "that action . . . will be taken for only a few days, after which stores and enthusiasm will begin to run low."

In fact, as Eisenhower himself acknowledged in his memoirs, the resistance was "of inestimable value in the [French] campaign. Without their great assistance, the liberation of France would have consumed a much longer time and meant greater losses to ourselves." In a letter to SOE head Colin Gubbins in May 1945, Eisenhower elaborated on the importance of the wartime accomplishments of resistance movements throughout occupied Europe. "In no previous war, and in no other theater during this war, have resistance forces been so closely harnessed to the main military effort," he wrote. "I consider that the disruption of enemy rail communications, the harassing of German road movements, and the continual strain placed on the German war economy and internal security services throughout occupied Europe by the organized forces of resistance played a very considerable part in our complete and final victory."

OSS director William Donovan also gave high marks to the French underground's contributions. In a letter to President Harry S. Truman, Donovan, whose intelligence and sabotage agents played an active role in the fighting in France, wrote that the battle for that country "showed as never before the extent of the assistance that an oppressed people, given supplies and leadership, can render its allies in the course of its liberation."

Yet despite these encomiums, considerable controversy remains about the value of the work of the resistance in France and elsewhere in Europe. "One may ask if an enterprise in which around 75,000 French men and women, some of whom were *résistants* while others were mere innocents caught up in savage German reprisals, perished in German concentration camps, and another 20,000 died in France, often after horrible torture, was worth the fairly trifling return in intelligence and 'action,'" the American military historian Douglas Porch has written. Porch maintained that "in the end, the Allies achieved victory by outproducing German factories and defeating German armies in the field. Sadly, one is driven toward the conclusion that the contribution of the Resistance to that victory . . . was minimal . . . [the effort] weighed little in the war's strategic balance."

Porch is hardly the only historian to contend that the impact of the resistance in France and elsewhere was greatly exaggerated. The skeptics' reaction was in part a response to the tidal wave of books published since the war celebrating the exploits of SOE, its officers, and the resistance as a whole and in some cases minimizing the enormous errors made by SOE.

Even more, however, such questioning was meant as a corrective to General de Gaulle's contention that French resistance had been widespread and had been largely responsible for the country's liberation. Though that claim was clearly false, de Gaulle persisted in making it after the war—a "necessary myth" that he hoped would heal the divisions in the nation and erase the shameful stain of its capitulation and official collaboration with Germany. In the view of one French historian, "de Gaulle had to convince the French that they had resisted. It was necessary that they disguise the truth from themselves."

Yet even though it's obviously true that the Allied armies were primarily responsible for liberating France, it's also true that the resistance movement, admittedly made up of only a small minority of French men and women, played an important contributing role when it was needed most: during and after the Allied invasion. As Julian Jackson has rightly noted, "If there had been no Resistance, France would still have been liberated, [but] if there were no Resistance, the Liberation would have cost the Allies significantly higher casualties." In Jackson's view, there was indeed "a Resistance myth which needed to be punctured, but that does not mean that the Resistance was a myth."

For those like Airey Neave, who actually had dealings with the resistance movements of occupied Europe during the war, the idea of historians examining their efforts as if they were figures on a balance sheet was distinctly offensive. "In recent years, attempts by professional historians in Britain to describe the actions and errors of men and women who fought the Nazis underground have assumed an unpleasant air of disdain," Neave wrote two decades after the conflict. "Academic writers have attempted to belittle their contribution to the war. That they would not have written in this vein had they taken part themselves is self-evident. No one who saw secret agents actually leave for occupied territory could afford such arrogance."

Others, such as Air Chief Marshal Arthur Tedder, the deputy commander of Supreme Headquarters Allied Expeditionary Force (SHAEF), have argued that the debate about the military contribution of the French resistance was beside the point. "While its military successes were undoubtedly worthwhile," Tedder wrote, "I believe that we ought to judge the Resistance in France on quite a different basis. . . . Its greatest victory was that it kept the flame of the French spirit burning throughout the dark years of the Occupation." In joining the resistance, French men and women were able to shed their sense of isolation and shame and gain a feeling of community and self-respect. In working to reclaim their country, they reclaimed themselves as well.

Eric Sevareid, one of the most astute and eloquent writers about the French wartime experience, made this point in his excellent autobiography *Not So Wild a Dream*. "A man whose army and country have suffered defeat is not a complete man afterwards; no matter how healthy his body, he is always a little sick," he wrote. "The conditions of defeat do not count. No matter if he fought bravely himself, no matter if his army never had a chance, no matter if he was betrayed by treasonable leaders—he remains a cripple. . . . It does not suffice that others restore his country. He must act again himself if he would recover.

"And this is, at bottom, why Frenchmen acted. This is why they never waited until invading troops insured their lives, but rose up in every village and city before we arrived, sometimes days before, and did things that were reckless, sometimes useless, but always magnificent and of imperishable memory. An Allied soldier would shake his head with incredulity to see a French farmer assault a German machine gun with a single grenade and a pistol. He would say, 'These Frogs are crazy,' not understanding why the farmer had to do this, even if he died in its doing."

CHAPTER 22

"A Tale of Two Cities"

Warsaw and Paris Rise Up

IN MIDSUMMER 1944, ALLIED FORCES WERE FAST APPROACHING two occupied capitals: Paris and Warsaw. For the one, the summer would end in the joy of liberation; for the other, it would end in a firestorm of death and destruction.

In late July, the people of Poland would have given anything to see the U.S. and British armies, now closing in on Paris, "standing at Warsaw's gates," a Polish resistance fighter observed. Instead, their centuries-old enemies, the Russians, were nearing the city.

Several weeks earlier, the Red Army, advancing along a thousand-mile front, had swept German troops out of eastern Poland, now claimed by Stalin. The Soviet troops had then pushed westward into territory that no one, not even Stalin, could contend was anything but Polish. Yet on July 22, the Kremlin announced the formation of the Committee of National Liberation—a small group of Polish communists handpicked by Stalin—and proclaimed it the legitimate interim civil government in Polish territories that the Soviets had captured from the Germans. The committee would set up shop in Lublin, the first major city in Poland to be freed from German control. It was possible, Stalin said, that in time it might become the nucleus of a new Polish government.

For members of the Polish Home Army, this was calamitous news. As they saw it, they had two options: do nothing and let Stalin take over all of Poland through his puppet government, or rise up against the Nazis and try to establish control themselves.

Since 1939, the Polish underground's primary goal had been to launch a national uprising when the moment was right. And in late July 1944, the Home Army's leaders believed that the time had finally arrived. The swift Soviet advance across Poland had made a nationwide uprising impossible, but there was still hope that the Home Army could at least drive the Germans out of Warsaw and take control of the capital before the Russians got there. Some 35,000 officers and troops of the Home Army were based in Warsaw, all eager to fight for Poland in this critical hour. "National dignity and pride required that the capital should be liberated by the Poles themselves," said a top underground official. "What kind of an army would it be, what sort of government, that, being in the capital, failed to take part in the battle for the liberation of the city?"

The Home Army's plans were based, however, on the assumption that Allied airlifts would bring in reinforcements. Living, as one underground official later put it, "in a world of illusion," the Home Army's leaders believed that Britain and the United States would rush arms and men to support the uprising once it broke out. Furthermore, they were convinced that their Western allies would pressure the Soviets to do the same.

In the last week of July, the situation grew increasingly urgent. Red Army patrols were spotted a few miles from Warsaw, and panicked German troops began to stream out of the city. On July 26, General Tadeusz Bór-Komorowski, the commander of the Home Army, radioed the Polish government in exile in London that "we are ready to fight for Warsaw at any moment. I will report the date and hour of the beginning of the fight." He asked that Polish fighter and bomber squadrons and the Polish 1st Parachute Brigade, which had been created and trained in Britain for precisely this moment, be sent to Warsaw as soon as possible. He also requested Allied bombing attacks on German airfields near the capital and the immediate dispatch of arms, ammunition, and other equipment.

Up to that point, neither the British chiefs of staff nor SOE had definitively informed the Home Army that it would not receive Allied aid in the event of an uprising. Only now did the British military make it clear that the hopes of the Polish resistance hadn't the slightest chance of being fulfilled. Focused as they were on postinvasion operations in Normandy, British commanders told the Polish government in exile that the Home Army's requests were "completely impossible." The Polish bomber squadrons in the RAF were currently engaged in raids over Germany, while Polish fighters were providing support for Allied troops in Normandy, as well as making strafing and bombing runs against enemy ground targets and escorting bombers and ship convoys. The Polish parachute brigade, meanwhile, had been put under the command of General Bernard Law Montgomery for deployment in western Europe.

Knowing all this, General Kazimierz Sosnkowski, who had replaced the late General Władysław Sikorski as commander in chief of the Polish armed forces, vehemently opposed the idea of an uprising, believing that without strong Allied assistance, it "could only lead to useless bloodshed." Early in July, Sosnkowski ordered the Home Army to cancel its plans. But during the last week of July, he left England to visit Polish troops in Italy; while he was gone, the rest of the Polish Cabinet sent a message authorizing Bór-Komorowski to "proclaim the insurrection at the moment which you will decide as most opportune."

Sosnkowski was not the only Cassandra to warn the Home Army of looming disaster. Jan Nowak, the young Home Army courier who had spent several months in London in 1944 and had flown back to Poland in time for the uprising, told Bór-Komorowski and his lieutenants to expect no help from the Allies. Based on Nowak's report at a July 29 meeting of Home Army leaders, a number of participants urged Bór-Komorowski to delay the uprising. Others felt differently. "We have no choice," one leader declared. "I ask you to imagine a man who has been gathering speed for five years in order to leap over a wall. He runs faster and faster, and then, one step before the obstacle, the command is given to stop! By then he is running so fast that he cannot stop: if he does not jump, he will hit the wall. Thus it is with us. In a day or two, Warsaw will be at the front."

His comment underscored the vast chasm of understanding between London and Warsaw. In Whitehall offices, coolness and rationality reigned. In Warsaw, there was only desperate passion. For five years, its people had suffered hunger, terror, and death. Now the chance for rebellion had come, and they would not be stayed by British logic. Throughout the city, men, women, and children began retrieving revolvers, rifles, grenades, and other arms from the places where they had been hidden since 1939. The weapons were cleaned and surreptitiously distributed to members of the Home Army.

At precisely 5 P.M. on August 1, thousands of windows and doors were flung open all over Warsaw, and the uprising began. From balconies, rooftops, and windows, underground soldiers cut down passing German troops with a cascade of rifle and small-arms fire. Other Poles lobbed grenades at Nazi headquarters and hurled Molotov cocktails at ammunition dumps and troop transports. In dozens of neighborhoods, ordinary citizens—housewives, workers, university professors, shopkeepers—dragged tables, chests, desks, and sofas into the street to build barricades against German tanks and troops. Long-hidden Polish flags were unfurled and draped from apartment windows.

By nightfall, virtually every visible trace of the German occupation had vanished. Warsaw residents had torn down German street and shop signs, posters, inscriptions, and flags. Portraits of Hitler and other prominent Nazis were affixed to the barricades so that Germans would have to fire at images of their own leaders.

During the first three days of the insurrection, the Home Army fighters, only some 2,500 of whom were well armed, gained control of most of Warsaw. In that critical first stage of the fight, however, they failed to take several key military targets, including German airfields and the bridges over the Vistula River. The insurgents were already overextended and in desperate need of assistance. But no aid of any kind came from the Western Allies or from the Red Army, several units of which were encamped on the outskirts of the capital.

The Nazis, meanwhile, were bringing up reinforcements and preparing to counterattack. The Hermann Göring Division, an elite unit of Luftwaffe troops, was being rushed from Italy, and two more SS divisions were also on their way. Their aim, according to the Reich's top leaders, was to teach the upstart Poles a final lesson. "Every inhabitant of Warsaw must be killed, and there shall be no taking of prisoners," Heinrich Himmler declared. Once his forces had carried out that task, they were to flatten whatever was left of Warsaw. "From the historical point of view, this insurrection is a blessing," the SS chief crowed to Hitler. "Warsaw will be eradicated. . . . That nation which for seven hundred years has stood in our way . . . shall no longer be a problem for our children or even for ourselves."

As Himmler's SS and police units surged into Warsaw, Home Army radio operators sent desperate appeals to London, requesting weapons and ammunition. With the hours ticking by, the resistance felt more and more ignored and cut off from the world—a sense of isolation that was only heightened by reports of new Allied advances on the Normandy front and the liberation of a growing number of French cities and towns.

Nonetheless, their appeals finally began to have some effect. In early August, Winston Churchill and Anthony Eden pressed the British chiefs of staff to come to the Home Army's assistance with "maximum effort." The reluctant RAF dispatched several supply flights from Italy, but its losses were heavy, causing future missions to be canceled. Air Chief Marshal Sir Douglas Evill, vice chief of the air staff, said that effective aerial support "could only be provided by Russian tactical aircraft, operating at short range." And those, of course, were nowhere to be seen.

In Warsaw, meanwhile, the SS and police units dispatched by Himmler went from house to house on a wild rampage of looting, rape, and murder. In one neighborhood after another, residents were herded into courtyards and streets to be executed by machine-gun fire. By the end of the day on August 5, more than 10,000 civilians had been slaughtered in one Warsaw neighborhood alone. Over the next several days, the orgy of killing swept throughout the city.

Having pushed the Home Army from Warsaw's outer districts, the Germans targeted the insurgents' center-city stronghold. The uprising was slowly dying. Its lightly armed troops were defending themselves against Nazi forces possessing armored cars, tanks, long-range artillery, dive-bombers, and other heavy weapons. By the middle of August, the Germans were shelling and bombing the city twenty-four hours a day.

No part of Warsaw was out of artillery range, and much of the downtown area was on fire. Bricks were falling like rain, blazing timbers flew through the air, dust and smoke blanketed everything. Sidewalks and streets were littered with bodies, and many more corpses were buried in the rubble of collapsed buildings.

Still the Poles fought on. The combat was most savage in Warsaw's Stare Miasto (Old Town), just north of the city center, with its narrow, winding cobblestone streets and tall, beautifully restored medieval houses. There, in extremely close quarters, Poles took on Nazis in hand-to-hand combat so intense that it reminded some Germans of the last days of the Battle of Stalingrad.

From the cellars of bombed-out buildings, Home Army radio operators continued to tap out urgent pleas for Western aid. At one point Bór-Komorowski sent a personal message to both Churchill and Roosevelt: "Confident of the part we have played in the war effort of the Anglo-Americans, we have the full right to address to you, Mr. President and Mr. Prime Minister, this ardent appeal for immediate help to be sent to wounded Warsaw."

Moved by the insurgents' passionate resistance and stunned by the Germans' barbarity, Churchill insisted that the RAF resume its supply missions. Over the next week, almost a hundred planes were dispatched to the Warsaw area. The flak was murderous, and the aircrew casualties were huge. It was almost impossible to make accurate drops over the rapidly shrinking areas of the city still held by the Home Army. Nonetheless, many Polish pilots clamored to fly there. The pilots of 303 Squadron, who had so bedazzled Britain with their heroics in the Battle of Britain, went so far as to send a blunt telegram to Queen Elizabeth, the wife of George VI and mother of Elizabeth II: "WHEN IN 1940, THE FATE OF GREAT BRITAIN WAS IN THE BALANCE, BELIEVE US, YOUR MAJESTY, WE POLISH AIRMEN NEVER THOUGHT OF ECONOMIZING OUR BLOOD OR OUR LIVES. . . . AT THAT TIME, OVER BURNING LONDON, THERE WAS NO DEARTH OF POLISH OR BRITISH AIRMEN. ARE THEY TO BE LACKING NOW OVER BURNING WARSAW? IS THE CITY TO PERISH ON THE EVE OF VICTORY, AFTER YEARNING FOR LIBERATION FOR FIVE YEARS?"

The queen never answered the telegram; in fact, she probably never received it. In any case, 303 Squadron was not sent to Warsaw.

On August 2, the day after the uprising began, Churchill had gone before Parliament to proclaim the rebellion "a hopeful moment for Poland." He went on, "The Russian armies now stand before the gates of Warsaw. They bring the liberation of Poland in their hands. They offer freedom, sovereignty and independence to the Poles. They ask that there should be a Poland friendly to Russia." But the Red Army never advanced as far as Warsaw's gates. Indeed, it halted its headlong advance less than a dozen miles from the heart of the Polish capital. There it waited—and did nothing.

Churchill, with Roosevelt's backing, asked Stalin to allow Allied bombers carrying supplies to Warsaw to land at Soviet airfields to refuel and rest. It was the very least the Soviets could do, Churchill thought, but the response was negative. The uprising was a foolhardy affair, the Kremlin said, and "the Soviet Government could not lend its hand to it." Averell Harriman, the U.S. ambassador to Moscow, reported to Roosevelt that the Soviets wanted the Warsaw uprising crushed and would brook no attempt by the Western Allies to support it. George Kennan, the chargé d'affaires at the U.S. embassy in Moscow, declared that what Stalin was really telling the United States and Britain was this: "We intend to have Poland lock, stock and barrel. . . . You are going to have no part in determining the affairs of Poland from here on out, and it is time you realized this."

Churchill was fast coming to the same conclusion. Indeed, according to his doctor, the prime minister was consumed during the uprising with fears of Soviet aggression. But he was also personally caught up in the epic David-and-Goliath struggle taking place in Warsaw. The Poles' passionate resistance had won his esteem, and the searing eyewitness accounts of German atrocities against Polish civilians left him enraged. More than fifty years later, Churchill's grandson, also named Winston Churchill, remarked, "My grandfather was beside himself in desperation to secure help for the Poles."

But Roosevelt did not share that sense of urgency. He refused to do anything more to aid the Poles after Stalin denied Allied aircraft access to Soviet airfields. When the prime minister sent the president a wrenching eyewitness account of Nazi mass killings in Warsaw, FDR coolly replied, "Thank you for the information in regard to the appalling situation of the Poles in Warsaw and the inhumane behavior of the Nazis. . . . I do not see that we can take any additional steps at the present time that promise results."

THROUGHOUT THE EARLY DAYS of August, residents of Paris huddled around their forbidden radio sets to listen to BBC reports of Warsaw's agony. Many were haunted by the same thought: Would a similar calamity befall them and their beautiful city, still intact after four years of German occupation?

After a long, bloody summer of slogging across Normandy's hedgerow country, Allied troops in the north had finally broken through in late July and were now slicing their way into the heart of France. Yet even though U.S. forces were closing in on Paris, General Dwight D. Eisenhower had no plans for its immediate liberation. In fact, he planned to bypass the city, which he considered of little strategic importance, and roll on with all possible speed toward Germany.

As General Charles de Gaulle, then ensconced in Algiers, saw it, Eisenhower's decision spelled disaster for both Paris and himself. Not only was the capital in danger; so was de Gaulle's hope of controlling all of liberated France. His main rivals, the French communists, dominated the resistance movement in many parts of the country, including Paris. He had already received word that they were preparing an insurrection there. It was vital, in de Gaulle's view, that the Western Allies reach Paris before the situation spiraled out of control.

Determined to get his way, he set out to outmaneuver Eisenhower, just as he earlier had outmaneuvered Roosevelt, who had wanted no role at all for him in the postwar governance of France. Before the D-Day invasion, the president had decided that U.S. military forces would administer France until elections could be held. To that end, dozens of army officers were currently enrolled in a two-month crash course in public administration and the French language at the University of Virginia.

By contrast, Eisenhower and most British officials believed that de Gaulle and his French Committee of National Liberation should act as the provisional government of France. Churchill, although still upset with de Gaulle over his behavior prior to D-Day, reluctantly agreed to allow him to return to France for a brief visit the week after the invasion. The prime minister was responding to heavy pressure from the British press and public, as well as to strong lobbying by Eisenhower. In effect, the Allied commander, who had been given considerable latitude by Roosevelt in governing liberated areas, was making an end run around Washington.

In his memo authorizing de Gaulle's visit to Bayeux, on the Normandy coast, Churchill wrote, "I suggest that he should drive slowly through the town, shake hands with a few people and then return, leaving any subsequent statement to be made here." As usual, de Gaulle had other ideas. At that point, most of the French knew him only as a spectral voice, to which they had listened over the BBC throughout the war years. "He was a ghost to those millions, an ideal," according to one observer. "Now he had to give himself flesh and blood [and] become a political reality."

When de Gaulle arrived in Bayeux on June 14, he was mobbed by huge crowds of cheering, sobbing townspeople wherever he went. After walking Bayeux's streets for hours, then addressing its population in the town square, he traveled to the nearby town of Isigny, where he did the same. When he returned to England that night, he left behind in Normandy one of his top aides, whom he had assigned to act as governor of the region. With Eisenhower's tacit support, de Gaulle was undermining Roosevelt's attempts to impose an Allied military administration on France. Whether Washington liked it or not, the French general was now in charge of the liberated areas of his country.

De Gaulle's biggest challenge, however, was to gain control of Paris—France's political, social, and economic epicenter—as soon as it was freed. Like Warsaw before the uprising there, Paris was a tinderbox, its residents eager to settle the score with the Germans and erase the humiliation of their country's capitulation. "On the barricades," one resistance leader proclaimed, "we must wipe out the shame of 1940." Such feelings were stoked by the French communists, who con-

General de Gaulle triumphantly returns to France eight days after D-Day.

trolled the unions and underground press in Paris, as well as two of its three major resistance organizations.

In mid-August, a series of communist-inspired strikes was launched in the capital; railway men, police officers, and postal and telegraph workers, among others, walked off the job, paralyzing the city. The communists called for an armed insurrection on August 18. At 7 A.M. that day, small bands of resistance fighters throughout Paris opened fire on German patrols. Other groups burst into public buildings, ousting the occupants and taking over. In a matter of hours, French flags were fluttering from windows and rooftops as far as the eye could see.

As in Warsaw, barricades sprang up all over the capital. At the Place du Palais-Royal, the actors of the Comédie Française, the national theater group of France, built their own huge obstruction, using sofas, bureaus, and other items of furniture from their theater's scenery storeroom. In the Paris police headquarters, now occupied by the resistance, the nuclear physicist Frédéric Joliot-Curie and his assistants from the Collège de France made Molotov cocktails out of a variety of materials, including sulfuric acid and potassium chlorate, that they had brought from their laboratory.

Although the uprising caught the Germans by surprise, it didn't take long for them to respond. Troops and tanks charged into the center of the city and the surrounding neighborhoods, wounding and killing hundreds. Among those battling the German forces were thousands of Gaullist resistance fighters. They had been ordered by de Gaulle not to engage in overt rebellion, but once the uprising began, they felt they had no choice but to join.

Faced with a fait accompli, de Gaulle traveled from Algiers to Eisenhower's headquarters in France to press him to launch an immediate Allied attack on Paris. If the supreme commander refused, de Gaulle said, he would withdraw the French 2nd Armored Division from Allied command and dispatch it to Paris on his own authority. Veterans of the fighting in North Africa, the 16,000 men in the division, under the command of General Philippe Leclerc, had arrived in Normandy just two weeks before to take part in the march on Paris.

But neither de Gaulle's appeals nor his threats made any headway with Eisenhower. To him, the capture of Paris—and the time and matériel, particularly gasoline, it would entail—would put at risk his overriding goal of reaching the Rhine River and crossing into Germany before the Wehrmacht could reorganize.

Yet Eisenhower changed his mind the very next day, thanks to the pleading of a young resistance leader named Roger Gallois, who came from Paris to present him and General Omar Bradley, the commander of U.S. ground forces in France, with alarming news. The resistance fighters in the capital were barely hanging on, Gallois said. If the Allies did not come to their aid immediately, hundreds of thousands of Parisians would lose their lives. Furthermore, the German commander in Paris, General Dietrich von Choltitz, was under orders from Hitler to destroy the city before surrendering it to the Allies. Although Choltitz didn't like the order, he thought he had no choice but to carry it out.

He sent word through Gallois that only the Allies' speedy arrival could stop him.

Both Eisenhower and Bradley were swayed by Gallois's report. On August 22, Eisenhower's Supreme Headquarters Allied Expeditionary Force (SHAEF) ordered the French 2nd Armored Division, supported by the U.S. 4th Infantry Division, to race toward Paris. Three days later, on the morning of August 25, Leclerc's troops reclaimed the delirious capital of their country. Throngs of Parisians embraced and kissed the soldiers as they marched and rode by. Glasses of champagne and cognac were handed to them. People climbed up onto the moving tanks, threw flowers and food, waved handkerchiefs and flags. High above the crowds, the great church bells of Paris—in Notre Dame, Sacré-Coeur, Sainte-Chapelle—rocked the city with their joyous peals. Not even gunfire from German snipers and sporadic duels between Allied and German tanks could dampen the celebration.

De Gaulle's triumphant entry into Paris later that afternoon foiled

Parisians celebrate the liberation of their city by Allied troops.

the communists' scheme to establish a government in the capital before he arrived. The following afternoon, in a carefully planned scenario, de Gaulle introduced himself to the people of Paris. After relighting the eternal flame at France's Tomb of the Unknown Soldier, the general, followed by hundreds of his men, marched down the Champs-Élysées to Notre Dame. From sidewalks, rooftops, windows, and balconies, hundreds of thousands of Parisians lustily cheered him.

That evening, the CBS war correspondent Larry LeSueur tried to capture for his American listeners the jubilant mood of the city. "Tonight," he said, "all Paris is dancing in the streets."

DURING THE LAST DAYS of August, while the people of Paris were still toasting their liberators with champagne, thousands of Polish resistance fighters and civilians slipped down manholes and disappeared into Warsaw's stinking, night-black sewers. The sewers had become the Poles' only means of escape from the shelled and bombed-out ruin that Stare Miasto had become. Against all odds, elements of the Home Army had held out for almost a month there. Now cut off from ammunition, food, and water, they were on the verge of annihilation, and Bór-Komorowski gave the order to evacuate. Even so, fighting continued in other parts of the city. In Berlin, Himmler remarked to his lieutenants, "For five weeks, we have been fighting the battle for Warsaw. This is the most bitter struggle of all we have had since the start of the war."

An RAF bomber pilot who flew over Warsaw during the uprising later remarked that if Dante could have seen the burning city, he would have had a realistic picture of his Inferno. Yet few people in the West had any idea of Warsaw's anguish. Except for some short news accounts and a few editorials, the drama, heroism, and tragedy of the uprising went largely unnoticed in Britain and the United States. The liberation of Paris and the Allied advance toward Germany were dominating newspaper and magazine headlines and radio broadcasts.

There was, however, one major exception. In a piece called "A Tale of Two Cities," *The Economist,* an influential British political and international affairs journal, provided a grim, "heartbreaking" comparison

Warsaw's Stare Miasto (Old Town) in ruins after the 1944 uprising.

between Warsaw's struggle for survival and the relatively easy liberation of Paris. Almost exactly five years earlier, *The Economist* pointed out, just three nations—France, Britain, and Poland—had gone to war against Germany. Since that time, the British, with help from their allies, prominently including the Poles, had managed to hold on to their freedom, and the French, with even more Allied help, were recovering theirs. But now the Poles were trying to drive out the Germans, and in this "vastly bloodier and more desperate battle," they have been "almost unsupported by their Allies, materially or even morally."

The Russians' refusal to provide military support or to allow their allies the use of their air bases was "intolerable," *The Economist* added. "The rising in Warsaw is a glorious contribution to the Allied cause, and it cannot be refused. Talks are now going on about aid to Warsaw. . . . In honor and expediency alike, they should have only one result, and that speedily. But, incredibly, the present prospect is said to be precisely the opposite. To our joy in victory, it seems that the Allies may have to add the ultimate shame of desertion."

The agitated Churchill was making a belated effort to forestall that outcome. He already had given in to Stalin on the question of eastern Poland; now the issue of a free and independent Poland was obviously in jeopardy. If Warsaw were ruined and the Home Army wiped out, Stalin would have a much easier time installing a regime of his own creation. The Kremlin, Churchill concluded, "did not mean to let the spirit of Poland rise again in Warsaw."

In a War Cabinet meeting, he discussed with its members the possibility of sending bombers to help Warsaw and having them land on Soviet airfields without permission. But in the end, the British government decided against taking unilateral action. It was, Churchill wrote after the war, another in the "terrible and even humbling submissions [that] must at times be made to the general aim."

Making one final push to aid the Poles, Churchill renewed his appeal to FDR to support a plan to send large-scale relief to Warsaw and also, if need be, to "gate-crash" Soviet airfields. The president, who was facing reelection in two months and wanted no hint of dissension within the alliance to harm his chances, remained unwilling to confront Stalin. Responding to Churchill's plea, Roosevelt said he had been informed by intelligence sources that the Germans were now in full control of Warsaw: "The problem of relief for the Poles . . . has therefore unfortunately been solved by delay and by German action, and there now appears to be nothing we can do to assist them." (FDR was wrong: the uprising would continue for another month.)

Six weeks after the insurrection began, Stalin, knowing that the Home Army was doomed, withdrew his objections to the use of Soviet bases by U.S. bombers. On September 18, more than a hundred B-17s dropped containers containing submachine and machine guns, pistols, grenades, medical equipment, and food rations. The supply mission was far too late; most of the containers drifted into parts of the city already reclaimed by the Nazis. "Had the containers been dropped in the first days, when two-thirds of the city was in our hands, they might have decided the outcome of the battle," Bór-Komorowski later remarked. Stalin, having scored propaganda points for his onetime gesture of supposed magnanimity, rejected requests from the British and U.S. military for repeat missions.

When the commanders of the Home Army launched their uprising on August 1, they believed they would have to hold out for only four or five days before help came, which was, as it happened, the exact scenario that unfolded in Paris. In Warsaw, however, the Home Army and the city's other residents held out—with no reinforcements at all—for sixty-three days. As SS troops pushed the Poles into smaller and smaller areas of the city, hope still flickered that outside aid would arrive in time to save what was left of them and Warsaw. But help did not come, and in early October, even hope died.

Food, water, and ammunition were gone. Disease was rampant. In the few districts still held by the Home Army, dozens of people were crammed into every basement and cellar, many on the verge of death. Faced with the prospect of total annihilation of the city's population, the Home Army leadership decided it had no choice but to capitulate. At 8 P.M. on October 2, Bór-Komorowski signed a surrender agreement at German headquarters. The following day, the Polish underground radio station sent a farewell message to London. The announcer's voice cracked with emotion as he said, "We have been free for two months. Today, once more we must go into captivity."

A staggering 200,000-plus people—about a quarter of the Warsaw residents who had survived the war to that point—had been killed in the uprising. All those left were ordered by the Germans to evacuate their ruined city. On the morning of October 5, the survivors emerged from their cellars and shelters, most of them soon to begin an existence in German POW, concentration, and labor camps. Leading the procession out of Warsaw were Bór-Komorowski and his Home Army troops. A phalanx of SS men awaited them several hundred yards away. As the Poles moved forward, Bór-Komorowski began singing the Polish national anthem. With tears in their eyes, his men and the civilians behind them joined in the hymn. Their voices swelling in intensity, they sang "Poland Has Not Yet Perished as Long as We Live" as they marched toward the waiting Germans.

Hitler did not massacre the survivors of Warsaw as he earlier had vowed to do, but he made life as hellish for them as possible. Many thousands died in German captivity before the war was over. Auschwitz was the destination for more than six thousand of the city's residents, mostly women and girls, a number of whom were Jews who had been hidden by Polish Christians in Warsaw and whose real identities remained secret. They were not sent to the gas chambers, but many succumbed from cold, starvation, disease, and physical abuse before the Soviets entered the camp in the spring of 1945.

More than twelve thousand other women from Warsaw ended up in the unspeakable squalor of the vastly overcrowded Ravensbrück camp, where, like at Auschwitz, there was virtually no food and the sanitary conditions were appalling. Hundreds of the women were pregnant, the result of rape by German soldiers during the uprising. When the babies were born, they were deliberately starved to death. Many of their mothers died as well.

Hitler, meanwhile, followed through on his pledge to destroy Warsaw. Nazi sappers divided the city into districts, each given a date for destruction. House by house, block by block, district by district, the remnants of the Polish capital were systematically and methodically burned and dynamited. All that was left when Russian troops finally "liberated" the city in January 1945 were ruins and the unburied dead.

PARIS, BY CONTRAST, was remarkably untouched when de Gaulle took control of it in late August. There was little mourning there—the uprising had claimed fewer than two thousand lives—and the city, its beauty unmarred by bombs, was once again open for both business and pleasure. The Allies took over hundreds of hotels for their own use, and within days a frenzied round of partying began. Most Parisians—and the French in general—had very little to eat, but there was a thriving black market in food, liquor, and wine for those who could afford the exorbitant prices. The city's best restaurants, which had served members of the Wehrmacht and Gestapo just a few days earlier, were now welcoming hordes of Allied officers and journalists.

De Gaulle, however, was not among them. His sole focus at the time was to consolidate his authority over Paris and the rest of the country and to mobilize its resources for the liberation of all of France and the final Allied assault on Germany. Within days, he had effectively disbanded the French resistance, bringing its units under the control of the regular French army and ordering SOE officers who had worked with resistance fighters to return to England.

The general spent much of the fall of 1944 touring France's main provincial centers and meeting their residents. In Besançon, a bustling city in the east of the country, Eric Sevareid stood in the middle of a huge crowd "jammed elbow to elbow" for two hours in a cold September rain, everyone patiently awaiting de Gaulle. Having lived in France for several years before the war, Sevareid was intimately familiar with the sourly cynical attitude of most of the French toward politics and politicians. But as he gazed at the faces around him, he observed "an intentness, an almost fanatical look of reverence such as I had never dreamed to see in this country."

Sevareid compared the assured, poised de Gaulle he saw that day with the rigid, unsmiling neophyte leader he had observed leading a meager parade of Free French troops in London on Bastille Day 1940. In those four years, Sevareid remarked, de Gaulle "had learned how to gesture, how to speak confidentially and colloquially to the people. He asked the people of Besançon to sing the *Marseillaise* with him, and then he walked slowly down the narrow streets, waving and touching the outstretched hands of hundreds. So it went, in every city and town he visited, the voice and myth becoming Gallic reality. I remembered how, in those other days, no Frenchman had seemed great to the French. Others had—Roosevelt, Churchill—but never one of their own. Now there was a great Frenchman too, and they accepted him as such."

So had much of the rest of the world. By early fall 1944, most of the Allied nations, including those in occupied Europe, had recognized de Gaulle and his committee as the provisional government of France. Roosevelt resisted for as long as he could, but, finding himself isolated on the issue, he finally gave in. On October 23, the United States recognized the general's committee. FDR made the announcement without first informing Churchill, who, despite growing misgivings, had loyally continued to follow Roosevelt's lead on matters involving de Gaulle. Caught flat-footed, the British government scrambled to issue its own announcement of recognition.

Three weeks later, Churchill made his first visit to liberated France. Given the extremely turbulent relationship between the prime minister and de Gaulle, both French and British officials feared the worst. "We all tremble for the result," said a British Foreign Office staffer. He and the others needn't have worried. As insufferable as de Gaulle had been in adversity, he was, as one historian wrote, "magnanimous in victory." On November 11—a bright, cold Armistice Day—the people of Paris and their leader gave Winston Churchill a welcome so warm and joyous that "it had to be seen to be believed," marveled Duff Cooper, the new British ambassador to France. "It was greater than anything I have ever known."

More than half a million Frenchmen lined the flag-bedecked Champs-Élysées and nearby streets as de Gaulle, Churchill, and the top officials of their governments strode down the wide thoroughfare to a dais half a mile away. Some in the massive crowd "were cheering; some were laughing; some were sobbing; all were delirious," remembered General Pug Ismay, who was among the officials in the procession. "All we heard was '*Vive Churchill!*' '*Vive de Gaulle!*' '*Vive l'Angleterre!*' '*Vive la France!*'" From the dais, the prime minister and general stood at attention as French and British troops paraded past. On de Gaulle's orders, a French band played a popular military march, "Le Père la Victoire" ("Father Victory"). "For you," de Gaulle said to the beaming Churchill. In his memoirs, de Gaulle noted, "It was only his due."

Caught up in the emotion of the day, both leaders put behind them, at least temporarily, their bitter antagonisms. For Churchill, it was a magical moment. From the day his beloved France had fallen in 1940, he had insisted to all naysayers, including Roosevelt and many in the prime minister's own government, that it would, like a phoenix, rise from the ashes one day. That day was now here, and he paid tribute to the Frenchman who had shared his belief and done so much to make it a reality. In a speech to French resistance leaders, Churchill described de Gaulle as the "incontestable leader" of France. "From time to time," the prime minister conceded, "I have had lively arguments with him about matters relating to this difficult war, but I am absolutely sure that you ought to rally round your leader and do your utmost to make France united and indivisible."

De Gaulle returned the favor, acknowledging the vast debt he and France owed to Churchill and the British. At a lunch honoring Churchill, he noted, "We would not have seen today if our old and gallant ally, England . . . had not deployed the extraordinary determination to win and the magnificent courage which saved the freedom of the world. . . . I do know that France . . . will not have forgotten in a thousand years what was accomplished in this war through the blood, sweat, and tears of the noble people whom the Right Honorable Winston Churchill is leading to the heights of one of the greatest glories in this world. We raise our glasses in honor of Winston Churchill . . . and England, our ally, in the past, the present, and the future."

Though clearly heartfelt at the moment, that concept of partnership and unity would be badly strained in the difficult months to come. And as the British and French officials toasted each other, a more immediate shadow darkened their celebration: Allied victory, which had seemed so tantalizingly close at the time of Paris's liberation, had slipped, for the moment, out of reach.

CHAPTER 23

"I Was a Stranger and You Took Me In"

Defeat at Arnhem

AFTER THE LIBERATION OF PARIS, THE ALLIED JUGGERNAUT continued its mad dash across France. In the north, the Twenty-first Army Group, commanded by General Bernard Montgomery, moved along a front sixty miles wide, covering 250 miles in four days and freeing a string of French cities and towns. Belgium's turn came next.

Montgomery's forces, which included British, Canadian, Polish, Belgian, Czech, and Dutch troops, began marching into Belgian cities in the first days of September. On September 3, the Welsh Guards liberated Brussels. After watching its delirious residents throwing flowers and bottles of beer into Allied jeeps and trucks, a British correspondent wrote, "The joy of Paris was a pallid thing compared to this extravaganza."

The next day, Antwerp was freed—a particular triumph because of its deepwater port, the second largest in Europe, which before the war had handled up to a thousand ships a month. Until Antwerp's liberation, only one other port, Cherbourg in northern France, had been available to unload the supplies needed to continue the Allied drive toward Germany. At that moment, all three Allied armies were running desperately short of just about everything, especially gasoline, which made Antwerp more important to the Allied cause than Paris or any other liberated city.

It was thanks to the Belgian resistance that Montgomery's troops were able to seize the huge thousand-acre port complex intact. Before his multinational force reached Antwerp, resistance fighters had overwhelmed the port's German garrison, preventing its soldiers from detonating the explosives they had defensively prepositioned throughout the facility. Resistance members also acted as guides for British tanks as they threaded their way around entrenched German positions outside Antwerp and entered the city along a less defended route.

By preserving the docks, warehouses, locks, sluice gates, and other machinery, the resistance had done its part to end the war as quickly as possible. Now it was up to Montgomery's forces to do theirs. Although the port was no longer under German control, it could not actually be used until Allied troops also controlled the forty-mile Scheldt River estuary, which linked Antwerp with the North Sea. Supply ships could not negotiate the estuary while its banks still bristled with German guns.

Montgomery had received numerous warnings—from the Royal Navy, Belgian resistance leaders, and Eisenhower himself—about the vital importance of clearing the enemy off the estuary's approaches. Admiral of the Fleet Andrew Cunningham, Britain's first sea lord, declared that Antwerp was "as much use as Timbuctu unless the German forts were silenced and the banks of the Scheldt River occupied."

To sweep the estuary clean would have been relatively easy at that point: the Germans were on the run, and their defenses were cracking. "At that moment, had they chosen to do so, the British could have driven onwards up the forty-mile coast . . . with nothing to stop them," Max Hastings wrote.

But Montgomery, who had given the British their first battlefield victory at El Alamein in 1942, decided against it. There was no need to hurry, he thought. With the Germans so close to defeat, the mopping up of the Antwerp defenses could be done at the Allies' leisure. Besides, his exhausted troops needed a couple of days to "refit, refuel, and rest" after their race across France. His decision turned out to be a strategic disaster and one of the worst Allied mistakes of the war.

At the time, Montgomery was focused on what he considered a far more pressing issue: how to cross the Rhine River, just eighty-five miles away, and become the first Allied commander to enter Germany.

On September 7, he informed London that he hoped to be in Berlin within three weeks, apparently not taking into account that a lunge like that would require food, gas, and other supplies that could arrive only through a major port such as Antwerp. Later, one of Montgomery's top generals admitted that he, like his boss, never stopped to think about the necessity of taking the estuary. "My excuse," the general said, "is that my eyes were fixed entirely on the Rhine, and everything else seemed of subsidiary importance."

To be fair, Montgomery and his subordinates were hardly the only Allied senior officers to feel that way. Drunk with their success in France, other field commanders and top SHAEF officers had also convinced themselves that the Germans, so close to collapse, were incapable of recovering. Victory was in their grasp, they were sure, perhaps as early as Christmas. On September 1, General Walter Bedell Smith, Eisenhower's chief of staff, told reporters that "militarily, the war is won."

Battered as they were, the Germans had other ideas: while the British celebrated in Antwerp, they were on the move. Just a few miles to the northwest, the German Fifteenth Army, an 80,000-man force that had been mauled at Normandy and lost most of its transport, had taken temporary refuge in the Pas de Calais area, not far from the southern bank of the Antwerp estuary. At that point, they easily could have been trapped by Montgomery's army. Realizing that the Allied drive had come to an abrupt halt, Fifteenth Army commanders swiftly dispatched their men across the waterway northwest of Antwerp. While some were left behind to reinforce the estuary's approaches, most escaped into Holland.

Not until September 13, nine days after Antwerp's liberation, did Montgomery assign the clearing of the port's approaches—a "low-priority mission," he called it—to the Canadian and Polish troops under his command. Because of the rapid buildup of German defenses, the Canadians and Poles didn't have enough men for the job, and the operation was scrubbed until more troops became available. Ultimately, it would take three months to get rid of the Germans—a task that could have been accomplished in a couple of days had it been done in early September. With Antwerp still closed, Cherbourg remained the only supply port for the entire Allied Expeditionary Force. Gasoline and other supplies grew increasingly scarce, threatening to shut down the Allies' headlong drive.

Always reluctant to admit error, Montgomery nonetheless admitted after the war that he had made a "bad mistake—I underestimated the difficulties of opening up the approaches to Antwerp." "Bad mistake" doesn't do justice to his fumbled handling of Antwerp, which set off a chain of events that ultimately prevented the Allies from smashing into Germany and ending the war in 1944.

As a result, many more people would die, soldiers and civilians alike. For the Netherlands, the consequences would be especially dire.

AFTER BELGIUM AND MOST of France had been freed, it appeared for a couple of glorious days that the Netherlands' hour had finally arrived, too. In early September, residents of Dutch towns and villages near the Belgian border watched with delight as panicked German troops streamed past, all of them heading east. "It was a flight of dirty, exhausted, silent scarecrows on foot and of sleeping men in trucks," an eyewitness recalled. "Hitler's drilled divisions had changed into a miserable horde of frightened, hunted men."

According to rumors, the Allies were swiftly approaching the Dutch border. On September 3, the day of Brussels' liberation, Dutch prime minister Pieter Gerbrandy went on the BBC to make it official: "I wish to give a warm welcome to our Allies on our native soil. . . . The hour of liberation has come." The BBC's Radio Orange reported that the town of Breda, just a few miles from the border, had been taken, and Eisenhower, although a bit more circumspect, affirmed that freedom for the Dutch was imminent: "The liberation that the Netherlands have awaited so long is now very near."

Thousands of people, their arms filled with flowers, gathered joyfully on the outskirts of Amsterdam, The Hague, and other major cities, waiting to welcome their liberators. Waving Dutch flags, the cheering throngs shouted "Long live the queen!" and sang the national anthem. The troops never arrived, however, and the crowds discarded their flowers and flags and drifted home. The BBC backed off its earlier broadcast, saying that there were no further official reports about an advance into the Netherlands. Breda was still in German hands: the British troops seen there, it turned out, had been a patrol that had crossed the Belgian-Dutch border by accident.

As the Netherlands' residents soon discovered, Montgomery had halted his troops just short of the border and was keeping them there, despite intelligence reports from the Dutch underground that Germany no longer had enough forces in the Netherlands to stop a swift Allied advance.

THE REPORTS THAT MONTGOMERY ignored had come from a fair-haired, bespectacled young officer in Dutch battle dress, who paid a call on Montgomery on September 7 at his headquarters in Brussels. He was Prince Bernhard, the thirty-three-year-old son-in-law of Queen Wilhelmina and commander in chief of the Dutch resistance forces.

The greeting that Bernhard received from Montgomery's aides was somewhat less than enthusiastic. In fact, it was downright patronizing. But the prince was accustomed to not being taken seriously; ever since his marriage to Princess Juliana, he had been engaged in a battle to win respect from those around him.

There were, as Bernhard himself acknowledged, a number of major strikes against him. He was German—and, even worse, he had joined the Nazi Party as a student. Before he was allowed to marry Juliana in 1937, an official inquiry had been held to determine his true political leanings. Finally, after he renounced his German citizenship and convinced Wilhelmina and the government that he was opposed to Hitler, the wedding was permitted to take place. "This is not the marriage of the Netherlands to Germany," Wilhelmina assured her people, "but simply the marriage of my daughter to the man she loves."

In the stuffy, straitlaced circles of the Dutch court, Bernhard had been an anomaly from the start. He had acquired the reputation of a daredevil and playboy who, in the words of his friend Erik Hazelhoff Roelfzema, "exuded an aura of action and adventure." Known for driving his Ferrari at high speeds, he came close to dying in a car accident in 1938.

When the Dutch queen and government had fled to England in 1940, Juliana and the couple's two small daughters had been dispatched to safety in Canada. Bernhard had remained behind with his mother-in-law in London, where he had trained with the RAF, won his wings, and flown with a Dutch squadron in bombing missions over occupied Europe. Impressing Wilhelmina with his newfound seriousness and sense of purpose, he became one of her key advisers. In 1943, she appointed him liaison officer between the Dutch military and the rejuvenated resistance forces at home. A year later, he was named commander in chief of the resistance. If it hadn't been for the war, Bernhard later said, "I would have been just another royal figurehead, lashed to the bow of the Ship of State."

Although considerable skepticism was aroused by his appointment, it melted away almost immediately. As London's *Daily Telegraph* later put it, Bernhard "played a vital and rather under-appreciated part in fusing the Dutch military and amorphous resistance factions into one force which eventually spearheaded the Allied advance into the Netherlands." Resistance members "adored him and listened to him," his chief of staff said, "and by bringing these brave but jealous, idealistic but egotistical men together, Prince Bernhard performed a near miracle."

In his meeting with Montgomery, Bernhard reported that, according to Dutch intelligence agents, the route through Holland and across Germany's vulnerable northern frontier was, for now, relatively defenseless. If Montgomery's army moved immediately, it could bulldoze its way into the Ruhr, the industrial heart of Germany, and bring about the Reich's defeat.

Several German military leaders later agreed with that assessment. If the Allies had mounted "a major thrust resulting in a breakthrough anywhere," Germany's collapse would soon follow, according to General Günther Blumentritt, chief of staff to Field Marshal Gerd von Rundstedt. Blumentritt was convinced that the Allies would indeed strike across Holland and into the Ruhr. Years after the war, Rundstedt said he had believed that the war would be over within two weeks.

During their meeting, Bernhard cautioned Montgomery that the window of opportunity for such a thrust would soon close. The Wehrmacht's retreat across Holland was slowing, reinforcements were on their way from Germany, and the German defenses at the Belgian-Dutch border were being rebuilt. But Montgomery rejected everything the prince told him, saying, "I don't think your resistance people can be of much use to us." It was obvious, Bernhard remarked, that "Montgomery didn't believe any of the messages coming from my agents in Holland."

As part of his *Englandspiel* campaign, the Abwehr's Hermann Giskes had planted seeds of distrust in the minds of British officials about the morale and security of the Dutch resistance. The general's reaction also reflected the overbearing, often hostile attitude that many high-ranking British officers had toward their smaller European allies. Montgomery, whose command included thousands of European troops, was particularly noted for his lack of knowledge of and regard for them. Once, during a visit to a Polish division in his army, he asked its commander whether Poles spoke to one another in Russian or German. He was stunned to learn that they had their own language. About the European forces under his command, he wrote, "I would rather not have them at all."

Bernhard got the impression that Montgomery and his staff "considered us a bunch of idiots for daring to question their military tactics. I was sick at heart because I knew that German strength would grow with each passing day. But nothing I said seemed to matter."

Before Bernhard left, Montgomery unbent enough to give him an inkling of what he was planning next. "I am just as eager to liberate the Netherlands as you are," he said, "but we intend to do it in another, even better way. . . . I am planning an airborne operation ahead of my troops."

In fact, Montgomery had already approached Eisenhower for approval of his plan, which called for U.S., British, and Polish paratroopers to seize a series of bridges and canals in Holland and establish bridgeheads for advancing Allied infantry forces, who would then cross the Rhine and enter Germany. The last bridge to be captured, by the British 1st Airborne Division, spanned the Rhine at the Dutch town of Arnhem.

The plan, called Operation Market Garden, presented many difficulties from the start. For one, the Allied offensive in Europe was literally running out of gas, thanks to Montgomery's failure to open the Antwerp port. The fuel that was left was being fought over by the various field commanders, all of them "obsessed with the idea that, with only a few more tons of supplies, they could rush right on and win the war." Montgomery, whom Churchill had just promoted to field marshal, told Eisenhower that the remaining resources should be his. He insisted that a bold thrust to the northeast, carried out by British forces and supported by U.S. troops, would have a much better chance than any other of breaking into Germany and bringing the war to a close.

Initially, Eisenhower thought the proposal absurd. "Monty, you're nuts," he said. "You can't do it." He ordered Montgomery to focus on opening Antwerp, but the British general kept pressing him, and Eisenhower began to weaken. A sharp political undercurrent underlay their wrangling. Several weeks before, General George Marshall, the U.S. Army chief of staff, had ordered Eisenhower to take direct operational command of all Allied land forces in Europe, replacing Montgomery, who had held that job since Normandy. It was time, Marshall felt, to underscore the United States' dominance on the European front, no matter how much Churchill, Montgomery, and the rest of the British might protest.

Montgomery was Britain's most popular military figure, and Britons, including Churchill, were incensed by his demotion. But no one was more upset than the prickly, arrogant field marshal himself. Highly critical of Eisenhower throughout the war, he never fully accepted the move and repeatedly challenged the supreme commander's authority for the duration.

Although Eisenhower, in turn, disliked Montgomery, he felt it important to placate him as much as possible. But there was another reason for his lessening opposition to Market Garden. The more he thought about it, the more intrigued he became by its audacity: perhaps it *could* resurrect the weakening Allied offensive, as Montgomery promised. Like his critic and rival, Eisenhower was seduced by the idea that the German military was so shattered that it would not—and could not—mount a stout defense of its homeland. On September 10, he signed off on the plan.

Market Garden was to be the greatest paratroop and glider-borne infantry operation ever staged behind enemy lines, far bigger and more complex than the airborne force that had landed in Normandy the night before D-Day. Months of planning had gone into that effort. For the operation in Holland, planners had been given just seven days to draw up a blueprint that, to have any chance of working, had to proceed like clockwork. Yet the chances of that were infinitesimal. The vast logistical difficulties of the airborne part of the operation all but precluded success. Equally daunting was the fact that the massive procession of tanks and ground forces assigned to relieve the paratroopers at the bridges would be forced to use just one highway—a narrow road that ran through marshy countryside, laced with dikes, for more than sixty miles. Before the war, the Dutch military had conducted an exercise using that same road for a simulated advance on Arnhem and decided it would lead to catastrophe. But the British never consulted the Dutch as they prepared for the operation. When Dutch generals learned of the route the British were planning to take, they tried to dissuade them, to no avail.

Dutch intelligence operatives, meanwhile, passed on news of a heightened German presence in the area around Arnhem. Instead of the scattered, weak units that the planners expected, two elite SS panzer divisions had reportedly been relocated near the paratroopers' landing spots. Equipped with heavy tanks, these units contained the best fighting troops in the German army, about to face off against British paratroop forces with no tanks or heavy weapons and very limited supplies of ammunition.

When other intelligence sources confirmed the Dutch report, Major Brian Urquhart, the chief intelligence officer for Market Garden, tried to impress on senior officers the gravity of the situation. They refused to believe him. "It was absolutely impossible to get them to face the realities," Urquhart said. "Their personal longing to get into the campaign before it ended completely blinded them." When he persisted, he was accused of being "hysterical and nervous" and was finally dismissed from the operation. The presence of tanks at Arnhem "was the one awkward fact that would not fit the desired pattern, so the best thing

*Brigadier
General John
Hackett (right)
with General
Bernard
Montgomery.*

was to sweep it under the carpet," the historian Ralph Bennett later wrote.

Also voicing dismay over the "light-heartedness and inexperience of our airborne planners" was Brigadier General John Hackett, the thirty-three-year-old commander of the British 4th Parachute Brigade, which would take part in the assault on the Arnhem bridge. Born and raised in Australia, the Oxford-educated Hackett, who was known as "Shan," had headed the brigade from its formation, leading it into battle in North Africa and Italy. He was known for his rapier wit, as well as for his "inability to suffer fools, especially senior officers." Though the young brigadier was extremely popular with his men, his superiors considered him "rather argumentative, with firm ideas" that often did not coincide with theirs.

One of those "firm ideas" was that Market Garden was a disaster about to happen. "After harrying a defeated enemy across western Europe, Allied commanders and staff tended to think they knew it all," Hackett observed. "Those of us who had some experiences of fighting against the German army . . . knew that however light their existing strength was, a real threat to an objective of vital importance would be met with a swift and violent response."

General Stanisław Sosabowski, the commander of the Polish 1st Parachute Brigade, which would also fight at Arnhem, agreed. After listening to the planners' optimistic views at one meeting, Sosabowski exclaimed, "But the Germans, the Germans! What about them?"

Sosabowski's force had been created to fight for the liberation of its country, and the fact that it was to be sent on this slapdash mission while the Home Army was still engaged in its doomed battle for Warsaw made the Polish general particularly irate. But his objections found no favor with British staff officers, who were known for ridiculing his heavy accent and "giggling like schoolboys" when he expressed his views. As Prince Bernhard noted, the British military "doesn't like being told by a bloody foreigner that they're wrong." Or, as it turned out, by anyone else.

SUNDAY, SEPTEMBER 17, was a beautiful day in Arnhem, with sunshine bathing the prosperous, peaceful little resort town and gilding the nearby Rhine. Many residents took a stroll that afternoon, intent on enjoying one of the last warm days of the waning summer. Known for its gracious hotels and well-kept houses and gardens, Arnhem had never looked lovelier, one resident thought. Then came the roar of approaching planes and the jaw-dropping sight of thousands of men dropping from the sky. For Arnhem and its people, nothing would ever be the same again.

Shan Hackett and his men were among the parachutists drifting down into that bucolic setting, which, as he feared, would soon become an inferno. Shortly before taking off from Britain, Hackett had told his staff and battalion commanders to forget all the optimistic talk bandied about by Market Garden's top commanders. Given the "German capability for a swift and violent response to any threat that really mattered," he said, his men should brace "for the hardest fighting and worst casualties" imaginable. His pessimism, already strong, had deepened when he learned that, instead of being dispatched to Arnhem at one time, the paratroopers were to be sent in waves over several days because of a shortage of cargo planes and gliders. Making matters worse, his own division was to be dropped several miles from the Arnhem bridge, its chief objective.

All of Hackett's dire predictions, along with those of Sosabowski and the Dutch, came true. The German panzer divisions were indeed dug in at Arnhem, and they quickly responded. Although the paratroopers would have suffered many losses by landing on or close to the Arnhem bridge, they would have seemed small compared to the number of casualties they actually took. Almost nothing worked according to plan. Within twelve hours of their arrival, the British had forfeited any chance of capturing the bridge. The only battle they faced now was the struggle to survive.

The advance of the British and U.S. ground forces, meanwhile, was extraordinarily slow. As predicted, heavy tanks and trucks bogged down in the Netherlands' soggy soil, and the only road on which the forces could travel was soon blocked by disabled vehicles. Allied infantry came under heavy fire from Germans on either side of the road, many of them Fifteenth Army soldiers who two weeks earlier had escaped from Belgium through the gap left by the British at the Antwerp estuary.

Another failure of this "epic cock-up," as one British officer described the operation, was the monumental breakdown of the British radio communication system, particularly that of the 1st Airborne, just as the battle began. Transmitters were lost or ceased to work, and, with no one knowing where anyone else was, there was no way to coordinate a systematic attack. Actually, although the British never took advantage of it, they did have a communications alternative: the Dutch phone system was working, and the resistance suggested its use to British commanders. But because of their suspicions of the Dutch, the British dismissed that suggestion. They also turned down offers by the resistance to act as guides and to provide information about the composition and location of German forces. "We were prepared to do anything, even sacrifice our lives if necessary," one resistance fighter later said. "Instead, we felt useless and unwanted. It was now increasingly clear that the British neither trusted us nor intended to use us."

In the view of Cornelius Ryan, who wrote *A Bridge Too Far,* the magisterial history of Market Garden, the British "had an outstanding force at their disposal whose contributions, had they been accepted, might well have altered the grim situation of the British 1st Airborne Division." The few Britons who did accept help from the Dutch, such as Major Derek Cooper of the Guards Armoured Division, were abundantly rewarded. Ordered to get through to the headquarters of the U.S. 82nd Airborne Division at the Nijmegen bridge, Cooper was guided there by resistance members, who, he said, were "absolutely invaluable."

In Arnhem itself, where the fighting was fiercest, dozens of Dutch civilians braved withering gunfire to retrieve dead British soldiers and carry the wounded to makeshift casualty stations in nearby houses and hotels. At one point, Arnhem resident Kate ter Horst—a "figure of truly heroic proportions," in the words of Shan Hackett—sheltered more than two hundred injured British paratroopers in her home. Other residents hid British officers in their houses and sheds to prevent their capture.

The fighting in Arnhem was savage and bloody, and the town, bombarded by German artillery fire, became a charnel house. Many if not most buildings were burned to the ground, and the bodies of soldiers and civilians were scattered everywhere. "Arnhem, one of the most scenic spots in the Netherlands," was now a "miniature Stalingrad," Ryan wrote.

On September 25, eight days after Market Garden began, most of the tattered remnants of the 1st Airborne were evacuated under cover of darkness, while many of the wounded surrendered to the Germans. Once they had secured their victory, the Germans treated their British prisoners with great consideration; one British officer called his captors "kind, chivalrous, even comforting." But they showed no such compassion to Dutch civilians, executing anyone they found who had aided the British. "It was pretty dismaying," a British captain said, "that while the Germans were giving us food, water, and cigarettes, on the other side of the square they were shooting out of hand Dutchmen whom they believed had helped us."

Overall, the number of Allied casualties from Market Garden totaled more than 17,000. Of the 10,000-man force at Arnhem, fewer than 3,000 escaped death, injury, or capture. Civilian casualties were estimated to be as high as 5,000. Amid the smoking ruins of the town, its surviving residents took refuge in cellars and other ad hoc shelters, struggling to live without gas, electricity, or water, and with very little food. A few days after the British surrender, the Germans ordered all Dutch civilians to leave Arnhem and nearby villages.

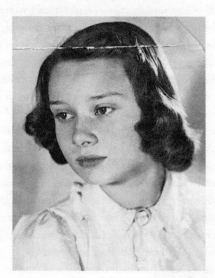

Audrey Hepburn in 1942.

Audrey Hepburn, who lived three miles outside Arnhem, watched the evacuation in horror. "I still feel sick when I remember the scenes," she said years later. "It was human misery at its starkest—masses of refugees on the move, some carrying their dead, babies born on the roadside, hundreds collapsing with hunger." The residents of Arnhem were not allowed back until the Allies finally liberated the area in April 1945.

IN ALL THE CHAOS and carnage of Arnhem, several hundred wounded British paratroopers managed to dodge death and evade capture. They were whisked away from hospitals, casualty stations, and battlefields by resistance members, who hid them in villages and towns several miles away. Audrey Hepburn's mother provided food for several, and Audrey herself delivered messages from the resistance to men in hiding.

One of those rescued was Shan Hackett, who, with his troops, had taken part in the brutal hand-to-hand combat in Arnhem. At the end, he was one of the few surviving members of his 1,000-man brigade, which, in his words, had been "the heart and center of my life" for two years. In less than a week, it had been virtually wiped out.

Hackett, who had been hit in the abdomen and leg by shrapnel from a mortar shell, was so gravely wounded that a German doctor who examined him said that nothing could be done and he should be allowed to die in peace. A South African surgeon from the 1st Airborne, who had also been taken prisoner, thought otherwise; he operated on the general and saved his life.

A few weeks later, when the German high command ordered the British wounded to be sent to prisoner-of-war camps, several members of the resistance smuggled Hackett out of a hospital near Arnhem. Weak, ashen-faced, and still in severe pain, he was taken to a tidy white house with a gable roof near the center of Ede, a bustling market town about twelve miles away. There he was put to bed in a tiny upstairs room with lace curtains, a white counterpane on the bed, and a needlepoint sampler of Sleeping Beauty on the wall.

His nurses and protectors were three middle-aged, unmarried sisters—Ann, Mien, and Cor de Nooij—who had never been involved in resistance activities before. But when asked by members of the underground to hide a wounded British officer, they had immediately agreed. "Thank God I now have something worthwhile to do!" one of them had exclaimed.

When Hackett, several days into his convalescence, looked out a window for the first time, he realized the extraordinary risk the sisters were taking in hiding him. On the street below, dozens of German officers and soldiers were "passing to and fro within speaking distance of me." Ede, which was less than three miles from the Rhine, had a sizable German presence: in addition to a large contingent of troops stationed there, it served as a rest center for soldiers on leave from the front. Indeed, many of the houses surrounding the de Nooij home had been requisitioned as German billets; the backyard adjoining the sisters' garden belonged to a house filled with German military police.

Yet hiding a British general under the noses of the enemy never seemed to faze the sisters or, for that matter, any of the members of their extended family who came to visit Hackett. He became particularly close to Johan Snoek, a son of one of the sisters and an ardent member of the resistance, and Johan's sister, Marie.

As the days and weeks passed, this hard-charging, hot-tempered brigadier, accustomed to giving orders and having his way, found himself enjoying the quiet, small comforts and rhythm of daily life in a household that "was now becoming my whole world." Once he was well enough to get out of bed, he would come down in the evening to the parlor, where the sisters, whom he now called "Aunt Ann," "Aunt Cor," and "Aunt Mien," gathered with other family members. While the "aunts" sewed and darned, Hackett played chess with Johan, read Shakespeare from an English-language version that Aunt Ann had found for him, or worked on his daily Dutch lesson with Marie. He was given his own mug for drinking tea and was gently teased for liking milk in it—the de Nooijs called it "kinderthee." At nine o'clock most nights, someone would take out a radio from its hiding place behind a cupboard and they would listen to Radio Orange on the BBC. Before they all said good night, a member of the family would read several chapters from the Bible. Those evenings, Hackett recalled, were suffused with "peace and industry and contentment. . . . It never occurred to me to give much thought to the strange and dangerous circumstances in which it was all happening. This was now my life; it had become for me the norm."

On November 5, Hackett celebrated his thirty-fourth birthday with the de Nooijs. "It would be a terrible thing," Marie told him, "for anyone to spend his birthday among strange people, far from home, and nothing be done about it all." Awakened at six on the morning of his birthday, he was told to leave his bedroom door open so that he could hear what was going on downstairs. In the parlor below, the family gathered around a small organ and sang, in English, all the verses of "God Save the King." They had wanted to celebrate in the evening, Aunt Ann told him, but that was when the Germans were out in force on the streets, and "it would not have been wise to arouse their curiosity with the British national anthem." A few hours later, all the de Nooijs came to his room, bearing coffee and a huge apple cake made from prewar flour that the aunts had been saving for a special occasion. Atop the cake was a small painted Union Jack that bore the words "Right or wrong, my country."

After the impromptu party was over and the family had left, Hackett broke down and cried. "Such loving kindness to a stranger in adversity, on whose behalf these people had already accepted so many dangers with such modesty and courage, was a thing beyond words then—and never to be forgotten afterwards," he later wrote. That evening he went downstairs to spend a couple of hours with those he now regarded as "my family."

For the four months he stayed with the de Nooij sisters, Hackett never stopped marveling at the willingness of these otherwise quiet, gentle women to defy the Germans. Early in his recuperation, he had trouble sleeping, which was made worse by the barking of a large Alsatian dog roaming around the backyard of the German military police office. When he mentioned the problem to Aunt Ann, she marched over to the house and confronted the head of the police detachment. "Someone in my house around the corner is very sick," she said. "This person cannot sleep because of that dog of yours and the awful noise it makes all night. Will you please have the goodness to see that at night it is kept locked up?" The startled German nodded, and the barking stopped.

As part of Hackett's recovery, one or another of the sisters would take him for walks in the early evening to build up his strength and stamina. These strolls past "the tidy gardens and prim, dignified little houses of the older part of town" were the highlight of his day, he recalled. Yet though he enjoyed the sight of "the steep gables, the snow on the ground, the gentle mist of winter twilight," he constantly worried about being stopped by one of the many Germans who brushed past him and whichever sister was accompanying him that night.

The women, however, never seemed to notice or care. Aunt Ann, in fact, seemed to go out of her way to court danger. During one evening stroll, she and Hackett walked to the post office in Ede so she could mail letters warning other town residents of the dangers of collaborating with the Germans. More than a dozen German soldiers were lounging in front of the building, smoking and talking. Hackett nearly fainted when Aunt Ann, her arm linked with his, pushed her way through the soldiers and deposited the letters in the mailbox outside. After apologizing to the Germans for disturbing them, she and Hackett continued their walk. Years afterward, he still had difficulty understanding how "this mild and unassuming woman" could be bold enough to "move straight into the eye of danger with someone at her side whose presence was her death warrant if he were discovered—even using him to help carry letters almost as lethal to her if she were found out."

Buoyed by the sisters' example, Hackett decided to join the resistance himself. He and Johan started their own underground newspaper, a single-sheet mimeographed weekly that they called *Pro Patria*. Boasting a circulation of two hundred, it focused on the news of the war and the situation in the Netherlands. Hackett, who was the paper's military correspondent, wrote a column under the headline "Notes on the War in the West." *Pro Patria* was published for nearly a month, until the Gestapo in the area began to take an interest in it and Hackett and Johan had to shut it down.

As cozy and sheltered as Hackett's life was with the de Nooij family, the harsh realities of war increasingly intruded. In the early days of the war, the family had laid in a store of food as a hedge against shortages, but those reserves were almost gone. They continually pressed on Hackett small luxuries—an occasional egg, a spoonful of jam—that they didn't allow themselves. He protested, but with no success. "When these ladies had once made up their minds about something, there was little more to be said," he noted.

When Hackett asked the sisters where they got the rations to feed him, they explained that the Dutch resistance, as part of their program to protect underdivers, had provided ration cards, either forged or stolen, for him. But by late 1944, there was no longer enough food to meet the monthly ration, and Hackett, like the Dutch, was never without a nagging feeling of hunger. Although the food shops were virtually empty, people stood in long lines to get whatever they could. In downtown Ede, central kitchens were organized to provide town residents with a daily half liter of cabbage and potato stew per person. "One or another of our household stood in that queue with a pot every day," Hackett said. "After a time, that, too, stopped." Even when they did get a bit of food, there was no gas to cook with. Coal also had disappeared. In the frigid winter of 1944–45, the only heat in the de Nooij house came from a wood-burning stove in the parlor, fed by several cords of wood that Hackett himself had chopped. To wash in the morning, he had to break a thin layer of ice in the water pitcher in his room.

In the months after Arnhem, the Germans' crackdown on the Dutch showed itself in other ways. German police and soldiers made more frequent visits to the houses in Ede, snatching up food, woolen goods, furniture, china, glassware, bicycles, skates, and anything else that struck their fancy. Farmers outside the town lost their cattle and workshops their machinery. When the Germans came calling on the de Nooij sisters, however, they repeatedly met with failure. On one occasion, Aunt Cor feigned a fit of hysterics. Hearing shrieks downstairs, Hackett looked outside his bedroom window to see two German soldiers "almost slinking away from the door. A cloud of defeat brooded over their heads."

But the sisters couldn't expect to get away with such theatrics forever. Even more worrisome, the Germans were stepping up their searches for underdivers in the area. Although the de Nooij house had not yet been a *razzia* target, a hiding place, fitted with a trap door, was constructed under the top-floor landing, and the sisters and Hackett conducted daily drills to get him hidden as quickly as possible in case of a search. Knowing that his presence was putting in mortal danger these people he had come to love, Hackett was haunted by the fear of what would happen to them and other Ede residents if he were captured: "the searches, the reprisals, the taking of hostages, and the summary punishments, all too dreadful to think of."

In January, "Bill" Wildeboer, the leader of the resistance in Ede, came to the house to tell Hackett that rumors had reached the Germans of a British paratroop general on the run somewhere near Arnhem. But

Wildeboer also brought with him the possibility of escape: he mentioned that dozens of other Arnhem survivors, hidden in nearby villages and farms, had already been spirited away to freedom. It was time for Hackett, now almost fully recovered from his wounds, to follow their example.

IN MID-SEPTEMBER 1944, U.S. troops had crossed the Belgian-Dutch border and liberated the three southernmost provinces in Holland. Though the rest of the country, which included all its major cities, remained in German hands, the freed territory provided a base for an organized rescue effort of the hundreds of paratroopers who remained in hiding around Arnhem. The effort, called Operation Pegasus, was put in motion by MI9.

Actually, a few evaders had gotten away soon after the Arnhem debacle, thanks to resistance members who had taken them individually down the Waal River to the liberated provinces in the south. After MI9 found out about the improvised escape line, Airey Neave was dispatched to southern Holland to orchestrate a much larger effort. As it happened, MI9 already had an operative in the occupied part of the country, who was to contact the hidden paratroopers and organize their evacuation. He was Dignus "Dick" Kragt, a British subject with an English mother and Dutch father, who had been parachuted into Holland in 1943 to set up an escape line for downed Allied fliers. The line, which ran from the Dutch town of Apeldoorn to Brussels, had been used for the rescue of more than a hundred airmen by the time of Market Garden.

A joint effort of the Dutch resistance, MI9, and the British and U.S. military, the first phase of Pegasus was launched on the night of October 22, 1944. Throughout the Arnhem area, small bands of British paratroopers, totaling 138 men, stole away from the farmhouses, barns, chicken runs, and other places that had served as their hideaways and quietly followed Dutch guides to a central meeting place. There they were loaded onto trucks and taken to a forest about three miles from the banks of the Waal. The Germans had stepped up their patrols in the area, and the walk to the river seemed an eternity to the paratroopers, especially when they reached the end of the forest and had to follow a drainage ditch across an open field.

Thanks in large part to the skill of their guides, they made it to the Waal without incident. After flashlight signals were exchanged across the river, the men were loaded onto rubber boats manned by soldiers from the U.S. 1st Airborne Division and ferried across. A few minutes later, the first paratroopers to reach the other shore were welcomed back to freedom by Airey Neave. Later that night, MI9 sent a message through the BBC to Bill Wildeboer in Ede: "Everything is well. All our thanks."

The success of Pegasus's first mass escape attempt, however, led to the failure of the second. Having learned of the operation, a newspaper reporter in London wrote a story about it. Now tipped off about the escapes, the Germans greatly strengthened their patrols on the Waal. Neave and his colleagues debated about whether to proceed with Pegasus II but finally decided to go ahead. On the night of November 18, another 150 Arnhem survivors headed for the river, but this time they were ambushed by German security police. Several in the party were killed or wounded, among them a number of resistance members. This time only five paratroopers made it to freedom.

Though the ambush caused the cancellation of any more large-scale rescues, the British and the Dutch resistance continued with individual attempts, in some cases using canoes down the Waal. In the early months of 1945, forty more paratroopers were ferried to freedom. The British were particularly eager to engineer the escape of Shan Hackett, who had been too badly injured to join the earlier escape parties.

Hackett's departure from Ede was set for January 30. On his last evening with the de Nooij family, they carried on with their usual evening routine—playing chess, reading, sewing, listening to the BBC—all the while trying to keep their emotions in check. Before saying good night, he told the family to listen closely to Radio Orange every night from February 7 onward. When they heard the message "The gray goose has gone," they would know that he was safe and free.

Later, as he packed his few belongings, Hackett looked around the spare little bedroom that had served as his refuge—at the lace curtains, the nightstand holding the English-language books that the sisters had found for him, the white counterpane on the bed, the needlepoint sampler of Sleeping Beauty on the wall. He was happy to be going home, of course, but that joy was balanced by a "heavy stone of sadness." Unlike most Britons, he had come to know firsthand what it meant to live in an enemy-occupied country, to understand and share the privations and dangers, the hopes and longings of the people imprisoned there. Having been part of that life, however briefly, he had forged a bond that would never be broken.

Hackett's mind kept returning to a verse from the book of Matthew in the Bible: "I was hungry and you gave me meat; I was thirsty and you gave me drink; I was a stranger and you took me in." The de Nooij family had done all that and more for him. In the process, they had bestowed on him something "rare and beautiful—an [example] of kindness and courage, of steadfast devotion and quiet selflessness." He had often seen bravery in battle. Now he also knew "the unconquerable strength of the gentle."

Early the next morning, Hackett, accompanied by Johan, bicycled away from Ede. He was dressed in one of Johan's old suits and carried false papers identifying him as Mijnheer van Halen. On his jacket he wore a button signifying that he was deaf, so that if stopped by a German patrol, he would have a legitimate reason for not answering any questions. Hidden in his small bag were copies of the three editions of *Pro Patria* and a letter from the de Nooij family to Queen Wilhelmina, which, Hackett later wrote, "expressed their loyalty, trust, and affection for her."

On their seven-day journey, he and Johan were passed from one safe house to another. For much of the way, they were accompanied by guides from the resistance. "It was like being a child again, led by the hand in a crowd," he observed. "I had neither the power to influence events nor the curiosity to inquire into their nature. I was content to be carried along."

In one farmhouse where they stopped, he changed into the tattered, blood-stained remnants of his old uniform, complete with paratrooper badge and battle ribbons. In his pockets, he placed his British Army identity card and his identity documents as Mijnheer van Halen: "I was still both of these, but I felt myself growing hourly more and more the first, less and less the second."

On the night of February 5, Hackett waited on the banks of the Waal many miles downstream from Ede. A heavy fog was swirling, and the wind blew in great gusts. Suddenly he saw a number of dark figures emerge from the fog. "Good luck," a woman whispered in English. A man shook Hackett's hand while another murmured in Dutch, "Good luck, Englishman." A second woman felt for his arm, then put a package into his hand. "Look, here are biscuits for your journey."

A boatman took him to a canoe, and the two embarked on a silent, tense journey down the Waal. Several hours later, as dawn was breaking, they tied up at the little river port of Lage Zwaluwe, in liberated Holland. After Hackett, shaking with cold, clambered out of the canoe, he heard a cheerful, British-accented voice say, "Hullo, Shan." It was Tony Crankshaw, an old friend and an officer with the 11th Hussars Regiment. "We've been expecting you," Crankshaw said. "Have a drop of brandy."

Led to a house filled with men in khaki battle dress and a great deal of tobacco smoke, Hackett collapsed in a chair, "surrounded once more by the familiar and comfortable jumble of the British army in the field." The next day, he was summoned to dinner at Montgomery's headquarters, where he was plied with oysters and wine, then put onto a plane to England.

One of the first things he did after arriving home was to place a call to the BBC in London. That night, as the de Nooij family listened to Radio Orange, they heard the message they had been eagerly awaiting: "The gray goose has gone."

IN THE AFTERMATH OF ARNHEM, the general in direct command of the operation, Frederick "Boy" Browning, was awarded a knighthood, an action that stunned U.S. lieutenant general James Gavin, the commander of the 82nd Airborne Division, which had also fought in Market Garden. Browning "lost three-quarters of his command and a battle" but "returned home a hero and was personally decorated by the

King," Gavin remarked. "There is no doubt that in our system he would have been summarily relieved and sent home in disgrace."

If Browning and his subordinates weren't to be blamed for the fiasco, they had to find a scapegoat. Their choice was General Stanisław Sosabowski, who had been skeptical about the operation from the outset. "The worst thing that a subordinate can do is to question orders and to be proved right," the historian Michael Peszke has noted. "Sosabowski's independent attitude and the fact that all his original warnings were proved correct made him the obvious target." The fact that he was a foreigner contributed, too. However specious the reasons for the attacks on him, Sosabowski, who had lost most of his troops at Arnhem, was relieved of his command.

It was only in October 1944, after the Market Garden disaster, that the battle to clear the estuary at Antwerp finally began. An assault that could have been won with minimal casualties instead took eighty-five days and cost the Allies a total of 30,000 men. The war on the western front, meanwhile, slipped into a stalemate. Reinforcing their defenses, the Germans dug in deep and held the line in the forested hills separating their homeland from the rest of western Europe. "Between our front and the Rhine," General Omar Bradley remarked, "a determined enemy held every foot of ground and would not yield. Each day, the weather grew colder, our troops more miserable. We were mired in a ghastly war of attrition."

The failure to liberate the Netherlands also meant that Hitler could launch his V2 campaign against London with impunity in September 1944. When the Germans lost France, they moved the V2 rockets based there to sites near The Hague and other Dutch cities, all of them within two hundred miles of London. The V2 launching areas remained in German hands throughout the winter, and Londoners continued to see their homes devastated by the new terror weapon. But they weren't the only ones who suffered; Brussels and Antwerp were also hit hard by the V2 rockets.

Antwerp was a particular target because of its port. On December 15, 1944, a V2 slammed through the roof of a crowded 1,200-seat movie theater in the city's downtown. For a week, rescuers used cranes and bulldozers to clear the rubble and reach the dead and injured. A rescue crew freed an American soldier, who stumbled out of the wreckage carrying two dead children in his arms. He had been sitting next to their mother, whose head had been severed in the blast. Nearly six hundred bodies were finally recovered, more than half of them Allied soldiers and sailors.

Overall, more than four thousand Belgians died in V2 attacks. In greater Antwerp alone, more than sixty-seven thousand buildings were destroyed, including two-thirds of all housing in the city.

PRIVATIONS AND SUFFERING WERE also in store for the Dutch, who had dared to side with the Allies in what they hoped would be the liberation of their country. When it appeared in September that all of the Netherlands would be freed, Radio Orange broadcast an order from SHAEF to Dutch railway officials to halt all rail service "in order to hinder enemy transport and troop concentrations." As one historian noted, "it was the most important act of defiance to the Nazis the Dutch were ever asked to make."

The order took everyone in Holland by surprise. It did the same to the Dutch Cabinet in London, which had not been informed of SHAEF's action. Prime Minister Pieter Gerbrandy, the only Dutch official who had seen and approved the order, was not concerned about its possible consequences. "Don't worry," he told an associate. "On Saturday, we shall be in Amsterdam."

That did not happen, of course, and the repercussions were frightful. More than 90 percent of the 30,000-man Dutch railway force had obeyed the summons to strike, a stoppage that halted not only the transport of German soldiers but also all food and coal supplies to Amsterdam, The Hague, and the Netherlands' other major cities. In retribution for the strike, the Germans embargoed all shipping on Dutch waterways—the only other method of moving food and fuel.

While the French and Belgians continued to celebrate their liberation, the Dutch, who had come so heartbreakingly close to freedom, were now facing famine.

CHAPTER 24

The Hunger Winter

The Netherlands'
Looming Destruction

ONLY FORTY MILES SEPARATE THE DUTCH CITY OF BREDA from The Hague. Nijmegen lies just fifty-five miles from Amsterdam. Yet in late 1944 and early 1945, those cities, despite the short distances between them, might as well have been on opposite sides of the moon.

Breda and Nijmegen were located in two of the three freed southern provinces of the Netherlands. The Hague, Amsterdam, and the country's other large population centers were all in the northwest, which was still occupied by the Germans. The area would remain under German control until May 4, 1945, only four days before the official end of the war in Europe. Once Market Garden failed, the Allies had no immediate interest in liberating the rest of Holland, and the Nazis were free to take their revenge on the Dutch for the pre-Arnhem railway strike and myriad other acts of resistance.

In the days immediately after their victory at Arnhem, the Germans blew up the ports of Rotterdam and other cities and flooded thousands of acres of farmland. The leaders of the railway strike were imprisoned, and several were killed. From then on, every act of Dutch rebellion, however small, was met with mass executions. When resistance fighters ambushed and severely wounded General Hanns Rauter, the ruthless head of the SS in Holland, more than three hundred Dutch citizens lost their lives in retaliation. In Amsterdam, twenty-nine young Dutchmen were executed on a garbage dump in the center of the city, and several buildings were set afire.

SS murder squads kept busy through the last months of the war, shooting hostages on street corners and in the central squares of Holland's major cities. "You saw them lying everywhere in groups of twenty—and they left them there as a warning," one Dutchman observed. Another wrote of the sprawled bodies he witnessed, "You look and feel punch drunk, shattered, and you don't know what is stirring in your soul."

Terror came to the Dutch countryside, too. After a skirmish between four German soldiers and a group of resistance fighters near the small town of Putten in central Holland, hundreds of German troops surrounded the village. Its male residents, more than six hundred in all, were dispatched to German concentration camps; fewer than fifty survived the war. The women and children were sent away, and most of Putten's houses were burned to the ground. When a young Dutchman traveled to the village to look for his parents, all he found was "smoking ruins and deadly silence."

As horrific as their terror tactics were, the Nazis tended to be selective in applying them. No one, however, escaped the extreme hunger caused by the German food embargo that was imposed after the railway strike. Before the embargo, a Dutch citizen's average daily calorie intake was 1,300, less than half of what constituted a normal diet. A month later, the average number of calories consumed per day had fallen to 900.

In this once prosperous country, communal kitchens were set up to feed millions of people with whatever food remained. City dwellers roamed the countryside by the thousands, begging or bartering with farmers. The misery, one observer wrote, was heartbreaking: "People with their feet torn, blood in their shoes. Some had no shoes and wrapped their feet in rags."

In one of the coldest and wettest winters of the century, there was almost no coal, gas, or electricity. To cook and warm themselves, the Dutch had to scavenge wood from anywhere they could find it. Trees in parks, woods, and along once leafy city avenues were cut down; in just three months, Amsterdam lost more than half of its estimated 42,000 trees. Bridge railings disappeared, as did wooden cross ties from tram lines. Abandoned houses were raided for their joists, beams, and staircases.

By the end of 1944, Holland's main cities were landscapes of desolation—their gardens bare; their parks, streets, and canals filled with mountains of garbage; their entry roads barricaded with heavy concrete walls. Dark and deathly silent, they were pervaded, one Rotterdam resident wrote, by "a quiet, oppressive apathy." There was no traffic, no industrial or commercial activity. Schools were closed because of the lack of heating. People took refuge in their homes, huddling together for warmth and avoiding any unnecessary activity so they could conserve their rapidly flagging energy. On New Year's Eve 1944, one Dutchman wondered whether 1945 was going to be "the year of liberation or the year of our death." Another observed in his diary, "For the first time in my life, I went to bed before midnight, glad that this black, disastrous year was over." Early 1945, however, proved even

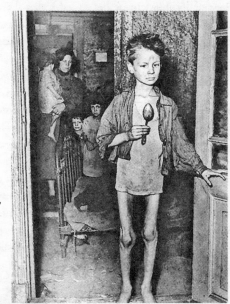

An emaciated boy in Amsterdam during Holland's Hunger Winter.

bleaker than the year before. In January, the daily food ration fell to 460 calories, then to 350 in February.

Outside Holland, the war was drawing to a close. On February 4, Roosevelt, Churchill, and Stalin met at Yalta. The Russians were less than fifty miles from Berlin. None of that mattered to the Dutch, whose only thought revolved around finding something to eat. People fought over anything remotely edible: when a bin full of gruel was accidentally spilled at a central kitchen in Rotterdam, bystanders scraped it off the street. Sugar beets and dried tulip bulbs became diet staples. Although they tasted dreadful and had no nutritional value, they filled the stomach and lessened to a small degree the always gnawing hunger. "By March, our faces were pale green, like imitation van Goghs," a Dutch painter recalled.

Near Arnhem, fifteen-year-old Audrey Hepburn and her mother subsisted on turnips, tulip bulbs, and nettles. "Everyone tried to cook grass," Hepburn later recalled, "[but] I couldn't stand it." In early 1945, she was five feet, six inches tall and weighed less than ninety pounds. The lack of food had made her so weak that she could barely walk, much less dance. Like many other Dutch citizens, she suffered from jaundice, anemia, edema, and other health problems stemming from her severe malnutrition.

But at least she was still alive. By March, thousands of her countrymen had died of starvation. The death rate climbed so swiftly that undertakers could not supply enough coffins for the dead or find enough men to bury them. Emaciated corpses were piled high in hospitals and churches throughout the country. A visitor to a cemetery in Rotterdam noted a row of "shrunken bodies lying next to each other. No flesh on their thighs or calves. Most had bent arms and legs, the hands clenched as if the poor devil was still asking for food."

In Amsterdam, a resident wrote, "My old, beautiful, and noble city is in a death struggle." So, too, was much of the rest of the country. But with a few prominent exceptions, no one outside seemed to care.

IN LATE DECEMBER 1944, Erik Hazelhoff Roelfzema was summoned to a meeting with Queen Wilhelmina in London. Two years earlier, he had joined a Dutch RAF squadron and was now flying bombing missions to Germany, but he remained close to the queen. When he was shown into the sitting room of her house in Chester Square, Wilhelmina looked up from the small armchair in which she was sitting. The young Dutchman knew immediately that something was wrong: "For the first time, in my experience, she did not rise to greet me." With great emotion, she exclaimed, "Have you heard? They're dropping dead in the streets!" Roelfzema, having no idea what she was talking about, stared at her in bewilderment. Gesturing impatiently, the queen repeated, "The people are dropping dead in the streets!" Her visitor shook his head; he was, as he wrote later, "living my self-satisfied life in the RAF," completely unaware that the Netherlands had become a "hell of hunger, terror, and death." The queen refused to accept his ignorance. "Don't you *know*?" she kept insisting. "Haven't you *heard*?" Shocked by her "horror and grief," Roelfzema "thanked God when our meeting ended."

His obliviousness to his country's plight was not unusual. Most people in London, not to mention the rest of the world, had no idea of the starvation ravaging Holland. Determined to focus international attention on what was happening in their nation, Wilhelmina and her prime minister, Pieter Gerbrandy, embarked on a campaign to persuade the Allies to liberate all of Holland immediately or, failing that, to provide direct aid to their starving people.

Gerbrandy, for his part, was responsible for what one historian called "one of the most impressive exercises in public relations in the history of World War II. The Netherlands was put on the map by him." In October 1944, Gerbrandy sought out dozens of British and foreign journalists in London to inform them of Holland's suffering. His efforts resulted in a stream of sympathetic stories, including one in *Newsweek* that began "'Famine, floods, cold, and darkness . . . a people starved, frozen, and drowned.' Pieter Gerbrandy said these were the winter prospects unless the Dutch are liberated this fall."

Having begun to alert the world, the diminutive prime minister now faced a far greater challenge: persuading the Allied commanders to change their strategy in order to save his country. His difficulty was compounded by the enormous psychological and cultural gulf that existed between the high-ranking Allied military commanders and the Dutch and other peoples of occupied Europe.

Allied officers working at General Eisenhower's headquarters in France seemed totally unaware of what the Europeans had to endure. They were ensconced in the five-star Trianon Palace Hotel at Versailles, a wooded area near Paris that had been the apex of French royal indulgence in the seventeenth century and was now transformed into an emblem of Allied privilege. They lived in a world of luxury and comfort, supplied with seemingly endless quantities of U.S. cigarettes and steaks, Scotch whisky, and French champagne. At the Trianon Palace, waiters in black tie served them meals on white linen tablecloths set with crystal stemware and gold-rimmed plates. One American general described hunting partridge near Versailles, with "all the farm hands for miles around acting as beaters."

It was at Versailles that Gerbrandy met with Eisenhower, having written the supreme Allied commander in December about a potential "calamity as has not been seen in Europe for centuries, if at all. . . . It is literally a matter of life and death." Eisenhower listened sympathetically to what Gerbrandy had to say, but he rejected the prime minister's appeal to change the Allied plans and liberate the rest of Holland. He explained that "military considerations and not political considerations" must dictate the Allied strategy. The best service the Allies could render to Holland, Eisenhower insisted, was to defeat Germany as soon as possible. It was the same rationale that had been given to those who, earlier in the war, had wanted help for Poland or assistance to save the remaining Jews of Europe from annihilation.

Perhaps Eisenhower was right. Certainly, many if not most military historians have agreed with his assessment. But it's not surprising that leading Jewish figures and Dutch and Polish officials, among others, were skeptical, believing that the "Defeat Germany first" mantra was a convenient excuse to avoid doing anything that U.S. and British leaders—political and military—had decided would not be to their own or their countries' advantage. In this case, the primary aim of the English-speaking allies was to appease the Soviets, who had been pressing hard for an advance from the west into Germany to relieve the pressure on the Red Army, which had swept across eastern Europe and was already deep into German territory. SHAEF's rationale was particularly hard for the Dutch to stomach, considering that the Allied fiasco at Arnhem had directly led to their country's calamitous situation.

In a letter smuggled out to the Dutch government in exile, a resident of Amsterdam declared that he and other citizens of Holland felt they were being needlessly and selfishly sacrificed. "The Allies are admired," he observed, "but they are also regarded as callous egotists." An Anglophile friend, he added, had exclaimed to him, "To let an ancient and civilized people like ours die without lifting a finger—my God, how can they do it?"

Queen Wilhelmina made exactly that point in personal appeals to Churchill and Roosevelt, who was of Dutch descent himself and who had welcomed Wilhelmina to the White House and his home in Hyde Park in 1942. "I felt as if I was addressing an old friend, so cordial were his feelings for the Netherlands," she noted about her visits with the president. Cordial they might have been—FDR assured her in 1944 that "I shall not forget the country of my origin"—but he also said there was nothing he could do, except to urge Eisenhower to "save food in Germany and keep it for use in Holland."

Churchill, for his part, was stricken by the tragedy taking place in Holland but told the queen he was no longer able to influence the Allied high command, adding "I must leave this to the generals." Earlier, he had told a friend that he was trying "to have Holland cleared up" but "it is not so easy as it used to be for me to get things done."

AS WRETCHED AS THEIR LIVES already were in early 1945, the Dutch were about to experience another jolt of suffering and death, this one directly inflicted by their allies. On the night of March 3, the RAF, alerted by the Dutch resistance, sent more than fifty aircraft over The Hague to bomb V2 launching sites in a forest not far from the city. On their first bombing run, the planes missed their targets and dropped incendiary and high-explosive bombs on several residential areas more than a mile away. Oblivious to their error, they returned for two more attacks on the same locations.

The raids caused the deaths of five hundred people and severe injuries to thousands more, as well as the destruction of more than three thousand houses, one of them the home of the Dutch resistance leader responsible for the tip about the launching sites. More than twelve thousand people were left homeless. Thanks to the bombardment, said a report smuggled out of Holland, "the temper of the civilian population has become violently anti-Ally."

In London, Dutch officials reacted with astonishment, then outrage. No one was more furious than Wilhelmina, who aimed her formidable temper at the British military leaders and Churchill himself. This time the prime minister offered no excuses; he was as angry as the queen about what he called "this slaughter of the Dutch." In a scorching memo to the RAF and the Air Ministry, Churchill demanded a "thorough explanation" of the botched raid. "Instead of attacking these points with precision and regularity," he said, "all that has been done is to scatter bombs about this unfortunate city without the slightest effect on their rocket sites, but much on innocent human lives and the sentiments of a friendly people."

To the Dutch, the British government expressed "deep regrets." According to an investigation, the officer who had briefed the aircrews on the mission had mixed up the vertical and horizontal coordinates of the target. The Foreign Office assured Dutch officials that the guilty officer had been court-martialed, although there is no evidence that anyone was actually punished.

The vehemence of Churchill's remarks reflected a boiling over of his long-simmering indignation at the huge civilian casualties caused over the previous two years by the RAF and U.S. Army Air Forces bombing campaigns in western Europe, particularly in France and the Low Countries. The raids had primarily been aimed at factories turning out German war matériel, V1 and V2 launching sites, and, as D-Day approached, the destruction of much of the French railway network, in order to prevent the transport of German troops to the Normandy beaches. The necessity of those raids was self-evident. In the case of the railways, by seriously damaging rail lines, repair yards, and bridges, the bombing reduced French rail traffic to 30 percent of its normal level, greatly impeding enemy movements. Clearly, collateral damage was inevitable.

What was not so easily understandable was the conspicuous lack of care that the Allied air forces took to limit that damage, often strewing bombs over city centers from high altitudes—bombs that fell nowhere near their military targets. As the raids intensified, so did the civilian casualties. In March and April 1945, four bombing runs on railway yards in Paris left more than 1,100 Frenchmen dead. On May 26, raids on railway facilities in ten French cities caused almost six thousand casualties. To many Frenchmen, "the Anglo-Americans seemed more capable of bombing France than liberating her," the historian Julian Jackson observed.

Not infrequently, the raids focused on targets of no real military value. When the U.S. Army Air Forces bombed the railway passenger station in the southern French town of Avignon, Francis Cammaerts "wanted to shriek out loud—how many of us are you going to kill before it's over!!! They simply didn't know a German troop train carrying tanks would never go through a passenger station!! Or anywhere near it."

In Belgium, the bombing runs were directed mostly at urban factories, with considerable collateral damage in nearby working-class neighborhoods. The killing of 209 children and 727 other civilians in an April 1943 raid on a factory complex in Antwerp prompted especially vigorous protests by the Belgian government in exile and resistance leaders.

Churchill had long complained to Allied military leaders about what he saw as their lack of concern in limiting civilian casualties. "Terrible things are being done," he wrote in May 1944. "This thing is getting much worse." The prime minister disputed the claim of SHAEF officials that they had chosen the best targets and warned that "you are piling up an awful load of hatred." He told his War Cabinet that he "had not fully realized that our use of air power would assume so cruel and remorseless a form." But his objections found no favor with SHAEF or Roosevelt, who wrote the prime minister that although he certainly hated the idea of inflicting severe casualties, he was "not prepared to impose any restriction on military action by responsible commanders."

WHILE THE DUTCH CONTINUED to suffer in late 1944, the Allies made several fruitless attempts to penetrate deep into Germany. Although Allied troops had first crossed the German border in October, every successive effort that year to advance more than a few miles into enemy territory ended in failure. In late December, the Germans, in a desperate attempt to regain the offensive, launched a massive assault against U.S. forces through the Ardennes Forest in Belgium. Known as the Battle of the Bulge, the German thrust resulted in the largest and most savage fight on the western front. The attack ultimately failed, but before it did, it claimed more than 100,000 U.S. casualties.

As costly as it was for the Allies, the Battle of the Bulge marked the beginning of the end for the Reich. In early 1945, the Allied armies resumed their march eastward and, by early March, began crossing the Rhine and pouring into the German heartland. Once again, their slog had become a gallop. In the words of Rick Atkinson, "The inner door to Germany had swung wide, never to be shut again."

Yet as Germany's war effort disintegrated, Hitler's recalcitrance increased. He not only refused to stop the fighting in his own country; he also threatened the wholesale destruction of western Holland, one of the last areas of western Europe the Germans still held. (The others were Norway, several French ports near the English Channel, and the Channel Islands.)

Under a Hitler directive, the Reich prepared for a final stand in what was now called "Fortress Holland," with German troops instructed to "fight to the last man and the last bullet." Orders were given to ready the demolition of all electrical plants, gasworks, bridges, railways, and, most deadly of all for Holland, its dikes. If the dikes were blown, the nation would be inundated by water within three weeks, resulting in a calamity of unimaginable proportions.

On April 17, the Germans gave the Dutch a glimpse of the frightful future in store for them when they destroyed a dike protecting a huge swath of fertile agricultural land in the north of the country. More than 50,000 acres were flooded, with dozens of farms and roads wiped out. More than twenty residents were shot as they tried to escape the floods.

In other parts of western Holland, the death toll from starvation continued to mount. At the same time, new life emerged in the form of the spring flowers that were blooming everywhere. "One has nothing to eat, misery has hit rock bottom, and one approaches the stage of indifference," an Amsterdam resident wrote. Yet "we buy flowers and we put them in our rooms, on the window sill. We salute them as they grow festively among the swollen bodies of edema victims, among the emaciated children and the garbage heaps in the parks."

With Holland's very existence threatened and thousands of its people dying, Churchill had finally had enough of SHAEF's excuses for not coming to the country's aid. In his efforts to force the Allied military to intervene, he received assistance from an unlikely source: Arthur Seyss-Inquart, the Nazi chief of the Netherlands. Realizing that the end of Nazi rule was near and hoping to save himself, Seyss-Inquart told Allied authorities in late March that he would allow them to provide aid to the Dutch.

Even so, it took four weeks for the Allies to prepare a relief operation. Before his sudden death on April 12 in Warm Springs, Georgia, Roosevelt had agreed to the plan but insisted that the Soviets must approve any negotiations with the Germans to allow emergency food supplies to be sent. Subsequent talks with the Soviets and then with Seyss-Inquart dragged on for days.

Finally Eisenhower himself balked at the delay. Having changed his mind about helping the Dutch, the supreme commander urged the Combined Chiefs of Staff to forget the red tape and allow him to launch the relief effort. At long last, they gave in, telling him in an April 24 cable that they had "decided to leave the matter in your hands."

ON APRIL 27, ELEVEN DAYS before the official end of the war in Europe and after more than twenty thousand Dutchmen had died of starvation, the Allies launched what the British called Operation Manna, a massive airlift of food to Holland. The initial drops were carried out by the RAF, whose pilots had been pressuring their superiors for days to speed up the process. At one airfield, RAF crews marched to their commander's office, chanting "The Dutch must get this food—the Dutch *will* get this food."

On the night of April 26, volunteers braved hail and rain to load more than six hundred tons of food—flour, corned beef, powdered eggs, coffee, tea, and chocolate, among other items—into 263 bombers based at various British airfields. The workers included crews just returned from bombing missions over Germany. One of them, a pilot, put a note in a food pack that read, "To the Dutch people. Don't worry about the war with Germany. It is nearly over. These trips for us are a change from bombing. We will often be bringing new food supplies. Keep your chins up. All the best. A RAF man." The following morning, the bombers set off on what, for them, was a unique mission: saving lives rather than ending them.

Thanks to thousands of posters tacked up all over Holland, the Dutch had been alerted to the food drop. As bombers swooped low over cities, towns, and the countryside, their astonished crews saw masses of cheering, waving people everywhere they looked—on red-tiled roofs, in fields, on country roads, and in city streets. "An old man on a bike waved so passionately that he almost fell off," recalled a British journalist aboard one of the planes.

In The Hague, a resident reported that he and his neighbors "ran outside with hats, shawls, flags, sheets, or anything else that we could wave at the planes which were thundering over our streets in an interminable stream. In a flash, our whole quiet street was filled with a cheering, crying, waving crowd, and the elated people were even dancing on their roofs." A Dutch official later observed, "The emotion and enthusiasm were so tremendous that one forgot one's hunger." Another declared, "If any emotions could still stir our blunted feelings, it was these generous gifts of those who were recognized as our friends in the moment of our greatest distress."

The airlift continued for more than a week, with some five hundred British and three hundred American aircraft dropping almost eight thousand tons of food. Dutch resistance fighters and former members of the Dutch army took charge of its distribution. For the most part,

the Germans lived up to their agreement not to interfere with the dropping and collection of supplies. Even more important, they ceased all
military operations in Holland. For the Dutch, the war—finally if
unofficially—had come to an end. "Fear was finished, and death has
fled," an Amsterdam resident wrote in his diary. Another noted, "We
are no longer isolated from the world. The Dutch prison door has been
rammed open."

On May 4, a little over a week after the airlift began, German commanders in Holland, northwest Germany, and Denmark officially surrendered to General Montgomery at his new forward headquarters
near the German city of Hamburg. Shortly before nine o'clock that
night, a stammering announcer on a Dutch resistance radio network
broke into regular programming to announce that Holland was now
officially free. "Long live the queen!" he exclaimed. "Long live victory!"

For the second time in ten days, the country went mad with joy. "I
saw people dancing in the street; they were jumping up and down," a
man in Rotterdam recalled. "Honorable burghers who would never
lose their composure, and certainly would never run, were now racing
around like boys, hugging each other, throwing their hats in the air."
In The Hague, a teacher who had spent the last two years in hiding
ventured outdoors for the first time to join his neighbors in their celebration. "It gives me a shock to see them," he wrote in his diary. "Some
are so thin that I hardly recognize them. . . . It is terrible—their pale,
emaciated faces—but the joy shines in all their eyes, the happiness for
our newborn freedom."

In another Hague neighborhood, residents abruptly stopped their
merrymaking when they heard over a radio in a nearby window the
strains of the "Wilhelmus," the Dutch national anthem, whose playing
had been forbidden by the Germans. Their voices trembling, a few
people began to sing, with others joining in: "Wilhelmus of Orange
am I, of Dutch blood; true to the fatherland am I till death." Composed in the sixteenth century during Holland's eighty-year fight for
freedom against the Spanish, the anthem, in the words of the Dutch
writer Henri van der Zee, "expressed the longing for freedom that our
forefathers had felt." The solemnity was shattered a few minutes later,

Dutch crowds greet their Canadian liberators with flowers in May 1944.

however, when a gramophone down the street started playing "When the Saints Go Marching In." Van der Zee, who as an eleven-year-old boy took part in the celebration, noted that the title of the jaunty American jazz tune reflected how the Dutch "felt about the Allies. I, who had never heard it before, listened spellbound while some people began to dance."

All over the country, there was dancing that spring, mostly to the music that the Canadian and British troops who now marched through the Dutch streets brought with them. Loudspeakers on street corners played "Moonlight Serenade," "Chattanooga Choo Choo," "The White Cliffs of Dover," and "Don't Fence Me In." The soldiers taught Dutch teenagers to dance the swing and handed out gum and candy to the children. In return, a British war correspondent wrote, "We have been kissed, cried on, hugged, thumped, screamed, and shouted at until we are bruised and exhausted. The Dutch have ransacked their gardens, and the rain of flowers which falls on the Allied vehicles is endless."

All the festivities, however, couldn't mask the fact that Holland was a terribly blighted country and thousands of its people were still dying. General Alexander Galloway, the head of the British military's relief effort, saw this for himself during a nationwide tour. "On first appearance," he reported to his government, "the condition of the people has proved unfortunately very deceptive. Allied soldiers were greeted with cheers and bunting, and made their progress through a smiling countryside. But it was deceptive because men and women who are slowly dying of starvation in their beds cannot walk gaily about the streets waving flags." He went on, "It is an empty country, inhabited by a hungry, and in the towns, a semi-starved population."

Two journalists who had just arrived in Amsterdam were besieged at their hotel by dozens of emaciated people begging for food. A doctor they interviewed told them that at least thirty thousand Amsterdam residents were close to death. In The Hague, the wife of the British ambassador to Holland, who had returned to his post after five years of war, reported to Winston Churchill about her visit to one of the hospitals there: "The babies were tragic. They looked like old men—or else they lay in a semiconscious state. . . . Most of the cases we saw had very distended stomachs, but no fat at all on their arms and legs. Many of them were bleeding at the feet." Her husband, Sir Nevile Bland, added his own postscript: "There is no possible doubt that undernourishment is universal and starvation and semi-starvation lamentably widespread."

Yet bad as the situation was, the worst was over. Thanks to a massive infusion of medical aid and food from Britain, Sweden, Switzerland, and other countries, hundreds of thousands of Dutch citizens were nursed back to health over the next weeks and months. One of them was Audrey Hepburn, who would live with the aftereffects of the war for the rest of her life. Never weighing more than 110 pounds, she continued to suffer from some of the health problems she had developed during the "hunger winter."

That, however, was all in the future: for now, she lost herself in the euphoria of the moment. Her older brother, Alexander, emerged from hiding as an underdiver, and her other brother, Ian, who had been forced into slave labor in Germany, walked 325 miles to return home to Arnhem. "We had almost given up when the doorbell rang, and it was Ian," she recalled. "We lost everything, of course—our houses, our possessions, our money. But we didn't give a hoot. We got through with our lives, which was all that mattered."

EVER SINCE SHAN HACKETT had returned to England in early February 1945, he had kept an anxious eye on the events unfolding in Holland. During that period, he had spent much of his time briefing the War Office and other government ministries about the events at Arnhem and the contributions of Dutch resistance fighters and other civilians in saving his life and those of other British soldiers. But his main focus was on seeking aid for the Dutch. In doing so, he was greatly troubled by what he saw as Whitehall's remoteness, its seeming unconcern, and its lack of understanding about what the Netherlands was enduring.

In late April, the British brigadier received the news for which he had been waiting: Ede and the area around it had been liberated. He arranged to hitch a ride on the next RAF cargo plane to Holland and hurried home to gather the bundles of goods he had collected for the de Nooij sisters and their family: "packets of tea (real tea!), coffee, sugar . . . tinned goods, clothes and other presents—and the precious letters from my family at home to my other family in Holland."

When he landed the next day at a military airfield in central Holland, a British Army car was waiting for him. In less than an hour, Hackett was back in Ede, his joyous mood matched by the weather. When he had left three months before, it had been gray and cold; now the sun shone, the trees were leafing out, and flowers were everywhere. But, as he noted, the changes went far deeper than the passing of winter: "A leaden pall of mourning had lain over this town when last I saw it. Now everything was as gay as a village wedding." Dutch flags were intertwined with Union Jacks in store windows. Streets he remembered as mostly deserted were now thronged with people of all ages, "looking about them as though they had never seen the place before, smiling, laughing, shouting to each other." Along the way, he recognized familiar landmarks: the church with its high steeple, the houses he had passed on his strolls with the aunts, and, most memorable, the post office where he and Aunt Ann had pushed past German soldiers to post her potentially deadly letters.

"With the certainty of a sleep walker," he directed the army driver down Ede's high street and into the narrow roadway where the de Nooij sisters lived. He got out with his packages and stared for a moment. Before him was the white fence "whose gate I had opened so often, the shape of whose latch I can still feel in my fingers. There was the little house, with the tidy curtains to the sitting room below." And there was Aunt Mien, standing in the doorway and smiling broadly. "There was no surprise upon her face as she came to meet me, only shining happiness, but she was in tears and laughter at the same time as I embraced her." Then the others crowded around—Aunt Ann, Aunt Cor, Marie, Johan—"everybody laughing and crying and talking at once."*

Rummaging among his packages, Hackett retrieved the coffee, and Aunt Mien whisked it away. Soon they were all sitting around the big kitchen table, savoring it—Hackett drank from the mug that had been his months before—and the little cakes he had also brought.

"Did you get my message?" he asked. "Did you hear about the gray goose?"

"Three times," Aunt Ann said. "We were so pleased and thankful." With a laugh, Aunt Mien chimed in, "We danced a jig around the table, every time! I do wish you could have seen us. Oh, dear!"

For Hackett, the rest of the day passed in a happy blur. Like a small boy, he embarked on a thorough exploration of the house. Everything was neat and spotless, just as before; the only difference he noticed was the location of the radio. Once hidden behind a cupboard, it now stood boldly on a table in the sitting room. On the upstairs landing, he lifted

---

* As General Galloway, the British relief official, indicated, the country towns had suffered less from the famine than the cities had; though much thinner than the last time Hackett had seen them, the de Nooijs were not starving or close to it, as so many other Dutchmen were.

the carpet and trap door beneath, then climbed into the hiding place the family had created to protect him from the Germans. In the barn, he picked up the ax and saws he had used to cut wood, feeling again "the sharp cold on my hands and smelling the fresh pine sawdust."

Within minutes, word of Hackett's return had spread through Ede, and a stream of visitors began to arrive: the pastor of the church where the family worshipped, the Dutch doctor who had treated his wounds, members of the resistance who had helped him escape. Later that evening, Hackett remembered, "we sat down together to supper again as one family and read the Bible again together after it. Everything was just as it had been before—but somehow a hundred times better."

At one point, he noticed Aunt Mien gazing intently at his smart paratrooper's uniform. "What is it?" he asked. "I had a wish," she answered, "and it came out all right. I couldn't tell you, of course. It wouldn't have come true then."

Hackett remembered how, during the family's modest little New Year's Eve celebration four months before, Aunt Mien had told everyone that they should wish "for what we wanted most in the New Year." As she had said that, she had stared intently at Hackett and the darned black jacket he was wearing. At that moment, "she looked up and our eyes met," he recalled. "With a tiny flash of an almost guilty smile, as though she had been found out, she looked away."

Now, on that lovely spring evening, he took her hand. "I think I knew what it was even then," he said. "It was to see me back here in uniform, with all that would mean."

"Yes, that was it." She smiled. "And here you are."

That night, Hackett slept soundly in the familiar little bedroom upstairs, with its treasured emblems of the love and security he had found in this house: the lace curtains, the neat white counterpane on the bed, and Sleeping Beauty on the wall.

CHAPTER 25

"There Was Never a Happier Day"

Coming Home

O N APRIL 26, 1945, QUEEN WILHELMINA RETURNED TO the Netherlands after nearly five years in exile. Her arrival was almost as unceremonious as her abrupt departure in May 1940. When she landed at an RAF airfield in the southern part of the country, there were no bands, no cheering crowds, no massed troops to greet her— just a small British honor guard and her son-in-law, Prince Bernhard.

Liberation for most of Holland lay more than a week away. The massive airdrop of food would begin the next day. But even if she could venture no farther than the three liberated Dutch provinces, the queen decided she had been gone long enough. Her tiny retinue consisted of only four people: her daughter, Princess Juliana, who had just returned from Canada, and three aides, all of whom were *Engelandvaarders*. They were Erik Hazelhoff Roelfzema and Peter Tazelaar, the dashing, hell-raising Dutchmen she had befriended when they had escaped to London in 1941. The third was a young secretary named Rie Stokvis.

The composition and small size of Wilhelmina's entourage were her idea. They were, she later said, "in keeping with the changed times. . . . [I wanted] to surround myself with people who had taken part in the resistance or had been *Engelandvaarders*—'nobles' in the best sense of the word." According to Roelfzema, "Rie, Peter, and I were symbols. We had . . . active war records, were young and of common birth.

Nothing about us recalled [the queen's] former, highly formalistic, and class-conscious surroundings, against which she nursed bitter resentment."

After Wilhelmina's RAF plane landed on that rainy spring morning, Roelfzema and Tazelaar jumped out onto the airfield's wet grass, helped push a mobile stairway into place for her and Juliana, then stood at attention on either side of the stairs. A few minutes later, the queen, dressed in a brown tweed suit, hat, and boots, appeared in the plane's door. Taking a deep breath, she paused for a moment and looked around. Roelfzema held out his hand to help her down the steep steps, but "she ignored it pointedly," he recalled. "Her first step back on Holland's soil after five years of exile, leaning on someone else? Unthinkable!"

For the next three weeks, Wilhelmina and her minuscule court took up residence in a small, stately manor house in the countryside, a few miles from the town of Breda near the Dutch border. The house, called Anneville, lay at the end of a row of beech trees. It had a curving drive, wide lawns, enormous shade trees, mounds of flowering rhododendrons, and a pond full of croaking frogs.

Although there was no official announcement of Wilhelmina's return, the news rapidly spread. On the night of her arrival, a guard hurried in to inform her that hundreds of people were heading up the drive. Ordering all the lights in the house to be turned on, the queen and Juliana stepped out onto the stone terrace in front. Five years before, as she had escaped from the Netherlands, she'd been deeply worried about what her people would think of her. Now she had her answer.

The gates were opened, and a huge throng of people—cheering, crying, singing, waving Dutch flags—pushed through, all of them there to welcome her home. Night after night, the scene repeated itself: countless men, women, and children walked or bicycled miles to Anneville, where they stood in line, sometimes for hours, to see and shake hands with Wilhelmina. "At first, Peter and I reveled in the processions, standing behind the Queen in our beribboned uniforms, gazing out impressively over the multitudes," Roelfzema wrote. "But it soon palled. It was all too intimate for bravado, too sad. From the eyes of the crowd, sunken deep in wan, pallid faces, something spoke that put our cheap vanity to shame, something between the people and their Queen from which we were excluded." One night, concerned that Wilhelmina was on the verge of exhaustion, Roelfzema told guards to hurry the people along. Overhearing him, the queen let him know what she thought of his order. "In my garden, nobody gets pushed around," she snapped. "Don't you *ever* forget it."

Although she retained her queenly demeanor, Wilhelmina refused to return to her former royal way of life. To her aides' dismay, she insisted on sharing her subjects' frugal existence, including the use of ration cards. Farmers in the area had filled Anneville's cellar with fruits, vegetables, and other scarce foodstuffs, but when strawberries were served to the queen on her first day there, she refused to touch them, saying, "I do not intend to eat anything that is not also available to the people." Fearing that such meager fare would damage her health, her aides occasionally tried to circumvent her wishes, once giving her steak for dinner. She took a bite, then stared suspiciously at Roelfzema. "Captain, this is steak," she declared. "Yes, Majesty," he replied. When she shot back, "Is everyone in Holland eating steak today?," he had to say no. She refused to eat any more, even when he pleaded that the farmers would be hurt if she kept refusing their gifts.

On May 4, Wilhelmina was working in her small study when Peter Tazelaar rushed in to tell her of the liberation of the Netherlands. She jumped up and vigorously shook Tazelaar's hand, all the while slapping his shoulder with her other hand. Soon she began the long, joyous process of reuniting with all her people.

Her first stop was The Hague. As despoiled as that once beautiful city had become, no one seemed to notice the ugliness that day; their sole focus was on their returning monarch. Tens of thousands of people, waving flags and carrying orange banners, flooded the streets, shouting, "Long live the queen!" As Wilhelmina's Packard convertible proceeded slowly along the route, the enormous crowds repeatedly broke through the police lines to cheer her. At times, her car was almost swallowed up in a sea of humanity.

According to one onlooker, this extraordinary outpouring of loyalty and affection was due to the mythical stature that Wilhelmina had assumed during the war, "synonymous as it was with liberty, pride, and Holland's national heritage." But the reason was more complex than that. Though Wilhelmina had indeed enhanced the country's monarchy and history by her resolute defiance of its occupiers, she had also succeeded in another, more personal goal: to break out of her hated royal "cage" and become one of the people. During her years in London, their sacrifices and suffering had become hers, a fact they recognized. As *Time* magazine later noted, "War brought Queen Wilhelmina a sense of comradeship with her people that she had never known. They respected her before. They loved her now."

LIKE HOLLAND, NORWAY WAS occupied by German troops until the last days of the war. Not until V-E Day, on May 8, 1945, did the 365,000 German soldiers in the country begin to lay down their arms in earnest.

In the end, Norway's fate, like that of the other occupied nations, was largely determined by its geography. Since it was of little political or strategic importance to any of the Big Three, it remained a sideshow throughout the war. Though there was a major disadvantage in that—Allied troops would not be available to liberate the country by force of arms—it also meant that Norway was not turned into a battlefield and could regain its independence without interference from its major partners.

But those benefits were not on the minds of Norwegian leaders as the war neared its conclusion. They feared that German military commanders would follow Hitler's order to make a final, bloody stand in "Fortress Norway"—the same directive he had given for Holland and other areas of western Europe still occupied by the Reich. When the Norwegians asked Churchill what the Allies would do if the Germans continued to resist, the British prime minister could only promise that the war would continue until Norway was free. But he also insisted that at that point the Allies had neither troops nor ships available to ensure liberation.

In early April, however, General Eisenhower and his SHAEF subordinates began to focus on the "dangers of leaving large German forces uncontrolled in Norway" for weeks after the war's end. Eisenhower urged the British War Office to consider sending the remnants of its 1st Airborne Division, which had been virtually wiped out in Operation Market Garden, to Norway as a disarmament force.

The British agreed, even though the division, at most, could muster only a few thousand men, less than one-tenth the size of the German military presence in Norway. Its head was to be General Sir Andrew Thorne, a canny, Eton-educated veteran of Dunkirk who currently commanded British troops in Scotland. Well aware that his tiny force would be no match for Norway's occupiers, Thorne resorted to a kind of sleight of hand to ensure the Germans' capitulation. As part of his plan, he formed a partnership with Milorg, Norway's military resistance movement, whose 40,000 members had been training for more than a year to take part in their country's eventual liberation.

Thanks in part to a strengthened relationship with SOE, Milorg was much more vigorous than in the war's early years. In the autumn of 1944, SOE agents began arriving in Norway to prepare for its liberation, organizing and training Milorg units to take over Norwegian factories, power plants, and transportation lines when the Germans surrendered.

On May 3, 1945, General Franz Böhme, the commander of German forces in Norway, declared that his troops were prepared to make a last-ditch stand if ordered by the Reich's military leaders. When Berlin capitulated five days later, Böhme reluctantly followed its lead, although there was considerable concern that soldiers and their commanders in the field would refuse to obey the order.

With most of his disarmament troops still en route to Norway, General Thorne assigned Milorg forces to come out in the open and begin occupying important factories and military installations, as well as perform other peacekeeping tasks, such as maintaining order between Norwegians and their former occupiers. Thanks to the close cooperation between the British military and Milorg, not to mention the discipline of the German troops and the Norwegian people, there were no serious outbreaks of violence on the way to Norway's liberation. It also helped that the Germans never knew how small Thorne's force really was; he managed to convince them, with no corroborating evidence, that it was large enough to crush them if they dared resist.

With their country's freedom assured, Norwegians in London began returning home. After taking part in the V-E Day celebrations in London, Crown Prince Olav and many members of the government in exile boarded British destroyers for Oslo. King Haakon stayed in London for another month. Shortly before his departure on the evening of June 4, he made a final visit to the BBC, where he recorded a message of thanks to the British people for their hospitality and support for his country.

On June 7, 1945, five years to the day he had left Norway, Haakon sailed into Oslofjord aboard the cruiser HMS *Norfolk*. Surrounding the *Norfolk* was a bevy of other British warships, including the *Devonshire,* which had borne the reluctant monarch to Britain in 1940. As the royal convoy steamed up the fjord, it was met by another, more impromptu escort: a massive flotilla of Norwegian fishing boats and other small craft.

At the stroke of noon, Haakon, wearing a blue admiral's uniform, stepped ashore in front of Oslo's City Hall, decorated with red-blue-and-white Norwegian flags and a huge "Welcome Home" banner. Cheers and shouts of *"Kongen leve!"* ("Long live the king!") erupted from tens of thousands of normally taciturn Norwegians, all jammed together in streets, on rooftops, and even in the rigging of ships. That night, an estimated 130,000 people paraded past the royal palace in a spontaneous salute to the man who had inspired them through the dark wartime years. "It is safe to say," a Norwegian historian remarked, "that there was never a happier day in the annals of the nation."

A few weeks later, Haakon wrote to his nephew, George VI, about the sorry state of the royal palace after five years of occupation: "Everything was out of place. My own room was pretty straight, but all the furniture and pictures are mixed up. The worst for me is that Quisling has remade Aunt Maud's room and changed it so much that I don't believe I can get it back in anything that can remind me of her time."

Even as he mourned the loss of cherished items that recalled his late wife, the king himself bore little resemblance to the unsure, unvalued ruler he had been when Queen Maud was alive. Thanks to his courage and resolution throughout the war, the man who had once thought of himself as an outsider in Norway now had an influence and authority unimaginable in the first thirty-six years of his reign. His extraordinary popularity was reflected in the fact that Norway was the only occupied nation whose resistance movement used the initials of its head of state—H7—as its symbol of defiance to the Germans. Haakon was, the *New York Times* later wrote, "the most beloved personage in the nation's history."

In 1947, when Haakon turned seventy-five, the people of Norway demonstrated their deep affection for him in a tangible way. Although he had never lived in kingly splendor, the ex–naval officer, with his deep love of the sea, had once requested a royal perquisite from the Norwegian government: a sailboat. The government leaders had turned him down. Now his subjects made up for that miserliness as millions chipped in to buy him a yacht for his birthday.

He christened it the *Norge*.

FOR BELGIUM'S MONARCH, there would be no crowds, no bands or flag-waving, no shouts of "Long live the king!" Unlike Wilhelmina and Haakon, Leopold III was not viewed as an indispensable symbol of unity and stability in his country. In fact, quite the opposite. The forty-three-year-old king was now seen as an emblem of the discord roiling postwar Belgian society, particularly between the nation's two main ethnic groups, the Dutch-speaking Flemish and the French-speaking Walloons. The announcement of Leopold's imminent return from German captivity in May 1945 set off a firestorm of controversy in Belgium, which had been liberated nine months before. In a cable to the State Department, the U.S. ambassador to Belgium reported that "the situation was full of dynamite."

For the first four years of the conflict, Leopold had been confined to the royal palace at Laeken, a suburb of Brussels. In June 1944, Heinrich Himmler had ordered that he and his family be deported to Germany. Imprisoned in a medieval fort in Saxony until March 1945, the royal family was then transferred to a wooden chalet, surrounded by a twelve-foot barbed-wire fence, near the Austrian town of Salzburg. Two months later, they were freed by U.S. troops.

However well intentioned, Leopold's decision to remain behind in Belgium after its defeat in 1940 had hurt him badly. He had made the

decision in emulation of his adored father, King Albert, who had refused to leave Belgium and its army during World War I. Fortunately
for both Albert and Belgium, the Germans had never conquered the
entire country, and the king had remained on its soil and continued to
rule. Albert's son, by contrast, had no chance to lead his people in wartime, since the Germans kept him apart from them for the duration.

As the war continued, criticism began to mount both inside and
outside Belgium about Leopold's public silence, particularly his failure
to encourage his countrymen to stand up to the Germans, as Wilhelmina and Haakon were doing from London over the BBC. The fact
that they were considerably freer than he to challenge the Nazis did not
seem to matter to his critics. "Were he to openly back the Allied cause,"
the Dutch historian James H. Huizinga noted, "Hitler would soon
make short shrift of him."

There was also scathing criticism of a meeting between Leopold and
Hitler at Berchtesgaden in November 1940. The king had reportedly
gone to Germany to ask the Führer to release all Belgian prisoners of
war, as well as to provide more food for his country and a guarantee of
Belgium's "national integrity"; all his requests had been ignored. Following the trip, neither Leopold nor anyone around him had publicly
revealed the reasons for it, and rumors had swirled that he was cooperating with the Germans.

Yet another black mark against Leopold was his marriage in September 1941 to Mary Lilian Baels, the twenty-four-year-old, London-
born daughter of a high-ranking Belgian government official. Not only
was Baels a commoner, but, even worse in the eyes of many Walloons,
she was of Flemish origin. According to Belgian tradition, in order to
avoid offending either the Walloons or the Flemish, the country's monarch was expected to seek out a foreign bride of royal blood. As a princess from Sweden, Leopold's first wife, Astrid, had met all the
qualifications; in addition, she had been greatly loved by the Belgian
public until her death. The idea of Leopold seeking personal happiness
in the middle of the war, especially with someone considered so unsuitable, was an affront to many.

Among those most upset were members of the Belgian government
in exile in London, especially Prime Minister Hubert Pierlot and For-

eign Minister Paul-Henri Spaak, who had done so much to blacken Leopold's name after Belgium's surrender in May 1940. After they themselves had been attacked for urging Belgian peace negotiations with Germany, Pierlot and Spaak had been compelled to proclaim their wholehearted support of the British war effort and their allegiance to the king, who was as popular among Belgians in 1940 as they were unpopular.

By 1943, the situation was far different. Leopold's passive attitude, which some Belgians saw as pro-German, had alienated many of his countrymen. In addition, his and his advisers' failure to communicate with the government in exile had only exacerbated the already deep and bitter resentment each side felt for the other. In November 1943, Pierlot and Spaak wrote to Leopold, demanding that he get rid of courtiers whom the London government considered anti-Ally, as well as issue a public statement condemning Belgian collaborators. They also insisted that he deny a recent flurry of rumors that he intended to set up a dictatorship in Belgium after the war. "It is not difficult to imagine the king's feelings," James Huizinga wrote. "The impudence of these men who had already besmirched his honor three years earlier and who were now offering new insults" was, in his mind, an affront not to be borne.

In his curt response, Leopold promised only to respect and obey the Belgian constitution. A year later, the king sent another letter to Pierlot and Spaak—a scathing denunciation of them for "gratuitously covering [Belgium's] sovereign and its national flag with obloquy" and thus doing the country "incalculable harm." The two officials must not be allowed to wield any authority in liberated Belgium, Leopold went on, until they acknowledged their error and "made full and solemn reparation."

Once again, war had been declared between the king of Belgium and the top leaders of the government. But that hostility did not become a major issue until after Belgium's liberation and the holding of a national election in February 1945. Although Spaak remained as foreign minister, Pierlot was replaced as prime minister by the leader of the country's Socialist Party, which opposed bringing Leopold back as king. The entire nation, it turned out, was also bitterly divided over

Leopold, with the divisions mirroring traditional splits between the left and right—Socialists and Walloons generally against him, the Catholic-dominated Christian Democratic Party and the Flemish supporting him.

When Leopold was freed in May 1945, the executive council of the Socialist Party called for his abdication, while the Catholic members of parliament demanded his swift return home. In a letter to the king, the rector of the Free University of Brussels wrote of his concern that riots would break out in French-speaking Wallonia if Leopold returned: "The question is not if the accusations against you are right or not . . . but that you are no more a symbol of Belgian unity."

When Belgian officials met with Leopold after he was freed, he pledged to meet the earlier demands made by Pierlot and Spaak: to get rid of the top members of his civilian and military staffs, strictly adhere to the country's constitution, and mete out prompt punishment to collaborators. But that wasn't enough for most members of the Socialist-dominated government; led by Spaak, they vowed to resign if the king returned, adding that if violence broke out, they would do nothing to maintain law and order.

Having made it politically impossible for Leopold to assume his throne again, the government then passed a law making it unconstitutional for him to resume his duties until and unless invited by parliament. Although Leopold refused to abdicate, he accepted temporary defeat, agreeing to live with his family in exile in Switzerland. His younger brother, Charles, was appointed prince regent.

In a farewell message to his people, Leopold lamented that "I have not had the happiness which you have known of being present at the liberation. Alone among the Belgians who underwent the sufferings of captivity and exile, I have not been allowed the joy of returning to my home and my fatherland."

Yet, difficult as the situation was for him, Leopold, like his fellow Belgians and the other citizens of western Europe, was able to live in freedom for the rest of his life. The same could not be said for the residents of Czechoslovakia and Poland, who were destined to exchange their wartime existence under one brutal tyranny for postwar life under another.

CHAPTER 26

"Why Are You Crying, Young Man?"

The West Turns Its Back on Poland and Czechoslovakia

O F THE MANY INTRIGUING "WHAT-IFS" OF WORLD WAR II, one of the most tantalizing is what might have happened to Czechoslovakia if General George Patton's Third Army had been allowed to liberate Prague during the waning days of the war, as its commander so badly wanted to do. If Patton had marched into the Czech capital, would the "Iron Curtain of the next half-century have had a very different shape," as the American writer Caleb Crain has speculated?

Winston Churchill, who was having second thoughts about assigning Czechoslovakia to the Soviet sphere of influence, obviously believed so. In an appeal to the new U.S. president, Harry Truman, on April 30, 1945, Churchill wrote, "In our view, the liberation of Prague and as much as possible of the territory of western Czechoslovakia by U.S. troops might make the whole difference to the postwar situation in Czechoslovakia, and might well influence that in nearby countries."

The only Western Allied force to reach eastern Europe during the war, the U.S. Third Army breached the western border of Czechoslovakia in late April 1945 and drove the Wehrmacht out of the country's three westernmost cities and towns, including the medieval city of Plzeň. The Americans had an easy time vanquishing the demoralized Germans, and Patton was anxious to continue his advance. One obstacle stood in his way: Eisenhower had ordered him to go no farther than Plzeň to avoid angering the Russians.

At that point, the Red Army had not advanced as far into Czechoslovakia as it had into Poland. Although Czech president Edvard Beneš had signed a treaty of cooperation with Stalin in 1943, his country was still regarded as a sovereign, independent nation. If Washington had agreed to Patton's advance, Prague would likely have fallen into the Western Allies' hands like a ripe pear; Patton's forces were only forty miles away, with the roads to the city wide open. The Soviets, by contrast, were at least 120 miles from the capital.

Edward Stettinius, Truman's secretary of state, agreed with Churchill that Czechoslovakia should be denied to the Russians and urged the president to authorize Patton's advance. Truman, however, had been in office for only two weeks at that point, and he left the decision up to George Marshall. The army chief of staff, in turn, kicked the request back to Eisenhower, who said no.

While all this diplomatic and military buck-passing was under way, the residents of Prague erupted in joy at the news of the Americans' presence nearby. Convinced that Patton's army was on its way to free them, they eagerly responded to an appeal from a Czech resistance radio station on May 5 to rise up against their occupiers and help the supposedly still advancing Allies rid Prague of the enemy.

As in the case of the Warsaw rebellion, the Germans fought back hard, determined to subdue the lightly armed upstart Czechs. Wehrmacht units beat back resistance fighters in bloody street battles, while SS units herded civilians out of their homes and mowed them down with machine-gun fire. Prague's fourteenth-century City Hall was set afire, as were several other landmark buildings.

There was, however, a major difference, in addition to sheer size, between the Prague and Warsaw uprisings: Western Allied troops were close enough to Prague to give the resisters immediate assistance. After hearing about the rebellion from U.S. intelligence agents who had been in the city, Patton pleaded with General Omar Bradley, his immediate superior, for permission to march to Prague as quickly as possible. "For God's sake, Brad, those patriots need our help!" Patton exclaimed. "We have no time to lose!" To ensure that Bradley would not be held responsible for his unauthorized advance, Patton offered to act as if he were doing it on his own and would "only report back from a phone booth when the Third Army was actually inside Prague."

Bradley, however, insisted on leaving the decision up to Eisenhower, who again turned thumbs down. Under no circumstances, he said, was Patton to take Prague. His sole concern, as always, was to end the war as rapidly as possible, and he saw no strategic benefit in capturing the Czech capital. In Eisenhower's view, all that would come of its liberation, beyond more U.S. casualties, would be problems with the Russians. George Marshall agreed: "Personally, and aside from all logistic, tactical or strategical implications, I would be loath to hazard American lives for purely political purposes." In fact, although Marshall didn't acknowledge it, "political purposes"—in this case, not antagonizing the Russians—were behind the U.S. decision not to liberate Prague.

Eisenhower did not mention to Bradley or Patton that he had already consulted the Soviets about the possibility of dispatching U.S. forces to the Czech capital—an idea they had predictably quashed. "We Communists realized that if we let American troops enter Prague, they would be our liberators," a Czech communist official observed years later. "The result would be an important political shift in the bourgeoisie's favor."

On May 9, the day after the European war ended and four days after the Prague uprising began, the Red Army reached the outskirts of the city. Prague was now in Soviet hands, as the rest of the country soon would be. "No [Czech] citizen could misunderstand the implication of the military strategy of those last days," the Czech diplomat Josef Korbel noted. "The West was not interested in Czechoslovak democracy; its fate was left to the Communist and Soviet forces. This realization had a shattering effect on the morale and psychology of the Czech people: after six years of agony, as in a kind of nightmare, they watched something like Munich happening once again." Sir Orme Sargent, an undersecretary in the British Foreign Office, told a friend that, as a result of the Americans' failure to push on to Prague, "the Russians would be glorified, and with them the Czech Communists. . . . Czechoslovakia was now definitely lost to the West."

When the Red Army finally did enter the city, its forces appeared to Prague's despairing residents to be just as predatory as the departing Germans. Acting more like conquerors than liberators, they treated the Czechs, their supposed friends and allies, in much the same ruthless manner they were now treating the citizens of the collapsed Third Reich. Eyewitness accounts reported widespread rape and drunkenness, wholesale looting, and wanton destruction of property.

Patton's army, meanwhile, remained in place forty miles away.

FROM THE EASTERN SLOVAKIAN town of Košice, Edvard Beneš watched apprehensively as the events in Prague unfolded. No one was more anxious to see it liberated by the Third Army than Beneš, who, in a voice quivering with emotion, blurted out, "Thank God, thank God," when he heard that Patton had entered Czechoslovakia. He immediately sent a telegram of congratulations and welcome to the U.S. general.

Like the British government, Beneš had developed severe misgivings about his cozy relationship with Stalin, which he was now beginning to view as a pact with the devil. Under the treaty he had signed with the Soviets in December 1943, he had agreed to return to Czechoslovakia by way of Moscow, as Stalin had ordered. When he and his government did so in March 1945, the Soviet leader demanded that the Czech communists who had spent the war in Moscow be appointed to head most of the major ministries in Czechoslovakia's postwar government. Beneš, with some hesitation, finally agreed. Only two top posts were not given to the communists: justice and foreign affairs, which continued to be headed by Jan Masaryk.

Clearly worried about the future, Beneš had gone to Moscow "without enthusiasm and with grave doubts," recalled František Moravec, Czechoslovakia's intelligence chief. Indeed, the Czech president had tried to postpone the trip on grounds of illness. But Stalin had insisted, and now, according to Moravec, Beneš "was moving toward the inevitable destiny which he had prepared for himself and his people by his decision to put his faith into Soviet hands."

Throughout the war, Beneš had shown little reluctance in following Stalin's orders and indeed had dismayed those around him with his obsequiousness to the Russians. "Beneš dealt with the Soviet Union," Josef Korbel observed, "not as one of the inescapable influences of Europe with which Czechoslovakia would necessarily have to reckon, but as an influence so compelling to the re-emergence and preservation of Czechoslovakia as to justify the bargaining away of his country's central values. . . . As at Munich, he had done little to stand up for the independence of his country."

Echoing that view, Moravec recalled how, late in the war, a Soviet diplomat, who in fact was an NKVD agent, had told him that Moravec would be appointed to a high post in the postwar Czech government provided he turned over information to the Soviets about the British intelligence service and about Beneš's postwar political and military plans. "I asked him how he could dare to ask me to spy on my President and my British colleagues, whose hospitality I had been accepting for six years of war," Moravec said. When the intelligence chief informed Beneš about the encounter, the president seemed shocked but told him that the Soviets "should not be judged by western standards of morality."

Not long afterward, Moravec realized that Beneš had also been approached by Soviet agents with certain demands. At one point, the president mentioned to Moravec the problem of "fascist elements in our army." The astonished Moravec later noted that "we had no fascists in our army. But Beneš repeated that the problem had to be faced: the fascists must be silenced and removed. The communists obviously had given him a list of persons they considered politically unreliable . . . i.e. all those who did not happen to be Communists."

In early April, Beneš and his now communist-dominated government, following in the Red Army's wake, finally arrived in Prague. Having defied the Soviets, Moravec was no longer part of Beneš's administration: he had been fired as head of Czech intelligence and deputy chief of staff of the country's armed forces. From then on, he wrote in his memoirs, "I was treated by the Communists as Enemy No. 1."

Yet even as the communists began to consolidate their political gains, the country regained several of the freedoms it had enjoyed before the war. Newspapers were allowed to publish varying points of view; bookstores sold books from the West, both current and classic, that had been banned under the Nazis; and movie theaters showed American and British films. Most Czechs basked in this seeming return to prewar normality. But some, like Moravec, who had seen firsthand what the Soviets were capable of, knew that a sword of Damocles was poised over their country, hanging by the thinnest of threads.

FOR THE POLES, the sword had already fallen. Their future had been settled at the Yalta Conference in February 1945, three months before V-E Day. At that point, thanks to the Soviets' military successes, Stalin unquestionably held the initiative in eastern Europe. By the time he sat down with Roosevelt and Churchill in early February 1945, the Red Army had swept the Germans out of most of Poland, Hungary, and Yugoslavia and was in effective control of Bulgaria and Romania. Soviet troops had marched into Austria and were advancing deep into Germany, with units on the Oder River, just forty-five miles east of Berlin.

The question of Poland dominated Yalta, taking up more time and causing more friction than any other subject on the agenda. Nonetheless, the discussions were an exercise in futility. As much as Churchill tried to convince himself otherwise, Stalin was determined to rule all of Poland, insisting that the puppet government he had established in the summer of 1944 would take control of the country after the war. He was supported by Roosevelt, who said that "coming from America," he had a "distant view on the Polish question" and made plain that his interest in it was essentially limited to its effect on his own political future.

Stalin was also aided by Roosevelt's apparent lack of concern about leaving the Soviet Union as the Continent's dominant military and political power. FDR made matters worse, in Churchill's view, by telling Stalin that he planned to pull all U.S. troops out of Europe, even including Germany, after two years. To thwart Soviet dominance, Churchill "fought like a tiger" at the summit to make sure that France's postwar role in Europe would be as strong as possible. By doing so, he thought, both Britain and France could serve—to some extent, at least—as counterweights to Russia. Under heavy pressure from the prime minister, Roosevelt and Stalin reluctantly agreed to make France one of the occupying powers in Germany.

However, when the discussion turned to the question of creating an independent government in Poland, Churchill, who had repeatedly promised the Poles in London that he would win back their freedom, did not put up the same kind of fight as he had for France. True, early in the discussion, he did make the argument that "Poland must be mistress in her own house and captain of her own soul." Stalin, however, had no patience with such high-flown rhetoric and would not change his mind. Faced with Stalin's intransigence and receiving no support from Roosevelt, Churchill gave in. He and Roosevelt accepted the Soviet puppet government, albeit one with some democratic window dressing. Under the agreement, the government would be enlarged to include several leaders from "Polish émigré circles," and free elections to create a permanent government would be held as soon as possible. Roosevelt and Churchill decided to take Stalin's word that the voting would be free of coercion, even though the Soviets had never allowed this type of election in their own country.

The announcement that the two Western Allied leaders had handed over the governance of their nation to a Soviet-controlled communist regime came as a stunning blow to the Poles in Britain. Before climbing into his RAF bomber on the night of February 13, a young Polish navigator, who had been a prisoner in one of Stalin's gulags early in the war, sat down in despair to write a letter to a friend. "Just think, I and so many others knocked about the world, fleeing like criminals, starving, hiding in forests—all only in order to fight for . . . what?" He was flying that night, he wrote. "It's the proper thing to do, they say, although anger and despair are in our hearts. . . . If the Germans get me now, I won't even know what I am dying for. For Poland? For Britain? Or for Russia?" Ten days later, he was killed on another mission over Germany.

Calling the Yalta Agreement a basic violation of the principles for which the Allies had fought, the Polish government in exile refused to accept it. The government, however, did instruct Polish forces to continue fighting, to "keep peace, dignity and solidarity, as well as to maintain brotherhood in arms with the soldiers of Great Britain, Canada, the United States, and France."

In Italy, General Władysław Anders and the men of the Polish II Corps, most of whom had also been in Stalin's gulags early in the war, were inclined at first to lay down their arms. But II Corps, which had captured Monte Cassino in May 1944 and was regarded as among the best units in the British Eighth Army, was playing far too important a role in the Italian campaign to be taken out of the line now, and the British high command denied Anders permission to withdraw his men. The Polish general obeyed, ordering them to fight on. The Polish corps continued its drive, storming and capturing Bologna a few weeks before the end of the war.

In their sweep north, as the men of II Corps liberated one small Italian town after another, they were surrounded by crowds of smiling, cheering people, many of them shouting "Viva Polonia!" Women threw flowers, men handed them glasses of wine, girls hugged and kissed them. For the Poles, it was a bittersweet time. "On the one hand, I was happy that I could bring freedom to these people," recalled one soldier. "But on the other hand, I was disappointed that this was not a Polish street that I was walking on, that I wasn't bringing freedom to my people and my nation, that this was not the fulfillment of our dreams."

Scarcely a month after the Yalta Agreement was signed, reports reached London of mass Soviet arrests of Poles in Kraków and other major cities. Thousands of Poles had already been shipped off to Soviet gulags, while others, mostly Home Army officers and men, were being accused by the NKVD of spying for Britain and the London Poles, whom the Soviets called "fascists." The Home Army troops "are starved, beaten, and tortured," according to one account from the Polish underground. "There are many deaths."

In late March 1945, sixteen prominent leaders of the Polish resistance disappeared after being invited to a meeting with Soviet military commanders. A number of the missing men would have been prime candidates for top positions in a broadly based postwar Polish government. For the next six weeks, Soviets ignored repeated British inquiries about them; finally, they revealed that the Poles had been arrested.

The leaders were later tried and sentenced to long terms in prison. Four of them died there.

Notwithstanding all these reports of savage Soviet treatment of the Poles, Britain and the United States, still chasing the chimera of "Allied unity," withdrew formal recognition from the Polish government in exile on July 5, 1945, and bestowed it on the communist government in Warsaw. "The Poles," Max Hastings noted, "ended the war as they began it, human sacrifices to the reality of power."

NEAR THE END OF JULY, Britain held its first general election since 1935. When the votes were tallied, Winston Churchill, so inspirational in wartime, had been unceremoniously turned out of office by weary, war-sick voters, who decided that they preferred the Labour Party to manage their crippled economy. The voters seemed to think that Churchill, having spent almost his entire life dealing with foreign affairs, wasn't up to these new domestic challenges. They had a point. Now seventy years old, he was still looking mainly abroad, more and more concerned about the growing Soviet dominion over most of central and eastern Europe. The situation in Poland especially haunted him. From his unaccustomed seat on the minority benches of Parliament, he urged the British not to turn their backs on the Poles, failing to mention the large part he had played in settling Poland's fate.

The new prime minister, Clement Attlee, and his Labour government felt rather differently about what was going on. If anything, they were more eager than Churchill had been to establish and maintain good relations with Moscow and were disinclined to do anything that might jeopardize the achievement of that objective.

At that point, more than 200,000 members of the Polish military were in Britain—veterans of campaigns on virtually every front in Europe and the Middle East, from Norway and Libya to France and Italy. Attlee's government called the presence of the Poles a "source of increasing political embarrassment" and pressured them hard to return to communist Poland.

Britain's military brass, particularly the RAF's high command, responded to that pressure with fury. While understanding the rationale behind the government's "cold and dispassionate attitude," the RAF made it very clear that it would not turn its back on the "strongest, the most loyal and faithful, and the most persistent European ally of all." An Air Ministry report declared in January 1946 that Polish pilots "are part and parcel of the RAF, they have fought with us during the whole of the war, they were with us in the Battle of Britain, and with us from D-Day onwards to 'the kill' in Germany. To contemplate, in their tragic hour, anything short of sympathetic and generous treatment is unthinkable."

The report was an indication of how dramatically the RAF's attitude toward the Poles had changed. When the Polish pilots had arrived in England six years earlier, they had been greeted with disdain by many of their British counterparts. But their courage, tenacity, and flying skill—and, above all, their crucial role in winning the Battle of Britain—had swept away most of the prejudice and replaced it with gratitude and friendship.

Although the Attlee government rejected the RAF's appeal for preferential treatment for the 14,000 Polish pilots and ground crew in Britain, it halted its campaign to try to force the Poles to go back to their homeland. The vast majority of Polish military personnel had already made it clear that they would not return to a Soviet-controlled Poland, even though most had homes and families there. Resigned to the reality that a great many Poles would be staying in Britain, the British government set up an organization called the Polish Resettlement Corps, which offered them temporary jobs until they could find permanent work.

Having been given asylum, tens of thousands of Poles settled in postwar Britain, trying to rebuild their shattered lives. General Anders remained there, as did General Stanisław Sosabowski, the commander of the Polish parachute brigade at Arnhem, and Count Edward Raczyński, Poland's ambassador to Britain, who praised Britain's magnanimity in his memoirs. "Throughout our history, there is no country in which Polish exiles have enjoyed more generous and imaginative help," Raczyński wrote. "The kindly attitude of the authorities has been matched by the friendliness of the British people."

Raczyński, however, had far greater advantages than other Poles.

He had lived for many years in London and knew it well. He also had money and high social status; most of his compatriots did not. In the bleak austerity of postwar Britain, they struggled to find jobs and decent places to live. They also grappled with the grim probability of a lifetime of permanent exile, of never seeing their country or loved ones again.

Adding to their pain was the Attlee government's attempt to draw a veil over Poland's many contributions to the Allied triumph. On June 8, 1946, Britain staged an elaborate victory parade, inviting armed forces from more than thirty Allied nations to London in a joyous celebration of their collective effort. Czechs and Norwegians marched down the Mall that day, as did Chinese and Dutch, Frenchmen and Iranians, Belgians and Australians, Canadians and South Africans, Sikhs and Arabs, and many, many more. However, the Poles—the fourth largest contributor of manpower to the Allied effort in Europe—were nowhere to be seen. They had been deliberately and specifically barred by the government, for fear of offending Stalin.

While church bells pealed and crowds cheered, Polish war veterans stood on the sidewalks and watched. One Polish pilot looked on in silence as the parade passed. Then he turned to walk away. An old woman standing next to him looked at him quizzically. "Why are you crying, young man?" she asked.

AS THE POLISH EXILES worked to make new lives for themselves in Britain and elsewhere, about 30,000 of their countrymen—15 percent of the Poles in Great Britain at the end of the war—did decide to return home. Their yearning for their country and families proved stronger than their fear of a future under Soviet control. Among those who went back was Marian Rejewski, the young Polish cryptographer who had first broken the Enigma cipher. Rejewski had spent the last two years of the war in Britain but never had been allowed inside Bletchley Park. At the end of the war, virtually no one working there had any idea how much they owed him and the other Polish cryptographers who had made Ultra possible.

Until November 1942, Rejewski and six Polish colleagues had continued their work with Major Gustave Bertrand and his French code-breaking team at a secluded château in the countryside of Provence. But the threat of detection by the Vichy French and Germans was steadily mounting. In early November, mobile direction-finding teams, in vans and trucks with circular antennas on their roofs, began sniffing around the area.

On the morning of November 8, Bertrand and his team learned of the Allied invasion of North Africa; three days later, they were told of the German takeover of Vichy France. Within hours of hearing that latest news, all the cryptographers fled the château, and the Poles went into hiding in Cannes on the French Riviera coast. The escape plan Bertrand had devised for them, which involved fleeing over the Pyrenees to Spain in two separate groups, proved ramshackle and badly executed. The guides for the first group abandoned their five charges just as they set out for the Pyrenees. The Poles were forced to proceed on their own, with no one to turn to for help or advice. They eventually found another guide, who betrayed them to the Gestapo. All five were sent to German concentration camps, where two died before the war's end. "Any one of these men might have purchased their freedom by telling their captors that Enigma had been broken but, none did," noted Dr. Reginald Jones, the British government's chief science warfare adviser. "Their loyalty to their allies matched their brilliance in cryptography."

The second group—consisting of Rejewski and Henryk Zygalski, another of the three young cryptographers who had first cracked Enigma—also encountered terrible difficulties. As they crossed the Pyrenees at the end of January 1943, their guide pulled a gun on them and demanded all their money and possessions. They made it to Spain on their own and were promptly taken into custody by Spanish police, who threw them in jail. They were kept in a series of Spanish prisons until early May, when the Polish Red Cross finally secured their release. In August, they were taken by ship to Britain.

"What a windfall for the English!" Bertrand, who had stayed behind in France, exclaimed when he learned of Rejewski's and Zygalski's arrival on British shores. To him, it seemed logical that the higher-ups at Bletchley Park would welcome the Poles with open arms. Instead, apparently for "security reasons," they were barred from Bletchley and assigned to a small Polish cryptography unit in Boxmoor, a small town near London. There they were put to work breaking low-level codes of SS forces in occupied European countries. "Setting them to work on [those codes] was like using racehorses to pull wagons," Alan Stripp, who worked in Bletchley Park's Japanese code section, later observed.

According to Stripp, bringing Rejewski and Zygalski to Bletchley Park would not only have served as an acknowledgment of what they had contributed to Ultra, it would also have greatly benefited the English code breakers, who were still struggling with the complexities of the Enigma naval cipher. "We cannot exclude the fact that the Poles' perfect knowledge of the machine and the habits of the German signalmen would have been very helpful if not decisive," Stripp remarked. "The extraordinary expertise we would have gained was largely put aside by British intelligence."

Dispirited by the British cold-shouldering, Rejewski drafted a note to the Polish government in exile in late 1944 suggesting that the British be urged "to cooperate with the Polish [cryptographers] as loyally as the Poles worked and continue to work with them." The exile government did ask MI6 to come to the aid of Rejewski and Zygalski, but there was no response. Evidently, British security and intelligence officials thought the Poles had been tainted by their two-year stay in Vichy France and five-month confinement in Spanish prisons. In fact, neither cryptographer had ever fallen into German or Vichy French hands, but the British refused to reconsider.

The Poles' cause was weakened by the fact that neither Dillwyn Knox nor Alastair Denniston—the two Bletchley Park officials who had worked most closely with them in the early days of the war—were around to plead their case. Denniston had been replaced as head of Bletchley in late 1942, and Knox, who had been the Poles' greatest advocate, had died of cancer three months before Rejewski and Zygalski came to Britain.

With both gone, no one at Bletchley Park seemed to have any memory of the crucial events of 1939 and 1940: the British and French visit to Warsaw, the Poles' gifts of two Enigma machines and the techniques they had used to break the early codes, and the day-to-day cooperation between the British and the Polish-French code-breaking centers in France. The breaking of Enigma had become a British monopoly, and the Polish cryptographers and their crucial contributions were thrust into the shadows. "It is clear that of the many people who worked on Enigma, very few ever knew about the Polish contribution," Stripp noted. "The 'need to know' principle extended to that as to many other matters there."

As it happened, some newcomers to Bletchley did receive a garbled version of the Poles' involvement. Reginald Jones, who spent some time there, was told by Bletchley's deputy chief that the Poles had somehow stolen an Enigma machine and presented it to the British. "Such a theft, of course, would have been a tremendous coup of the cloak-and-dagger variety, but it would not by itself have been a cryptography feat," Jones observed. Gordon Welchman, who would eventually head the Bletchley section responsible for work on Enigma, was told the same thing when he was hired.

In Jones's memoir of the war, published in 1978, he repeated the story of the Polish theft of the German cipher machine. Welchman did the same in his 1982 book about his work at Bletchley Park. Both men were greatly chagrined when they later learned what the Poles had really done. "The credit I gave them was utterly inadequate," Jones said, and he tried to make amends in another, later memoir. Welchman, for his part, wrote a long paper, entitled "From Polish Bomba to British Bombe: The Birth of Ultra," shortly before he died in 1985. The paper began, "Until just before the Second World War, a small Polish team of three mathematician-cryptologists, headed by the brilliant Marian Rejewski, had been happily breaking the German military cipher machine, the Enigma, for many years." Later in the paper, Welchman wrote that Britain's Ultra operation "would never have gotten off the ground" if it hadn't been for the Poles' prior work. As generous as Welchman's tribute was, it didn't do much to change the conventional wisdom that the British had been responsible for breaking Enigma. It also came too late to have any impact on Marian Rejewski's life.

At the end of the war, Rejewski was depressed and in poor health, suffering from the rheumatoid arthritis he had contracted during his imprisonment in Spain. He rejected the idea of staying in Britain: there was nothing to keep him there, and he was desperate to reunite with his wife and two children, whom he had last seen six years earlier.

With his family, Rejewski settled down in his hometown of Bydgoszcz, a city in northern Poland. As was true for anybody who had lived in Britain during the war, he was under constant secret police surveillance from the day he returned. In the view of Poland's communist government, any previous contact with the West was equated with "fascism." Thousands of Poles who had fought the Germans in every way possible, whether as resistance fighters or as members of the returning Polish armed forces, had already been arrested and imprisoned. Some had been tortured and killed. For years after his return, Rejewski's mail was opened and his phone tapped. His friends and acquaintances were regularly quizzed about him, especially about what he had done during the war. Yet security officials never discovered his connection with the breaking of Enigma, and he remained free.

Rejewski did everything he could to avoid attracting the authorities' attention. He never became involved in any political or social activity, nor did he pursue high-level mathematics jobs. Instead, he worked at a string of low-level positions, including as an office clerk. A Polish historian called Rejewski's postwar life "nothing but depressing." His daughter said he had "a barren existence" until his death in 1980.

Not until the twenty-first century did the British government finally acknowledge officially that the Poles had indeed played a role in breaking Enigma. On July 12, 2001, a monument commemorating their contribution was installed on the grounds of Bletchley Park. Even so, it hardly did justice to the seminal nature of their work. The monument's inscription says only that the Poles' efforts "greatly assisted the Bletchley Park code breakers and contributed to the Allied victory in World War II."

In 2014, Sir Iain Lobban, the head of Government Communications Headquarters (GCHQ), Britain's signals intelligence agency, equated the breaking of Enigma to a relay race in which the baton had been passed by the Poles but the team "as a whole won the medal." True enough—but the fact remains that, to this day, the first runners on that relay team have been denied their full share of the credit and glory of the race's triumphant end.

CHAPTER 27

"A Collective Fault"

The Shadow of Collaboration

NEAR THE END OF THE WAR IN EUROPE, A TEENAGE BOY in Holland rhapsodized about what life would be like in peacetime. "There will be food again, and gas, light and water," he said. "Trains and trams will be running; our men will come back from forced labor in Germany; our prisoners of war and students will return. I will be able to go out whenever I want. I don't have to be afraid when a car comes into the street or when the doorbell rings at night. There will be papers, cinemas, dancing, and cars; families will be reunited."

But as the boy would soon discover, the grim reality of postwar life in Holland and the rest of scarred, impoverished Europe would bear scant resemblance to his comforting daydream. Bombed-out buildings still littered the landscape; in France alone, more than 1.5 million buildings had been destroyed, almost twice as many as in World War I. Throughout the Continent, rail lines, bridges, dikes, docks, and ports were in ruins. Once fertile farmland was flooded, cities were landscapes of desolation, and postal, telephone, and other vital services were largely nonexistent. There were food shortages everywhere; the same was true for coal and other fuel. During the bitter winters of 1945 and 1946 (two of the coldest in Europe on record), most homes, offices, and schools went unheated. In the words of the American journalist Theodore White, the countries of Europe were "as close to destitution as a modern civilization can get."

As the nations struggled to survive and then to begin the massive job of rebuilding, they were forced, too, to confront their wartime pasts, to acknowledge that although many of their citizens had defied the Germans, many others had collaborated with them. Like so many war-related questions, the issue of collaboration involved layer upon layer of complexity, including how the term should be defined. For some people, including the British writer and ex–MI6 operative Malcolm Muggeridge, the definition was obvious: "Under the German occupation, everyone who did not go underground or abroad was in some degree a collaborator and could be plausibly accused as such." Such black-and-white views tended to be held by people who lived in unoccupied countries such as Britain and the United States and, as a result, had no idea of the ambiguities of life under the Germans. In Britain and America, "we [regarded ourselves] as the good guys," the British novelist Paul Watkins remarked. "We did not have to think about what it might have been like to live as collaborators. Anyone who collaborated was weak and deserved to die along with the rest of the bad guys."

Those who thought of collaboration in simplistic terms could not comprehend the reality of trying to survive in an uncivilized, unstable environment in which the norms of society had broken down. "If you wanted to eat and go to school, you had to collaborate," Watkins observed. "The only other choice was to vanish into the hills or risk being sent to a concentration camp. If you wanted to hold on to some semblance of your former life, your only choice was to do as you were told."

Sir Isaiah Berlin, the noted Russo-British philosopher and historian, was more understanding of human nature in his thoughts on collaboration: in order to make it through the war, an individual might be forced to have dealings with the Germans, but "you did not have to be cozy with them." The historian Stanley Hoffmann, who lived through the German occupation in France, had another, more complex definition. Hoffmann divided collaboration into two categories: involuntary, in which one reluctantly recognizes the necessity of cooperation in order to survive, and voluntary, in which one exploits the necessity and actively abets the enemy for his or her own gain.

However collaboration was defined, those believed guilty of it were subjected to violent retribution at the end of the war. In every occupied country, resistance members and other citizens rose up against suspected informers and collaborators "with as much fury and disregard for personal civil liberties as collaborators had moved against resistance workers during the war," one historian noted. Such vengeance was particularly ferocious in France, where it was known as the *épuration sauvage* ("savage purification").

In the days and weeks following the war, thousands of French citizens were killed by their own countrymen, many of whom were members of communist and other resistance-linked groups. Estimates of the number of these summary executions vary widely, ranging from 6,000 to 40,000. Calling the period "one of the more squalid episodes in France's history," Malcolm Muggeridge observed that some of the killings, supposedly committed in the name of justice, were later revealed to be "the working off of private grudges and envies."

A hostile crowd surrounds Frenchwomen accused of sexual involvement with Germans. Stripped down to their undergarments, they likely all suffered the fate of the woman on the left—having their heads shaven.

Throughout the occupied countries, women accused of "horizontal collaboration"—sexual involvement with Germans—also suffered widespread public wrath. Their heads shaven, they were paraded like cattle through the streets of countless cities and towns, often stripped naked, sometimes beaten and/or tarred and feathered. Crowds of onlookers jeered and spat on them.

British and U.S. soldiers and war correspondents who witnessed these postwar spasms of violence were shocked at such intolerance by supposedly civilized Europeans. Most knew little of the brutality to which much of captive Europe had been subjected, especially in the last year of the occupation, or of the Continent's lawless wartime environment, in which once cherished personal and political rights had not existed for at least half a decade. "People who did not live under German domination . . . will find it difficult to understand that every moral law, convention, or restriction on impulses simply disappeared," a Polish resistance member wrote after the war. "Nothing remained but the desperation of an animal caught in a trap. We fought back by every conceivable means in a naked struggle to survive against an enemy determined to destroy us."

Shocking as the pan-European campaign of vengeance was, it burned itself out in a matter of months, with personal vigilantism finally giving way to the institution of official trials for collaborators. "In the circumstances of 1945, it is remarkable that the rule of law was re-established at all," noted the British historian Tony Judt, who wrote a magisterial history of postwar Europe. "Never before, after all, had an entire continent sought to define a new set of crimes on such a scale and bring criminals to something resembling justice."

In judging collaborators, France was faced with a particularly difficult dilemma. Its own government had been guilty of collaboration, as had a large percentage of its business, industrial, cultural, and social elites. Faced with such high-level cooperation with the enemy, the French courts were relatively selective in those they chose to charge and punish. By the end of 1945, only about 90,000 persons had been investigated or arrested. A little more than half—half of one percent of the population—were convicted of wartime offenses. Of that number, slightly more than 18,000 were given prison sentences, while the rest received other sanctions, including fines. Meanwhile, a relative handful of prominent Frenchmen were executed for war crimes, among them Vichy prime minister Pierre Laval. Marshal Pétain was also condemned to death, but because of his age and feebleness, the sentence was commuted to life imprisonment.

In other European countries, the net was thrown far more widely. That was especially true in Norway, whose postwar government arrested and tried about 2 percent of the nation's population, including all 55,000 members of Vidkun Quisling's pro-Nazi organization, Nasjonal Samling. About a third of that number—17,000—received prison terms.

In the Netherlands, nearly 100,000 people—slightly more than 1 percent of the population—were jailed for war crimes. More than half were members of the Dutch Nazi Party, many of whom had fought for the Germans in a Dutch unit of the Waffen-SS. In Belgium, 56,000 persons—slightly more than one half of one percent of the country's citizens—were sent to prison.

In Czechoslovakia and Poland, the question of how many citizens collaborated is virtually impossible to answer. Although there were certainly those who cooperated with the Germans in both countries, the Czech and Polish communist governments more often than not used the charge of fascist collaboration to rid themselves of perceived political opponents, including many Czechs and Poles who had served in the resistance or fought with the Western Allies during the war.

IN THEIR PROSECUTION OF those who had aided the enemy, the former captive countries paid virtually no attention to one type of collaboration—assisting the Nazis in the mass killing of Europe's Jews. Of the estimated 7 million civilians who died during the war in Poland, France, Holland, Belgium, Luxembourg, Norway, and Czechoslovakia,* nearly half—3.3 million—were Jews, most of whom were murdered in Nazi concentration and extermination camps.

* Polish civilian fatalities—5.6 million—accounted for a staggering 81 percent of the seven nations' death toll.

As Germany began to implement the Final Solution in late 1941, it relied heavily on the cooperation of local bureaucrats in the occupied countries. The most enthusiastic collaboration was found in France, where the Vichy government not only obeyed Nazi directives but did far more than the Germans asked. Indeed, just two months after France's capitulation in 1940, Vichy had already introduced anti-Jewish policies in its territory without receiving orders from Berlin to do so. Of the 76,000 Jews deported from France to the death camps, more than 90 percent were rounded up by French police.

In other occupied countries, the local aid was not quite as egregious. None of the other nations had indigenous governments like the one in Vichy; they were governed instead by German military or civilian administrations. Still, as a Dutch historian noted, the Germans "needed and received local administrative help in their efforts to isolate Jews in . . . society and to deport them to extermination camps." In all these nations, the Nazis had earlier weeded out civil servants and policemen who were not regarded as sufficiently pro-German; they then relied on the rest to do the Nazis' bidding, which most did with alacrity. Among the duties of the local police and militia was to take part in the Nazi roundups and deportation of Jews.

Cooperation extended far beyond local bureaucracies and police forces. Many ordinary citizens collaborated as well, informing on and otherwise betraying Jews who not infrequently were their neighbors, friends, or acquaintances. In some cases, local people actually took part in the murder of Jews. Perhaps the most notorious example occurred in the Polish village of Jedwabne in July 1941, where a group of Poles from the area, urged on by German forces, killed more than three hundred Jews.

ALTHOUGH IT'S BEEN MORE than seventy years since the end of the war, the issues of national and individual responsibility regarding the Holocaust remain highly fraught. How much should the former occupied nations be held accountable for the small minority of their citizens who helped the Nazis in carrying out the Final Solution? Or, indeed, for the vast majority of their people who did little or nothing to help the Jews?

Such widespread apathy has a number of explanations. Before and during the war, anti-Jewish prejudice was strong throughout the Continent, as it was in countries such as Britain and the United States. Even in western Europe, where Jews were more integrated into society than in the east, they were, more often than not, still considered outsiders. "During the first two years of the Occupation [in France], the prevailing sentiment towards Jews ranged from indifference to hostility," noted Julian Jackson. "People had more pressing concerns on their minds."

The same held true for the populations of other occupied countries. In Holland, "fear and their own worries held [the Dutch] back," remarked the writer Elsa van der Laaken, who was a child in The Hague during the war. There was "fear of losing a job or of being imprisoned. . . . People were self-centered now. Your own family and home came first, then you might see what you could do for others without endangering yourself and your family."

After the war, most of the governments and peoples of Europe chose to consign to oblivion their indifference to the fate of the Jews. Such deliberate forgetfulness was particularly obvious in France. More than twenty years after the war's end, revisionist historians and filmmakers finally began revealing the reality of the country's wartime experience. A particular landmark was *The Sorrow and the Pity,* a 1969 documentary by Marcel Ophuls that, through a series of filmed interviews, closely examined the Nazi occupation of France, shattering the idea of a country united in resistance and underscoring Vichy's collaboration with the Nazis, especially in the Jewish deportations. The film, which was extraordinarily controversial in France, was banned from French television and most movie theaters. It ended up playing in a single theater on the Left Bank in Paris.

Another twenty-four years would pass before France formally acknowledged its wartime complicity in the persecution and deportation of its Jews. "The criminal madness of the occupier was supported by the French people and by the French State," French president Jacques Chirac declared in July 1995. "It is undeniable that this was a collective fault."

Chirac's assertion was certainly true. Yet it's also important to keep in mind the complexity of the times and the harrowing moral choices that the French and other occupied Europeans had to make—a point made by Marcel Ophuls himself. In 2000, Ophuls said that his purpose in making *The Sorrow and the Pity* was not to condemn the French or "give a 'message' about how they behaved." He went on, "This would be pompous, stupid and prosecutorial—to make a statement about a country that had been defeated and had to live under these conditions for four years. I did not set this up to [indict] France for being a collaborator. In times of great crisis, we make decisions of life and death. It's a lot to ask of people to become heroes. You shouldn't expect it of yourself and others."

Nonetheless, when it came to saving Jews, many thousands of Europeans were, in fact, heroes. They may have comprised only a tiny minority of their countries' populations, but because of their efforts, nearly half a million Jews were able to survive the war.

A case in point is France, which, despite its actively collaborationist government, ended the conflict with about three-quarters of its Jews—some 225,000—still alive. As the historian Julian Jackson saw it, the success of the efforts to save French Jews "required the solidarity, passive or active, formal or informal, of the French people." Jackson added that, for decades before the war, "the Jews of France had looked to the State to protect them, if necessary, from sudden anti-Semitic outbursts of civil society. In the Occupation, it was civil society that helped protect Jews from the state."

In the effort to rescue Jews, no country faced more daunting challenges than Poland. It was extremely difficult to spirit anyone into or out of the tightly sealed ghettos in which many if not most of Poland's 3 million Jews were confined during the war. In addition, Poland was the only country in occupied Europe whose citizens and their families faced immediate execution if caught trying to help Jews.

Yet Poland was also the only nation whose resistance movement created a formal department for Jewish rescue. Known as Żegota, the group managed to find hiding places for thousands of Jews outside the ghettos. It also provided them with money, forged identification papers, food, and medical care. Of the 50,000 Polish Jews who escaped the Holocaust, "every [one of them] did so only because gentile Poles risked their lives to save them," declared Lucjan Blit, a Jewish socialist leader from Poland who spent much of the war in London. Echoing that view, the British-American historian Walter Laqueur, whose Jewish parents died in the Holocaust, wrote of the Poles, "It is not surprising that there was so little help, but that there was so much." The same could be said of the rest of occupied Europe as well.

CHAPTER 28

"The World Could Not Possibly Be the Same"

Planning for the Future

B Y THE LATE 1940S, MOST EUROPEANS WERE ANXIOUS TO PUT the war and its recriminations behind them. As Tony Judt noted, "Silence over Europe's recent past was the necessary condition for the construction of a European future." But planning for that future was complicated by the vast difference in experiences and outlook between those who had spent the war in London and those who had been trapped at home. Between the two groups, there existed a mutual failure to grasp what the other had endured.

Writing about the postwar disillusionment of himself and his fellow *Engelandvaarders*, Erik Hazelhoff Roelfzema noted that "through the entire war, one dream never left us, not for single day: our homecoming to Holland as we remembered it. We did come home, but the memory was crushed by reality and the dream exploded. Our country lay before us unrecognizable, emaciated like a wretch from a concentration camp. We couldn't cope, we turned away as from a leper, sickened and uncomfortable. We felt more at ease with our Allied buddies, with whom we had fought the war in freedom, than with our old friends who carried the mark of the Occupation." When he first visited his parents at their home in an affluent suburb of The Hague, Roelfzema recalled, he felt like "a creature from another, foreign world." When he left after a few hours, "we said goodbye like strangers."

For their part, Europeans who had stayed behind, particularly those who had risked their lives in resistance work, greatly resented their compatriots from London, who, in their view, had lived out the war in comfort and safety, with none of the daily tension, terror, and privations of occupation.

These deep fissures between compatriots were exacerbated by sharply divided visions of their nations' postwar futures. Many Europeans who had fought in the resistance, for example, were determined to re-create the sense of community that they had experienced during the war, which transcended traditional social and economic divisions. "The fearsome dangers had left no room for petty distinctions of social background, class and religion," wrote Roelfzema, who had been part of the Dutch underground early in the conflict. "We stood together, we fell together, we died together, brothers and sisters in the classic sense." What meaning could the war have if it did not result in radical changes in society, leading to a more just and equal world?

But the great majority of Roelfzema's countrymen, along with most other Europeans, did not share that view. Exhausted after the chaos and trauma of occupation, they wanted nothing more than to re-create the normalcy of their prewar world, craving little else but peace, order, a roof over one's head, enough to eat, and the other necessities of everyday life. They were encouraged in that view by members of their countries' governments in exile, many if not most of which were kept in power when they returned home. Representing continuity, these elder statesmen, Tony Judt observed, were "skeptical, pragmatic practitioners of the art of the possible. They reflected the mood of their constituents."

Not surprisingly, perhaps, one of those most disappointed by the return to the status quo was Queen Wilhelmina, who, throughout her long exile in London, had yearned to return to a Netherlands transformed, like herself, by the war. Her hopes had been strengthened by wartime reports from *Engelandvaarders* that the Dutch were disillusioned with the intrigues and divisions of the prewar political system and wanted drastic social and economic changes, in which they hoped the queen would play an important role.

When Wilhelmina returned to Holland, however, she discovered that few of her subjects felt, as she did, that the old political and government establishment should go. When national elections were held in May 1946, all of the prewar parties except the Dutch Nazis were returned to Parliament in roughly their former strength—a scenario that was replicated throughout western Europe.

Although the queen reluctantly backed down in her campaign for change, she defied efforts by government and court officials to force her back into the royal "cage" that had separated her from her countrymen before the war. She did so even as she took up residence again in the stately, cavernous Noordeinde Palace in The Hague, which, in Roelfzema's words, "was the embodiment of everything she hated about her former life." She once said to him in a facetious tone tinged with bitterness, "You and your RAF, you missed your targets often enough. So why couldn't you have dropped just one little bomb by mistake on this old place?"

Once reinstalled in the palace, she insisted on maintaining the informal style she had adopted in London and Anneville, breaking as many rules of protocol and tradition as she could and urging her staff to do the same. "She set the tone, we performed accordingly," said Roelfzema, who continued as Wilhelmina's military aide for several months after the war. "I whizzed around the endless corridors of Noordeinde Palace on a motorbike, scandalizing all the lackeys and missing the occasional ancient retainer by a hair."

Working hard to strengthen her bonds with her people, the queen, during the lean, impoverished postwar years, refused to turn on heat or electricity in her palaces as long as most Dutchmen had to do without them. She also rode a bicycle around the devastated countryside to meet and encourage farmers and others who had lost their homes and land.

Although Wilhelmina never again recognized the restrictions of class and rank, her subjects, to her distress, could not bring themselves to do the same. "To every Hollander, she offered her hand in equality, but Hollanders continued to bow," Roelfzema said. "After reigning for half a century, she had, in the end, become too democratic for them. She never gave up, but she could not break down the barrier."

Nonetheless, for all the disappointments of her later years, Wilhelmina could take pride in all that she had accomplished. During the war, she had, in fact, achieved her greatest childhood ambition—to perform great deeds, as her ancestors William the Silent and William of Orange had done. Hers, however, had been achieved not on the battle-field but during her London exile. In a never-to-be-repeated moment, a modern monarch of the Netherlands had been given the chance to exercise real leadership, and Wilhelmina had made the most of it. As was true of Winston Churchill and the British, World War II had been her "finest hour." She had stopped her defeatist government from ca-pitulating, kept Holland in the fight, and inspired and united her peo-ple, thus gaining "a victory that assured her a place in Dutch history second to none," Louis de Jong wrote. In doing so, she greatly strength-ened the House of Orange. Thanks to the queen, the Dutch monarchy had become not merely a "stabilizing element" in the country, *Time* magazine observed in May 1946, but "the spokesman for all elements of the people."

In September 1948, Wilhelmina, citing ill health and advancing age, abdicated after fifty-eight years on the throne, giving way to her daugh-ter, Juliana. She retreated to a small house in a suburb of The Hague, spending much of her time painting and babysitting her three grand-daughters. But before she died in 1962, she had the satisfaction of seeing her country finally making some of the dramatic social changes she had championed more than a decade earlier, including the so-called depil-larization of Dutch life—a breakdown of the rigid, centuries-old divi-sions between various segments of society, including that between Catholics and Protestants.

In much of the rest of western Europe, the stasis of the immediate postwar era also began to crumble, giving way to profound economic and social shifts. "The idea that the world as it was before the war could simply be restored . . . was surely an illusion," the Dutch writer Ian Buruma remarked. "It was an illusion held by governments as much as by individual people. . . . But the world could not possibly be the same. Too much had happened, too much had changed."

Indeed, one of the earliest harbingers of change occurred even be-fore the war ended, involving, of all countries, conservative France. On March 23, 1944, Charles de Gaulle's provisional government granted voting rights to women, a decision reflecting the enormous role that women had played in the French resistance. As it happened, twelve members of the provisional assembly approving the measure were women, the first in French parliamentary history. In 1946, just months after women's suffrage was inscribed in the country's new constitution, Frenchwomen cast their first-ever ballots in national elections.

In the rest of Europe, too, former resistance members began having an impact on their countries' political life, even though, in many cases, it took much longer than they had originally hoped. A considerable number eventually moved into positions of prominence and responsibility in their local and national governments, helping to enact significant social and economic reforms that in just a few years would culminate in the modern welfare state and change the face of western Europe.

Such an extraordinary transformation would not have been possible without the economic bounty of the United States, a country whose president had wanted nothing more to do with Europe once the war was over. At the Yalta Conference in early 1945, Franklin Roosevelt had made it clear that he had little interest in further close collaboration or partnership with the United States' Western Allies, whose empires and global influence were fast disintegrating. Serenely confident of his own country's power, he envisioned the Soviet Union as its main ally in dealing with postwar international problems.

The onset of the Cold War, however, put an end to that notion, as well as to Roosevelt's plan for a speedy U.S. withdrawal from European affairs. Having spent much of the war pacifying the Soviets, the U.S. government—now led by Roosevelt's successor, Harry Truman—launched a campaign to contain them. To do so, Washington realized it must not only maintain but increase its wartime involvement with Europe. Specifically, the Truman administration realized it must take urgent steps to assist European countries if total economic collapse and the spread of communism were to be warded off. "It is now obvious that we grossly underestimated the destruction to the European economy by the war," said U.S. undersecretary of state William Clayton after a fact-finding tour across the Continent in the spring of 1947. "Millions of people in cities are slowly starving."

In June 1947, George Marshall, now Truman's secretary of state, outlined what came to be known as the Marshall Plan, a far-reaching program to jump-start the economic recovery and reconstruction of Europe. For the countries of western Europe, the Marshall Plan offered resurrection from disaster. For their wartime eastern allies, Czechoslovakia and Poland, it spelled calamity.

ON JULY 12, 1947, U.S. officials held a meeting in Paris to discuss and explain the workings of the Marshall Plan. Every country in Europe was invited, and almost all agreed to attend. Among those who accepted were Poland, now in thrall to the Soviets, and Czechoslovakia, which, despite a strong communist presence, still retained traces of democracy. As Andrei Zhdanov, a top Soviet official and one of Stalin's closest allies, dourly put it, the Soviets had achieved a "complete victory of the working class over the bourgeoisie in every East European land except Czechoslovakia, where the power contest still remains undecided."

Indeed, in the summer of 1947, it appeared that the pendulum might be swinging away from Soviet influence. Ever since the war's end, the Czechs had grown increasingly disenchanted with their communist-dominated government. The strong-arm tactics of the state police had alienated many citizens, and farmers were up in arms over talk of collectivization. Workers, meanwhile, opposed communist demands for increased output without higher wages in return. With national elections scheduled for May 1948, it became increasingly clear that the communists would probably fail to achieve their goal of a majority in the country's parliament.

Such an outcome was unacceptable to Stalin, who had had enough of this facade of democracy. He used the United States' invitation to join the Marshall Plan as an opportunity to show the Czechs who was in charge. Summoning Jan Masaryk, the Czech foreign minister, and Masaryk's Polish counterpart to Moscow, the Soviet leader ordered them to reject the United States' offer of economic aid. Both did so.

To Masaryk, it was obvious that democracy was nearing its end in his homeland. "I left for Moscow as the minister of foreign affairs of a sovereign state," he remarked when he returned to Prague. "I am returning as Stalin's stooge." When a friend asked him how Stalin had treated him, he replied, "Oh, he's very gracious. Of course, he'd kill me if he could. But, still, very gracious."

It was an agonizing time for Masaryk. His friends in Britain and the United States chided him for not standing up to Stalin. If he and his noncommunist colleagues had insisted on accepting Marshall Plan aid, his critics said, they would have won the overwhelming backing of their fellow citizens, making it considerably harder for the Soviets to crack down on the country.

But in Masaryk's view, such resistance would have had little or no effect unless the Czechs had strong support from the United States and Britain. A few months earlier, he had traveled to Washington to urge the president and his administration to supply that support. But neither Harry Truman nor his secretary of state, Dean Acheson, would see him. The message was clear: the United States had written Czechoslovakia off. "What happened in Washington broke Jan's heart," said Marcia Davenport, an American novelist and close friend of his. Caught in the middle of a maelstrom, Masaryk alternated between despair and frantic gaiety. "He was straining to stay in a position that assaulted and offended everything he inherently was," Davenport remarked.

In Czechoslovakia, meanwhile, the conflict between the communists and their democratic opponents grew increasingly bitter. The situation came to a head in February 1948, when Václav Nosek, the communist interior minister and head of the state security police, illegally removed all noncommunist senior police officials from the force. The remaining noncommunist ministers in Edvard Beneš's cabinet resigned in protest after Nosek refused to reinstate the men he had purged.

The noncommunists assumed that Beneš would refuse to accept their resignations and keep them in a caretaker government, thus forcing the dissolution of parliament and the calling of an immediate national election. "Facing an implacable foe, the democratic leaders still put their trust in constitutional procedures," Josef Korbel wrote. "It was the decent procedure of decent men in such a situation, but the tragic weakness of such a process is that the enemy is often not burdened with any such regard for decency."

Beneš did nothing to help the ministers' cause, despite his insistence to Korbel a month earlier that he would never permit the communists to take over the government: "They have found out for themselves that I enjoy a certain authority in the nation. . . . They have come to realize that they cannot go against me." Beneš assured Korbel that "I shall not move from my place and I shall defend our democracy till my last breath. They know it, and therefore there will be no coup." But just as at Munich, his brave words meant nothing when the time came for action. Announcing that he planned to remain above the political fray, he refused to confront the communists, who immediately took advantage of the leadership vacuum.

On February 25, 1948, the masquerade of democracy in Czechoslovakia finally ended. The communists seized control of the government, and Beneš, once again surrendering his country to the demands of a foreign power, obediently signed the list of new ministers put before him. He remained as figurehead president and Jan Masaryk as figurehead foreign minister.

Two weeks later, Masaryk's body, clad in blue silk pajamas, was found at sunrise, lying spread-eagled in the courtyard of the Ministry of Foreign Affairs, just below his apartment. The Czech communists insisted that he had committed suicide. Nearly half a century later, the Prague police ruled that Masaryk had been murdered. "What died with him," wrote one historian, "was the liberty of his country."

The shocking twin deaths of Masaryk and Czech democracy reinforced the West's fear of an imminent spread of communism and galvanized it into action. Vigorous measures were taken to keep communists out of power in the governments of France and Italy. In Washington, Congress, which had been dithering over legislation authorizing the Marshall Plan, approved it immediately. Less than a month after Masaryk's murder, Truman signed the bill into law, granting an initial $5 billion in economic and technical assistance to sixteen European nations, including Britain, France, the Netherlands, Belgium, Norway, and Luxembourg. Another $8 billion was spent during the four years of the plan's existence; when it ended in 1952, the economy of every participant country had easily surpassed its prewar level.

The Marshall Plan marked a definitive parting of the ways between the two halves of the Continent. While western Europe embarked on several decades of unprecedented growth and prosperity, Czechoslovakia, Poland, and the rest of eastern Europe sank further into poverty and repression.

IN 1949, THE CZECHOSLOVAKIA coup and worries about further Soviet aggression gave rise to another historic event—the groundbreaking decision of the United States, Britain, and Canada to join the countries of western Europe in creating the North Atlantic Treaty Organization (NATO), which promised a collective defense by all member nations in the event of an armed attack on any one member.

The impetus for NATO came the previous year, just days after Masaryk's death, when Britain, for the first time in its history, committed itself to a peacetime European defense confederation. Its partners were France, Belgium, the Netherlands, and Luxembourg. On the day the European treaty was announced, President Truman expressed strong support in a speech to Congress, remarking that the "determination of the free countries of Europe to protect themselves will be matched by an equal determination on our part to help them do so."

Truman's declaration marked an extraordinary about-face for the United States, whose policy since its founding had been a determination to stay away from European military commitments. Indeed, at Yalta, Franklin Roosevelt had announced that all U.S. troops would be pulled out of Europe, including occupied Germany, within two years. But the Cold War had put an end to such aloofness; with the stroke of a pen in Brussels, the United States became a permanent and leading force in keeping peace in Europe.

Just as dramatic was the change of heart of the European nations, most of which had zealously guarded their neutrality before the outbreak of World War II. Norway was the most striking example. Arguably the least prepared of all the nations invaded by the Germans, it had spent almost nothing on its defenses in the interwar years in the hope of remaining distant from any future conflict. The shock of defeat and calamity of war, however, swept away its ostrichlike attitude.

The trauma of 1940 "really made us grow up as a nation," noted one

Norwegian who had fought in the war. "Until then we had been part of nothing. We learned the lesson that we had to be prepared ourselves; we couldn't just leave it to others to fight our wars. We became determined not to be taken with our pants down anymore." Since the creation of NATO, Norway has remained one of its most stalwart members.

CHAPTER 29

"My Counsel to Europe . . . : Unite!"

Postwar Europe Bands Together

ON A RAINY SPRING MORNING IN 1942, REPRESENTATIVES of the exile governments of Belgium, the Netherlands, and Luxembourg gathered for a meeting in central London. At that point, the outlook for the Allies seemed as dreary as London's weather: the Soviets appeared to be close to defeat, while the United States and Britain were still reeling from Japanese drubbings in the Pacific. But the Belgians, Dutch, and Luxembourgians had not met to bemoan the present. Trusting that the Allies would triumph in the end, they were there to plan their postwar future.

The three small nations had much in common. Clustered together on the rim of western Europe, they were precariously situated between two major European powers, France and Germany. Like Norway, they had put their faith in neutrality before the war. Like Norway, too, the German invasion of their countries had disabused them of the belief in every country for itself.

Only by banding together politically and economically, they believed, could their nations regain control of their destinies. And if all of western Europe formed such close ties, perhaps it could once again establish a measure of influence and sorely needed security. In pursuing these goals, the three governments in exile decided they would lead the way.

In September 1944, after more negotiations, they signed what came to be known as the Benelux Treaty, which called for the elimination of all tariff duties on the exchange of imported goods between their countries and the establishment of common tariffs on imports from other nations. The treaty also laid the foundation for eventual free movement of workers, capital, and services. As a result of this groundbreaking pact, which took effect in 1948, Belgium, the Netherlands, and Luxembourg helped change the face and future of Europe. The first stirring of the movement toward European unification, the Benelux Treaty became a catalyst for the more revolutionary steps soon to follow.

Paul-Henri Spaak

Among the leaders of the effort was Paul-Henri Spaak, the foreign minister of Belgium. Known as "Mr. Europe" after the war, Spaak became one of the founding fathers of the European union movement, together with such pioneers as Jean Monnet and Robert Schuman of France. Thanks in no small part to Spaak's close involvement, his hometown of Brussels emerged as the movement's base—home to NATO and various other supranational organizations, including the current European Union.

It was a remarkable turnaround for a man who had started the war in disgrace. Spaak's list of sins then were many: he had falsely accused King Leopold of treason, urged Belgium to capitulate to Germany in late 1940, and initially refused to go to England to continue his country's war effort. Once the war ended, Spaak again became a lightning rod in his homeland when he launched his successful drive to keep Leopold from returning to the throne. In 1950, after the king had won a national plebiscite to resume his reign, Spaak and other leftist leaders orchestrated a new campaign to keep him out, which prompted a full-blown political crisis. A general strike was called, which quickly turned violent. Riots broke out in Brussels and other cities, and several people were killed. With Belgium on the brink of civil war, Leopold finally gave in and abdicated in favor of his son, Prince Baudouin.

Throughout his checkered political career, Spaak resembled a modern-day Dr. Jekyll and Mr. Hyde. Confrontational and divisive in his own nation, he worked tirelessly after the war to heal the divisions among the countries of Europe and to bring them together. His transformation from rabble-rouser to international statesman was due largely to his three years in London. Thanks to the many new contacts he made there, his previously parochial outlook on the world became much more cosmopolitan. He even tried to learn to speak English, which, prior to his arrival in Britain, he had adamantly refused to do. He never quite mastered it; his biographer wrote that he was "rather like one of the Marx brothers pretending to talk in a foreign language." Spaak himself once said, "I'm often told that I look like Winston Churchill and speak English like Charles Boyer, but I wish it were the other way around."

Wartime London, as it happened, was a perfect breeding ground for European cooperation. Spaak and officials from all over the Continent worked and socialized together in a way that would never have been possible without the war. Their long stay in the British capital gave them a certain distance from narrow national concerns and allowed them to form close personal and official bonds that bore extraordinary fruit once the conflict was over. "If the European Community is compared to a house, those years of cooperative exile in wartime London are part of the foundation," wrote the historian Robert W. Allen.

For Spaak, the integration of Europe became an obsession. In 1944, he took note of a final message that a member of the Belgian resistance had scrawled on the wall of her jail cell shortly before her execution by the Gestapo: "I have opened a door to you which none shall close." Spaak declared, "When we have won this war, we must unite Europe. We cannot afford any more civil wars among our nations or we will destroy our civilization."

Yet even as they worked for European union, Spaak and other movement leaders were eager to retain their close ties with the British. For all their wartime difficulties with Britain, the exile governments were keenly aware of how much they owed the country that had given them refuge when they most needed it. Indeed, many Europeans hoped, as one Dutch official put it, that "Britain, our closest wartime ally and friend, will not only participate in European cooperation but take the initiative." That could happen, however, only if Britain agreed to abandon its historical insularity and, in Spaak's words, "consent to think of itself as belonging to Europe."

The Europeans looked to Winston Churchill, the man responsible for welcoming them to England in 1940, to lead the charge in aligning his country with the Continent. Initially, there seemed grounds for optimism. A longtime advocate of what he called "a United States of Europe," Churchill had discussed various permutations of the idea with members of his government throughout the war. To Anthony Eden, he had envisioned a new Europe, guarded by an international police force, "in which the barriers between nations will be greatly minimized and unrestricted travel will be possible."

After he lost the 1945 general election, Churchill devoted much of his time and energy to just such a unification campaign. In a speech that "fired multitudes of Europeans with hope and excitement," he declared, "When the Nazi power was broken, I asked myself what was the best advice I could give my fellow citizens in our ravaged and exhausted continent. My counsel to Europe can be given in a single word: Unite!"

In 1949, Churchill's efforts helped lead to the creation of a multilateral organization called the Council of Europe, based in the French city of Strasbourg. Among its ten members were Britain and five of the countries it had sheltered during the war: France, the Benelux nations, and Norway. From the start, however, the council's raison d'être was unclear; with no power and no authority to act, it functioned, for the most part, as a debating society.

Spaak, who became the first president of the council's assembly, grew tired of presiding over "solemn charades" of votes that approved "grandiose schemes [of European integration] which had no chance of being implemented." In early 1950, he snapped, "I admire those who can remain calm in the face of the present state of a Europe . . . that for five years has been living in fear of the Russians and on the charity of the Americans. In the face of all this we remain impassive, as if history was standing still and as if we had decades at our disposal . . . to abandon selfish nationalistic viewpoints."

It soon became evident to Spaak and other movement activists that Churchill, for all his eloquence, was unwilling to do anything concrete to make European union a reality and, further, that he and the British had no interest in becoming an integral part of Europe themselves. Indeed, Churchill had made his feelings clear in a *Saturday Evening Post* article written before the war: "We are with Europe but not of it. We are linked, but not comprised. We are interested and associated, but not absorbed."

As Churchill saw it, Britain's postwar destiny lay in its empire and a close alliance with the United States. He had underscored that view in his vehement declaration to de Gaulle in 1944: "Every time we must choose between Europe and the open sea, we shall always choose the open sea. Every time I must choose between you and Roosevelt, I shall always choose Roosevelt."

His aloofness from European affairs had its roots, of course, in Britain's centuries-old discomfort with continental entanglements. But it also had much to do with his and his country's refusal to accept the fact that its days as a world power were over. It had been bankrupted by the war, and its empire was on the verge of slipping away. But, clinging to the memory of its days as one of the Allies' Big Three, it couldn't abide the idea of surrendering any part of its national sovereignty. In the words of the future German chancellor Willy Brandt, Britain was unable to meet European demands to "renounce the insularity of her past greatness" and join the Continent in an alliance.

Also explaining Britain's standoffishness was its attitude toward the war. To the Europeans, World War II was a cataclysm that must never happen again. To the British, who had suffered neither invasion nor occupation, it was one of the proudest periods of their country's history—a "moment of national reconciliation and rallying together, rather than a corrosive rent in the fabric of state and nation." As Max Hastings remarked, the British came to regard the war "as the last hurrah of their greatness, a historic achievement to set against many postwar failures and disappointments."

Whatever the reasons for Britain's reserve, the European leaders had had enough. Realizing that it would never take the lead in the unification movement, they assumed the initiative themselves in 1950. In the forefront was Jean Monnet, an innovative, far-seeing French political economist who had spent most of the war in Washington, D.C., where he had served as an economic adviser to FDR and been a key figure in the United States' enormously successful wartime industrial mobilization program.

In 1940, Monnet had wanted nothing to do with Charles de Gaulle, whom he considered an "apprentice dictator." He had second thoughts three years later and became armaments minister in the general's French Committee of National Liberation in Algiers. After the war, Monnet, as chairman of the French National Planning Board, was de Gaulle's chief adviser on the reconstruction of the French economy.

In the spring of 1950, Monnet joined forces with Robert Schuman, France's foreign minister and a former member of the French resistance, to introduce a revolutionary economic plan for Europe. Monnet's and Schuman's proposal called for the integration of the two key industries of France and Germany—coal and steel—under a jointly appointed central authority. According to what came to be known as the Schuman Plan, the two countries would surrender their rights to protect and subsidize those industries, making each powerless to use the materials for purely national interests, such as weapons production.

Up to that point, France had shown no interest in reconciling with its former enemy and occupier and rejected the idea of any treatment of Germany on an equal footing. By contrast, Monnet and Schuman were intent on binding the two countries into what Schuman called "an embrace so close that neither could draw back far enough to hit the other." The new coal and steel community, which would be open to other European nations, "represents the first concrete step towards a European federation, which is imperative for the preservation of peace," Schuman said. West Germany, led by its first postwar chancellor, Konrad Adenauer, agreed to participate in negotiations. For the first time in history, Germany was on the verge of allying itself on equal terms with its western European neighbors.

When it made the plan public, the French government purposefully administered snubs to both the United States and Britain, in retaliation for their rebuffs of de Gaulle and the Free French during the war. It informed the U.S. government just one day before it made the official announcement and gave the British no advance notice at all.

The French demanded that the countries interested in joining the community must give their answer by June 2, 1950, or remain outside. British leaders, almost all of whom were hostile to the idea, declined the invitation. Their rejection would cost them dearly in both political and economic terms. As the historians Antony Beevor and Artemis Cooper put it, "any British pretension to the leadership of the Continent was finished."

On April 18, 1951, six western European countries—France, West Germany, Italy, Belgium, the Netherlands, and Luxembourg—signed the treaty establishing the European Coal and Steel Community. With this historic rapprochement between France and Germany, Europe took a giant step away from centuries of ruinous nationalism.

IN 1958, PAUL-HENRI SPAAK emerged as a pivotal figure behind the next great landmark event of European union—the creation of the European Economic Community (Common Market). Among Spaak's contributions was the drafting of the EEC treaty, which called for gradual abolition of economic barriers among the six nations that made up the European Coal and Steel Community. When the EEC began accepting new members soon afterward, Britain once again declined the invitation.

Just two years later, however, the British began having second thoughts. Their empire was dissolving, and their economy was weak. Unlike most of the rest of western Europe, they were still recovering from wartime austerity: basic food rationing in Britain had lasted until 1954. In addition, their "special relationship" with the United States had not turned out to be the close-knit, equal partnership they had sought during and after the war. From the start, the United States had made clear who the dominant partner was. An example was the 1956 Suez crisis, when President Eisenhower used economic pressure to force Britain to halt an ill-advised invasion of Egypt by British, French, and Israeli troops.

In 1961, the British, attracted by the success of the EEC's free trade policy, applied for membership in the organization. "If we try to remain aloof, bearing in mind that this will be happening simultaneously with contraction of our overseas possessions, we shall run the risk of losing political influence and of ceasing to be able to exercise any real claim to be a world power," a British Cabinet committee had warned the year before.

Yet now that the British were ready, albeit reluctantly, to join forces with the rest of Europe, they found themselves stymied by the ally who had been their greatest wartime scourge: Charles de Gaulle. Immediately after the war, de Gaulle had headed France's provisional government but had resigned in early 1946 because of sharp differences with other political leaders. In 1958, he had returned to power as president of France, still smarting from the humiliations he had suffered at the hands of Churchill and Roosevelt. When Britain applied for EEC membership, de Gaulle, recalling Churchill's words that the British would always choose America over Europe, took great satisfaction in vetoing its application.

For the eleven years that de Gaulle governed France, Britain was blocked from the EEC. When it finally became a member in 1973, the community's rules regarding agriculture and various other key areas had been set in ways that the British regarded as detrimental to their interests. "The price of British overdependence on the U.S. was that the country . . . aggravated its estrangement from the European Community," wrote the BBC broadcaster Jeremy Paxman. "It has never caught up since. By the 1990s . . . it was left blowing in the wind." In 1992, the Maastricht Treaty gave a new name, the European Union, to the EEC, which then numbered twelve members. Although Britain remained part of the expanded club, many of its citizens remained as skeptical about the European Union as they had been about its predecessor.

When World War II began in 1939, the British foreign minister, Lord Halifax, had spoken of his nation as being "on the fringe of this mad continent." In June 2016, a majority of Britons, clearly still in full agreement with that assessment, voted to leave the European Union.

The EU certainly had given its detractors in Britain and elsewhere a host of reasons for their hostility. The steady creep of its powers and regulations, the repeated crises involving the Euro and other economic matters, its failure to cope with large-scale immigration, the bitter squabbling among member countries, have all been causes of great concern.

But the antagonism toward the EU does not diminish the fact that the achievements of the campaign for European integration campaign were—and are—remarkable, particularly in the first four decades after the war, when, as the British journalist and historian Neal Ascherson noted, "prosperity, security, health, and equality all increased" in western Europe. "What began as a drive to remove tariff barriers and to free commercial exchange," the *International Herald Tribune* observed, "ended by banishing war between European nations [and] enriching the Continent beyond measure."

European unification also brought into the Western fold the two members of the war's Grand Alliance whose lands had fallen under Soviet domination. On July 9, 1997, eight years after their communist regimes crumbled, Poland and the Czech Republic were admitted to NATO.* In 2004, Slovakia also became a NATO member, and in that same year, all three nations were admitted to the European Union.

The day after Poland entered NATO, President Bill Clinton spoke

---

* On January 1, 1993, Czechoslovakia was divided into two independent states: the Czech Republic and Slovakia.

to thousands of jubilant Poles in the center of Warsaw's Stare Miasto. "Never again will your fate be decided by others," he promised. "Never again will the birthright of freedom be denied you." Among the multitude who cheered Clinton's words was Polish foreign minister Bronisław Geremek, whose Jewish father had perished at Auschwitz. Geremek later recalled that moment as "an unforgettable one for the country, a day on which its independence—so frequently swept away by historical storms—finally gained the protection that came with security guarantees by a mighty alliance. A day on which Poland found itself once again among its traditional allies."

DESPITE THE ROCKY POSTWAR relationship between the British and the Continent, the memory of Britain's wartime ties with occupied Europe continued to burn bright. For the Britons and Europeans who had worked together in London, the legacy of that period was deep and enduring. "During the Second World War, the bonds of friendship between Britain and the Netherlands were tightened in such a way that it had a lasting effect on our postwar relationship," noted Frits Bolkestein, a former Dutch secretary of defense. "We in the Netherlands shall never forget what Britain meant to us in those difficult days."

Even though his icy official relations with the British government never thawed, Charles de Gaulle felt much the same way. After Winston Churchill was defeated in 1945, de Gaulle wrote, "The essential and ineffaceable fact remained that without him, my undertaking would have been in vain from the start." When Churchill died in 1965, de Gaulle was prominent among the mourners at his state funeral in London. The French president was joined by hundreds of former resistance fighters from throughout Europe.

The Europeans' admiration of Churchill was rivaled only by their affection for the BBC, which, as Tom Hickman put it, "bordered on idolatry." When postal service was restored in Europe at the end of the war, a tidal wave of letters of thanks poured into Bush House, four thousand letters from France in the first month alone. Long afterward, Alan Bullock remembered, "it was almost embarrassing to go to Europe because there was so much fuss made about the BBC."

In the postwar years, a number of Europe's state-run broadcasting networks adopted the BBC as their model, including France's Radio-diffusion Française. In its inaugural transmission in October 1944, Ra-diodiffusion declared, "During the long dark four years, the BBC was a torch in the darkness and the embodiment of the promise of libera-tion. The world was in agony; but the BBC played its life-giving music. The world was submerged in lies; but the BBC proclaimed the truth. This tradition of truth and honor will be continued here."

On a personal level, former European exiles credited their experi-ences in wartime London and the array of friends they'd made there with expanding their horizons and making them feel part of a larger world. Joachim Rønneberg, the leader of the Norwegian commandoes who blew up the Norsk Hydro plant in 1943, observed many years after the war, "The British made me feel as if I had two homelands. When I was stationed in Britain, we talked of going home on a mission, but in Norway, we talked of going home to Britain to relax or to take on new assignments."

Many Britons, for their part, strove to maintain the close relation-ships they had formed with the Europeans who had been an integral part of their wartime lives. They included SOE's Francis Cammaerts, who stayed in close touch with the French citizens who had sheltered and otherwise cared for him during the perilous period before and im-mediately after D-Day. He and they shared, he said, a "love which was neither physical nor intellectual. It was eternal. Nothing can take it away."

Late in his life, Cammaerts and his wife settled in a small village in the Drôme Valley, an area in southeastern France where he had spent much of his time organizing resistance groups. From his house, he could see in the distance the Vercors plateau, where so many of his maquis had lost their lives in their gallant, doomed 1944 uprising against the Germans.

TO COMMEMORATE THEIR WARTIME bonds with the Europeans with whom they had worked, a number of British groups, military and ci-vilian, organized annual reunions, some of which continue to this day.

Among them was the RAF Escaping Society, sponsored and financed by the British Air Ministry as a way of paying tribute to the thousands of Europeans who had helped rescue downed Allied airmen. The society provided financial assistance to the families of escape network members who had been killed by the Germans and to members requiring medical treatment or who were otherwise in need. It also helped sponsor reunions between the airmen and those who had aided them.

Year after year, the most heavily attended gathering was the reunion celebrating the work of Andrée de Jongh's Comet line. De Jongh herself was revered by the hundreds of Britons and Americans whom she and her network had saved. Following the war, she was awarded high civilian honors by both nations: the U.S. Medal of Freedom and the British George Medal. She spent most of the rest of her life working as a nurse in leper hospitals in Africa.

Shortly after the war, the RAF demonstrated its appreciation for de Jongh in a unique way. Learning that her mother was on her deathbed in Belgium, Air Ministry officials ordered an RAF training flight from Africa to Britain to make an unscheduled stop in the Ethiopian capital of Addis Ababa, where de Jongh was then living. The plane picked her up and, in another unscheduled leg of the flight, took her to Brussels. After her mother's funeral, another RAF plane took her back to Ethiopia.

IN THE NETHERLANDS, the British veterans of Arnhem formed similar long-lasting ties with the civilians who had helped them survive after the debacle. The guiding force behind that effort was General John Hackett, who never forgot "the courage and compassion" displayed by the Dutch citizens who saved his life.

Shan Hackett's distinguished military career continued for more than two decades after the war. He was named deputy chief of the Imperial General Staff and later became commander in chief of British troops in Germany. He also served as commander of NATO's Northern Army Group, whose forces included the Dutch I Corps. After retiring from the army, Hackett was appointed head of King's College in

London. In the late 1960s, wearing a bowler and carrying an umbrella, he joined his students in marches demanding an increase in government student grants. In later years, he became a well-known television commentator and bestselling author.

Yet throughout Hackett's long and busy life, "the spiritual experience" he had had in Arnhem stayed uppermost in his mind. "This was a battle," he wrote, "but its significance as a human event transcends the military." He and his wife frequently traveled to Holland to visit the de Nooij sisters and the rest of their extended family. De Nooij family members, in turn, often stayed with the Hacketts in England. Hackett also became close lifelong friends with several other residents of Ede whom he had met during his convalescence there, including the resistance members who had spirited him out of the hospital at Arnhem and had later guided him to freedom.

In September 1994, half a century after Operation Market Garden, Hackett and other British survivors of the battle unveiled a stone monument near Arnhem dedicated "To the People of Gelderland," the Dutch province in which the fighting took place. Hackett wrote the inscription:

> 50 years ago British & Polish Airborne soldiers fought here against overwhelming odds to open the way into Germany and bring the war to an early end. Instead we brought death and destruction for which you have never blamed us. . . .
>
> You took us then into your homes as fugitives and friends, we took you forever into our hearts. This strong bond will continue long after we are all gone.

Shan Hackett died in 1997. By 2015, only a tiny handful of Arnhem survivors were still alive. But the prediction inscribed on the monument proved to be true: the attachment between the people of Arnhem and those who fought there has endured.

In 1945, the schoolchildren of Arnhem volunteered to care for the graves of the more than 1,700 Allied dead, mostly Britons and Poles, who were buried in the military cemetery in Oosterbeek, a wooded suburb of Arnhem where much of the fighting had occurred. Each child was assigned to one grave. In addition to laying flowers at the graves and keeping them tidy, the children wrote letters to the families of the men whose resting places they tended. Some developed close relationships with the families. When the children left school, they passed their responsibility on to a new generation of students, who did the same with the next generation. The tradition of "the flower children of Arnhem" still flourishes today.

Every year in September, thousands of people from around the world gather in Arnhem to commemorate what had happened there so many years before. During a simple yet powerful ceremony, the chil-

Schoolchildren in Arnhem place flowers on the graves of Allied troops killed during Operation Market Garden. The annual ceremony continues to this day.

dren of Arnhem circulate throughout the cemetery, solemnly laying flowers at the bases of the hundreds of white crosses and Stars of David.

According to Gerrit Pijpers, a Dutch organizer of the annual commemoration, many of the battle's survivors have asked over the years if they could have their ashes "buried here, next to their comrades." To them, Pijpers said, "this is home."

AUTHOR'S NOTE

I'M OFTEN ASKED HOW LONG IT TAKES ME TO WRITE A BOOK. THE length of time obviously varies, but on average it's two to three years. Not so with *Last Hope Island*. I first got the idea for it more than a decade ago and spent about a year doing research in the United Kingdom, on the Continent, and elsewhere. At that point, however, I decided to stop for a while. Focusing as it does on Britain and much of occupied Europe during World War II, the book was far more complex than any I had written till then. At that juncture in my writing career, it was, quite frankly, too much for me.

So I put it aside and went to work on other projects, including *Citizens of London* and *Those Angry Days*. *Last Hope Island,* though, kept drawing me back. It was such a compelling story—one that had never been fully told—with an enormous cast of the most colorful characters I've ever encountered. Another attraction was the light it sheds on the evolution of today's tumultuous relationship between Britain and the rest of Europe. So, after writing *Those Angry Days,* I plunged back into exploring the subject, and this time it worked. I did a considerable amount of new research, adding to the storehouse of material I had gathered years before. I also drew on research I had accumulated for other books I've written. For a writer, one of the advantages of focusing on a particular subject and time period in one's books, as I have done with Britain and World War II, is the ability to draw on a backlog of knowledge acquired over the years.

When it came time to organize the book, I decided to focus only on the six occupied European nations whose governments escaped to London in the spring and summer of 1940, along with General Charles de Gaulle and his Free French forces. In the spring of 1941, two additional

European countries—Greece and Yugoslavia—were invaded and occupied by Germany. The Greek government fled to Cairo and set up its wartime base there. The Yugoslav government, headed by King Peter II, arrived in London in June 1941. Although the British gave substantial aid to Yugoslav partisans resisting the Germans, the king and his government had little influence with the British or, for that matter, within their own country. For that reason, among others (including the sheer unwieldiness of adding another country to an already complex narrative), I decided to leave Yugoslavia out of the story.

FINALLY, I'D LIKE TO express my deep appreciation to the dozens of librarians and archivists who have so generously helped me in my research over the years. Some work for institutions that I consider to be special treasures, including the UK's National Archives at Kew, the Churchill Archives at Cambridge University, the Library of Congress in Washington, D.C., and the Franklin D. Roosevelt Library in Hyde Park, New York. Another jewel is Georgetown University's Lauinger Library, which has been my home away from home for the last two decades. With its extraordinary collection of books and other material on every aspect of international affairs you can think of, it has made my research incomparably easier.

A word of thanks as well to the many historians whose work I learned from and drew on in writing *Last Hope Island*. I'd like to single out Christopher Andrew, Asa Briggs, Max Hastings, François Kersaudy, David Stafford, and Robert and Isabelle Tombs.

Thanks to everyone at Random House, especially my brilliant editor, Susanna Porter, and her immensely talented associate, Priyanka Krishnan. A shout-out, too, to Henry Rosenbloom, Philip Gwyn Jones, Molly Slight, and the other folks at Scribe, who are publishing *Last Hope Island* in the UK and Australia. And to my longtime friend Gail Ross, for all that she's done for me and the other authors who are fortunate enough to have her as their agent.

And of course, to my husband, Stan Cloud, and my daughter, Carly. Words cannot express how much I owe to both of you. You are my everything.

Notes

INTRODUCTION

xiii "outgrown the continent": Jeremy Paxman, *The English: A Portrait of a People* (Woodstock, NY: Overlook Press, 2000), 32.
"sustained by a peculiar": Murrow manuscript, undated, Edward R. Murrow Papers, Tufts University.
"How horrible": *Times* (London), Sept. 28, 1938.

xv "After gallant France": Martin Gilbert, *Winston S. Churchill*, vol. 7: *Road to Victory, 1941–1945* (Boston: Houghton Mifflin, 1986), 1344.

xvi "It's impossible to explain": Tangye Lean, *Voices in the Darkness: The Story of the European Radio War* (London: Secker & Warburg, 1943), 149.
"Occupation had descended": Erik Hazelhoff Roelfzema, *Soldier of Orange* (London: Sphere, 1982), 38.
"drunk with happiness": Mrs. Robert Henrey, *The Incredible City* (London: J. M. Dent & Sons, 1944), 2.

xvii "Together, we have formed": M. Lisiewicz et al., eds., *Destiny Can Wait: The Air Force in the Second World War* (Nashville, TN: Battery Press, 1949), 343.
"No matter our": Erik Hazelhoff Roelfzema, *In Pursuit of Life* (Stroud, Gloucestershire, UK: Sutton, 2003), 110.

xviii "all those insane": Eve Curie, *Journey Among Warriors* (Garden City, NY: Doubleday, 1943), 481.

CHAPTER 1: "MAJESTY, WE ARE AT WAR!"

3 "full dress and orders": Florence Jaffray Harriman, *Mission to the North* (Philadelphia: J. B. Lippincott, 1941), 248.
"to defend them": Ibid.

5 "Majesty": François Kersaudy, *Norway 1940* (Lincoln: University of Nebraska Press, 1998), 68.
"If Hitler comes": Ibid., 12.

6 "It was very difficult": Richard Petrow, *The Bitter Years: The Invasion and Occupation of Denmark and Norway, April 1940–May 1945* (New York: William Morrow, 1974), 71.

"my old bathtubs": Kersaudy, *Norway 1940*, 11.

"so that Norwegian": Ibid.

"war was the kind": Sigrid Undset, *Return to the Future* (New York: Alfred A. Knopf, 1942), 6.

8   "I actually have plans": Julia Gelardi, *Born to Rule: Five Reigning Consorts, Granddaughters of Queen Victoria* (New York: St. Martin's Press, 2005), 64.

"so socialistic": Ibid., 148.

"Vermont offhandedly trying": John van der Kiste, *Northern Crowns: The Kings of Modern Scandinavia* (Stroud, Gloucestershire, UK: Sutton, 1996), 57.

9   "Herre Konge": Gelardi, *Born to Rule*, 148.

"You don't know me": van der Kiste, *Northern Crowns*, 87.

"All the small nations": Tim Greve, *Haakon VII of Norway: Founder of a New Monarchy* (London, Hurst, 1985), 117.

10   "We are fighting": Martin Gilbert, *Winston S. Churchill*, vol. 6: *Finest Hour, 1939–1941* (Boston: Houghton Mifflin, 1983), 106.

"at all costs": Kersaudy, *Norway 1940*, 49.

11   "how completely senseless": Halvdan Koht, *Norway: Neutral and Invaded* (New York: Macmillan, 1941), 68.

"the nation that bowed": Ibid., 71.

"Reservists and volunteers": Kersaudy, *Norway 1940*, 101.

12   "in flight again": Willy Brandt, *In Exile: Essays, Reflections and Letters, 1933–1947* (Philadelphia: University of Pennsylvania Press, 1971), 22.

"the cause of a free Norway": Ibid., 165.

13   "I think we all": Harriman, *Mission to the North*, 258–89.

14   "hysterical half-men": Undset, *Return to the Future*, 8.

"could not appoint": Kersaudy, *Norway 1940*, 103.

"The government is free": Ibid., 104.

"That instant": Ibid.

15   "this ridiculously small": Jack Adams, *The Doomed Expedition: The Norwegian Campaign of 1940* (London: Leo Cooper, 1989), 31.

"The king": Kersaudy, *Norway 1940*, 106.

16   "not good haters": C. J. Hambro, *I Saw It Happen in Norway* (New York: D. Appleton-Century, 1940), 150.

"an army of marauders": Undset, *Return to the Future*, 5.

"With weak": Hambro, *I Saw It Happen in Norway*, 161.

17   "The idea of": Lord Ismay, *The Memoirs of General Lord Ismay* (New York: Viking, 1960), 119.

"There were no": Kersaudy, *Norway 1940*, 141.

"elementary knowledge": Magne Skodvin, "Norwegian Neutrality and the Challenge of War," in *Britain and Norway in the Second World War*, ed. Patrick Salmon (London: HMSO, 1995), 16.

"We've been massacred!": Leland Stowe, *No Other Road to Freedom* (New York: Alfred A. Knopf, 1941), 143.

18   "one of the costliest": Philip Knightley, *The First Casualty* (New York: Harcourt Brace Jovanovich, 1975), 227.

"Always too late": John Keegan, ed., *Churchill's Generals* (London: Cassell, 2007), Kindle edition, loc. 632.

"So Norway is": Kersaudy, *Norway 1940*, 170.

"Please tell me": Ibid., 171.

18   **"You are killing"**: Ibid., 178.
19   **"blame should be attached"**: Ibid., 88.
       "The strict observance": Gilbert, *Finest Hour*, 228.
20   **"considering the prominent part"**: Winston S. Churchill, *The Second World War*,
       vol. 1: *The Gathering Storm* (Boston: Houghton Mifflin, 1948), 649–50.
       "You ask what": William Manchester, *The Last Lion: Winston Spencer Churchill:
       Alone, 1932–1940* (New York: Dell, 1988), 678.

CHAPTER 2: "A BOLD AND NOBLE WOMAN"

21   **"They have come"**: Erik Hazelhoff Roelfzema, *Soldier of Orange* (London:
       Sphere, 1982), 15.
       "had been expecting": Ibid.
22   **"who is so popular"**: Roger Keyes, *Outrageous Fortune: The Tragedy of Leopold III
       of the Belgians, 1901–1941* (London: Secker & Warburg, 1984), 212.
23   **"an immoral system"**: Roelfzema, *Soldier of Orange*, 16.
       "no one has given": "Worried Queen," *Time*, November 27, 1939.
       "I raise a fierce": H. R. H. Wilhelmina, Princess of the Netherlands, *Lonely but
       Not Alone* (London: Hutchinson, 1960), 151.
       "make a bold": Ibid., 37.
       "great deeds": Ibid., 50.
       "the cage": Ibid., 42.
       "any kind of initiative": Ibid.
       "If you are naughty": "Caged No More," *Time*, Dec. 7, 1962.
25   **"In certain respects"**: John Wheeler-Bennett, *Friends, Enemies, and Sovereigns*
       (London: Macmillan, 1976), 158.
       "I fear no man": Roelfzema, *Soldier of Orange*, 113.
       "By the spring": Wilhelmina, *Lonely but Not Alone*, 147–48.
26   **"Our little speck"**: Roelfzema, *Soldier of Orange*, 10.
27   **"a great new power"**: Keyes, *Outrageous Fortune*, 86–87.
       "bewildered and dazed": Lord Ismay, *The Memoirs of General Lord Ismay* (New
       York: Viking, 1960), 126.
28   **"Even if the troops"**: Ibid.
       "haggard and worn": Martin Gilbert, *Winston S. Churchill*, vol. 6: *Finest Hour,
       1939–1941* (Boston: Houghton Mifflin, 1983), 309.
       "was her": Sarah Bradford, *The Reluctant King: The Life and Reign of George VI,
       1895–1952* (New York: St. Martin's Press, 1989), 315.
29   **"calm and unruffled"**: Keyes, *Outrageous Fortune*, 209.
       "be the last man": Wilhelmina, *Lonely but Not Alone*, 154.
30   **"She was naturally"**: Bradford, *The Reluctant King*, 315.

CHAPTER 3: "A COMPLETE AND UTTER SHAMBLES"

31   **"When it is a question"**: "Leopold Goes to War," *Time*, May 20, 1940.
       "inviolability of Belgian territory": Roger Keyes, *Outrageous Fortune*, 4.
34   **"carry the conflict"**: Ibid., 109.
35   **"a determined"**: Brian Bond, *Britain, France and Belgium, 1939–1940* (London:
       Brassey's, 1990), 98.
       "fought like lions": Keyes, *Outrageous Fortune*, 190.

35 **"if the quality"**: Bond, *Britain, France and Belgium*, 98.
36 **"We have been defeated!"**: Winston S. Churchill, *The Second World War*, vol. 2: *Their Finest Hour* (Boston: Houghton Mifflin, 1949), 42.
37 **"It is so much"**: Janet Teissier du Cros, *Divided Loyalties* (New York: Alfred A. Knopf, 1964), xxiv.
    "The real truth": John C. Cairns, "A Nation of Shopkeepers in Search of a Suitable France: 1919–1940," *American Historical Review*, June 1974.
    "America is far away": Robert Tombs and Isabelle Tombs, *That Sweet Enemy: Britain and France: The History of a Love-Hate Relationship* (New York: Vintage, 2008), 502.
    "France wanted revenge": Margaret MacMillan, *Peacemakers: The Paris Conference of 1919 and Its Attempt to End War* (London: John Murray, 2001), 39.
39 **"the English have"**: Tombs and Tombs, *That Sweet Enemy*, 541.
40 **"A genuine alliance"**: Marc Bloch, *Strange Defeat* (Important Books, 2013), 68.
    "learners in the military arts": Martin S. Alexander, "Dunkirk in Military Operations, Myths, and Memories," in *Britain and France in Two World Wars: Truth, Myth and Memory*, ed. Robert Tombs and Emile Chabal (London: Bloomsbury Academic, 2013), 95.
41 **"Ever since 1907"**: François Kersaudy, *Churchill and de Gaulle* (New York: Atheneum, 1982), 26.
    "stew in her own juice": Tombs and Tombs, *That Sweet Enemy*, 571.
    "the French will look": Ibid., 528.
    "the finest in Europe": Kersaudy, *Churchill and de Gaulle*, 31.
    "utter dejection": Churchill, *Their Finest Hour*, 46.
    "There are none": Ibid.
    "one of the greatest": Ibid., 47.
42 **"Although there were"**: Bond, *Britain, France and Belgium*, 294.
    "We had only": Keyes, *Outrageous Fortune*, 107.
    "this inhuman monster": Charles Glass, *Americans in Paris: Life and Death Under Nazi Occupation* (New York: Penguin Books, 2009), 78.
    "a complete and utter shambles": Norman Gelb, *Scramble: A Narrative History of the Battle of Britain* (San Diego: Harcourt Brace Jovanovich, 1985), 10.
    "This is like": Tombs and Tombs, *That Sweet Enemy*, 553.
    "In all the history": Martin Gilbert, *Winston S. Churchill*, vol. 6: *Finest Hour, 1939–1941* (Boston: Houghton Mifflin, 1983), 377.
44 **"the Belgian Army might"**: Ibid., 362.
    "We don't care": Bond, *Britain, France and Belgium*, 92.
    "Belgian morale": Keyes, *Outrageous Fortune*, 214.
    "rotten to the core": Ibid., 324.
    "lesser breeds": Ibid., 187.
    "after assisting the BEF": Gilbert, *Finest Hour*, 407.
    "The Belgian Army has": Keyes, *Outrageous Fortune*, 338.
45 **"Never would King Albert"**: James H. Huizinga, *Mr. Europe: A Political Biography of Paul Henri Spaak* (New York: Praeger, 1961), 121.
    "an idle refugee monarch": Keyes, *Outrageous Fortune*, 308.
    "For the duration": Ibid., 360.
    "Defeat arouses": Irène Némirovsky, *Suite Française* (New York: Knopf, 2006), 313.
    "When one is fighting": Huizinga, *Mr. Europe*, 140.
46 **"good thing"**: Keyes, *Outrageous Fortune*, 372.

46 "the Belgian Army, virtually": Ibid., 187.
    "There has never": Bond, *Britain, France and Belgium*, 95.
47 "suddenly and unconditionally": Keyes, *Outrageous Fortune*, 362.
    "treating with the enemy": Ibid., 363.
    "for the space": Mollie Panter-Downes, *London War Notes: 1939–1945* (New York: Farrar, Straus and Giroux, 1971), 63–64.
48 "The king's capitulation": Keyes, *Outrageous Fortune*, 378.
    "savage and lying attacks": Ibid., 242.
    "the truth should not": Ibid.
    "vilification of a brave king": Ibid., 402.
    "fought very bravely": Churchill, *Their Finest Hour*, 95.
    "completed the full circle": Keyes, *Outrageous Fortune*, 399.
49 "Suddenly, without": Ibid., 402–3.
    "Seldom can a prime minister": William Manchester, *The Last Lion: Winston Spencer Churchill: Alone, 1932–1940* (New York: Dell, 1988), 677.
50 "the public interest": Keyes, *Outrageous Fortune*, 455.
    "a very grave injustice": Ibid., 459.
    "K.C. Clears King": Ibid.
51 "one of hideous complexity": John W. Wheeler-Bennett, *King George VI: His Life and Reign* (New York: St. Martin's Press, 1958), 452.
    "mon cher Bertie": Keyes, *Outrageous Fortune*, 310.
    "To act otherwise": Ibid., 309.
    "mixed up": Wheeler-Bennett, *King George VI*, 455.

CHAPTER 4: "WE SHALL CONQUER TOGETHER—
OR WE SHALL DIE TOGETHER"

52 "One feels": Richard Collier, *1940: The World in Flames* (London: Hamish Hamilton, 1979), 98.
53 "on further reflection": François Kersaudy, *Norway 1940*, 221.
    "unable to help": Ibid.
54 "God save Norway!": Halvdan Koht, *Norway: Neutral and Invaded* (New York: Macmillan, 1941), 126.
    "I am so afraid": Kersaudy, *Norway 1940*, 223.
    "extremely depressed": Ibid.
55 "So long as": Richard M. Ketchum, *The Borrowed Years, 1938–1941: America on the Way to War* (New York: Random House, 1989), 352.
    "We shall fight": Winston S. Churchill, *The Second World War*, vol. 2: *Their Finest Hour* (Boston: Houghton Mifflin, 1949), 118.
    "So you are admitting": Robert Tombs and Isabelle Tombs, *That Sweet Enemy: Britain and France: The History of a Love-Hate Relationship* (New York: Vintage, 2008), 555.
56 "the gutless collapse": Martin S. Alexander, "Dunkirk in Military Operations, Myths, and Memories," in *Britain and France in Two World Wars: Truth, Myth and Memory*, ed. Robert Tombs and Emile Chabal (London: Bloomsbury Academic, 2013), 98.
    "the French war effort": John C. Cairns, "A Nation of Shopkeepers in Search of a Suitable France: 1919–1940," *American Historical Review*, June 1974.
    "as bravely as": Julian Jackson, *France: The Dark Years, 1940–1944* (Oxford: Oxford University Press, 2001), 98.

56 **"very few British divisions"**: P.M.H. Bell, *A Certain Eventuality: Britain and the Fall of France* (London: Saxon House, 1974), 68.

"our contribution": Lord Ismay, *The Memoirs of General Lord Ismay* (New York: Viking, 1960), 142.

"on and on and on": Sir Edward Spears, *Assignment to Catastrophe*, vol. 2: *The Fall of France, June 1940* (London: Heinemann, 1954), 150.

"then in the provinces": Ibid., 206.

57 **"her neck wrung"**: Churchill, *Their Finest Hour*, 213.

"This was the great": Tombs and Tombs, *That Sweet Enemy*, 559.

"like an anthill": Jackson, *France: The Dark Years*, 100.

"all the ugliness": Charles Glass, *Americans in Paris: Life and Death Under Nazi Occupation* (New York: Penguin Books, 2009), 79.

58 **"too few arms"**: Robert Tombs, "Two Great Peoples," in *Britain and France in Two World Wars: Truth, Myth and Memory*, ed. Robert Tombs and Emile Chabal (London: Bloomsbury Academic, 2013), 10.

"shamefully feeble": Tombs and Tombs, *That Sweet Enemy*, 567.

"deliberate estrangement": Ibid., 568.

"The fact is": Ibid., 600.

"They seemed almost happy": Eric Sevareid, *Not So Wild a Dream* (New York: Atheneum, 1976), 155.

59 **"not to remain"**: Alexander, "Dunkirk in Military Operations," 107.

"a lifetime steeped": Spears, *The Fall of France, June 1940*, 48.

"Now we are all alone": Brian Bond, *Britain, France and Belgium, 1939–1940* (London: Brassey's, 1990), 117.

"Personally I feel": Alexander, "Dunkirk in Military Operations," 107.

"Certainly everything": Sir Alexander Cadogan, *The Diaries of Sir Alexander Cadogan, 1938–1945*, ed. David Dilks (New York: Putnam, 1971), 308.

60 **"Never"**: Churchill, *Their Finest Hour*, 145.

"One had to be": James H. Huizinga, *Mr. Europe: A Political Biography of Paul Henri Spaak* (New York: Praeger, 1961), 154.

"It would be difficult": Mollie Panter-Downes, *London War Notes: 1939–1945* (New York: Farrar, Straus and Giroux, 1971), 70.

"every crank in the world": Bell, *A Certain Eventuality*, 93.

61 **"All we knew"**: M. Lisiewicz et al., eds., *Destiny Can Wait: The Polish Air Force in the Second World War* (Nashville, TN: Battery Press, 1949), 35.

"Tell your army": Jan Ciechanowski, *Defeat in Victory* (Garden City, NY: Doubleday, 1947), 15.

62 **"to make every effort"**: Martin Gilbert, *Winston S. Churchill*, vol. 6: *Finest Hour, 1939–1941* (Boston: Houghton Mifflin, 1983), 574.

"The war is over": Lewis White, "The 1940 Evacuation," *On All Fronts: Czechs and Slovaks in World War II*, ed. Lewis White (Boulder, CO: East European Monographs, 1991), 71.

63 **"than an army"**: Lise Lindbaek, *Norway's New Saga of the Sea: The Story of Her Merchant Marine in World War II* (New York: Exposition Press, 1969), 33.

64 **"France has thrown in"**: Huizinga, *Mr. Europe*, 157.

"if you give up": Roger Keyes, *Outrageous Fortune: The Tragedy of Leopold III of the Belgians, 1901–1941* (London: Secker & Warburg, 1984), 417.

"You must have": Ibid., 382–83.

"virtually the entire": Ibid., 383.

65   **"were in too deep"**: Huizinga, *Mr. Europe,* 150.
      "lacked all social vices": Harold Callender, "General de Gaulle—The Legend and the Man," *New York Times Magazine,* July 9, 1944.
66   **"an improbable creature"**: Lord Moran, *Churchill at War, 1940–45* (New York: Carroll & Graf, 2002), 98.
      "the character of ": Dorothy Shipley White, *Seeds of Discord: De Gaulle, Free France, and the Allies* (Syracuse, NY: Syracuse University Press, 1964), 27.
      "brilliance and talent": Jean Lacouture, *De Gaulle: The Rebel, 1890–1944* (New York: W. W. Norton, 1990), 70.
67   **"an 'undisciplined act' "**: Ibid., 174.
      "your presence at my side": Charles de Gaulle, *The Complete War Memoirs* (New York: Carroll & Graf, 1998), 54.
      "rally French opinion": Spears, *The Fall of France, June 1940,* 312.
      "gaping faces": Ibid., 322.
      "in a hideously difficult position": Ismay, *The Memoirs of General Lord Ismay,* 356.
68   **"a loser"**: Lacouture, *De Gaulle,* 286.
      "I can't tell you": Ibid.
      "You are alone!": François Kersaudy, *Churchill and de Gaulle* (New York: Atheneum, 1982), 83.
69   **"Gen. de Gaulle"**: Ibid.
      "an act of faith": Ibid.
      "I was nothing": de Gaulle, *The Complete War Memoirs,* 83.
      "magnificently absurd": Janet Teissier du Cros, *Divided Loyalties* (New York: Alfred A. Knopf, 1964), 98.
      "I have neither": de Gaulle, *The Complete War Memoirs,* 83.

CHAPTER 5: "SOMETHING CALLED HEAVY WATER"

71   **"mixture between"**: Harold Macmillan, *The Blast of War: 1939–1945* (New York: Harper & Row, 1967), 80.
72   **"The country"**: Denis Brian, *The Curies: A Biography of the Most Controversial Family in Science* (Hoboken, NJ: John Wiley & Sons, 2005), 277.
73   **"not wish to"**: Per F. Dahl, *Heavy Water and the Wartime Race for Nuclear Energy* (Bristol, UK: Institute of Physics Publishing, 1999), 107.
75   **"For twenty generations"**: John Nesbitt, "Passing Parade," radio program, date unknown.
76   **"those mad Howards"**: William D. Bayles, "The Incredible Earl of Suffolk," *Saturday Evening Post,* Nov. 28, 1942.
      "Jack was a rebel": Ibid.
      "I don't see how": John Bartleson, Jr., "The Earl of Suffolk and the Holy Trinity," *The Disposaleer,* Feb. 1994.
77   **"won over completely"**: James Owen, *Danger UXB: The Heroic Story of the WWII Bomb Disposal Teams* (London: Abacus, 2010), 65.
      "The single thought": Bayles, "The Incredible Earl of Suffolk."
78   **"an unkempt pirate"**: Brian, *The Curies,* 292.
79   **"a young man"**: Macmillan, *The Blast of War,* 78.
      "something called heavy water": Ibid.
      "a truly Elizabethan character": Ibid., 79.
      "I have had": Ibid., 81.

79   "**may prove to be**": Owen, *Danger UXB,* 71.

80   "**I remember the spring**": Spencer R. Weart, *Scientists in Power* (Cambridge, MA: Harvard University Press, 1979), 170.
     "**Had the British**": Ibid., 176.
     "**If von Halban**": Ibid., 179.

81   "**some of them**": Owen, *Danger UXB,* 71.

CHAPTER 6: "THEY ARE BETTER THAN ANY OF US"

82   "**whole fury and might**": Winston S. Churchill, *The Second World War,* vol. 2: *Their Finest Hour* (Boston: Houghton Mifflin, 1949), 225.

83   "**with blond hair**": Virginia Cowles, *Looking for Trouble* (New York: Harper, 1941), 406.
     "**the infiltration of foreign pilots**": Alan Brown, *Airmen in Exile: The Allied Air Forces in the Second World War* (Stroud, UK: Sutton, 2000), 204.

84   "**was some one hundred**": Flying Officer Geoffrey Marsh, "The Collaboration with the English: Squadron 303, Kosciuszko," *Skrzydła,* Sept. 1–14, 1941.
     "**a rung or two**": Adam Zamoyski, *The Forgotten Few: The Polish Air Force in the Second World War* (London: Hippocrene, 1995), 58.
     "**All I knew**": John A. Kent, *One of the Few* (London: Kimber, 1971), 100.

85   "**The country was poised**": Josef Korbel, *Twentieth-Century Czechoslovakia: The Meanings of Its History* (New York: Columbia University Press, 1977), 136.
     "**We are only**": Ibid., 141.
     "**We have reached**": Ronald Clark, *Battle for Britain: Sixteen Weeks That Changed the Course of History* (New York: Franklin Watts, 1966), 114.

86   "**air units in this country**": UK Air Ministry, report on Polish Air Force, March 29, 1940, AIR 2/4213, National Archives, London.
     "**My mind was still**": Zamoyski, *The Forgotten Few,* 57.

87   "**I'm not having**": Richard Collier, *Eagle Day: The Battle of Britain* (New York: Dutton, 1966), 22.

88   "**We had to reverse**": Jan Zumbach, *On Wings of War* (London: André Deutsch, 1975), 66.
     "**we were not**": Ibid., 65.

89   "**They were a complete**": Zamoyski, *The Forgotten Few,* 79.

90   "**Most people who went**": *Daily Telegraph,* July 25, 2000.
     "**Some I couldn't remember**": Norman Gelb, *Scramble: A Narrative History of the Battle of Britain* (San Diego: Harcourt Brace Jovanovich, 1985), 219.
     "**intense struggle**": Churchill, *Their Finest Hour,* 325.
     "**On virtually every occasion**": Richard Hough and Denis Richards, *The Battle of Britain* (London: Hodder & Stoughton, 1989), 221.

91   "**Magnificent fighting**": Ferić, diary, undated, Polish Institute and Sikorski Museum, London.

92   "**You use the air**": Ibid.
     "**absolute tigers**": Edward Raczyński, *In Allied London* (London: Weidenfeld & Nicolson, 1963), 70.
     "**They are fantastic**": Rosme Curtis, *Winged Tenacity: The Polish Air Force, 1918–1944* (London: Kingston Hill, 1944), 7.
     "**their understanding and handling**": Zamoyski, *The Forgotten Few,* 93.
     "**Whereas British pilots**": Ibid., 94.

93   **"When they go tearing"**: Ibid., 90.

"**one of the decisive**": Churchill, *Their Finest Hour,* 332.

"**Even though**": Stephen Bungay, *The Most Dangerous Enemy: A History of the Battle of Britain* (London: Aurum Press, 2000), 346.

94   **"I am a Pole"**: Lynne Olson and Stanley Cloud, *A Question of Honor: The Kosciuszko Squadron: Forgotten Heroes of World War II* (New York: Alfred A. Knopf, 2003), 156.

"**Had it not been**": Zamoyski, *The Forgotten Few,* 97.

"**If Poland had not stood**": Speech by Queen Elizabeth II to Polish Sejm and Senate, Warsaw, March 26, 1996.

"**in the knowledge**": Alexander Hess, "We Were in the Battle of Britain," *On All Fronts: Czechs and Slovaks in World War II,* ed. Lewis White (Boulder, CO: East European Monographs, 1991), 95.

95   **"stripes of streets"**: Ibid.

"**the stern face**": Ibid., 99.

"**I am British!**": Ibid.

"**A feeling of deepest gratitude**": Ibid., 101.

"**Never in the field**": Churchill, *Their Finest Hour,* 340.

96   **"rare combination of steel nerves"**: William D. Bayles, "The Incredible Earl of Suffolk," *Saturday Evening Post,* Nov. 28, 1942.

"**He had us all**": Ibid.

97   **"Charles Henry George Howard"**: Ibid.

CHAPTER 7: "MY GOD, THIS IS A LOVELY PLACE TO BE!"

98   **"You walk"**: Quentin Reynolds, *A London Diary* (New York: Random House, 1941), 65.

"**swimming in the full tide**": Charles Ritchie, *The Siren Years: A Canadian Diplomat Abroad, 1937–1945* (Toronto: Macmillan, 1974), 59.

100   **"The Queen"**: Henri van der Zee, *The Hunger Winter: Occupied Holland, 1944–45* (London: Jill Norman and Hobhouse, 1982), 92.

101   **"as he had heard"**: John W. Wheeler-Bennett, *King George VI: His Life and Reign* (New York: St. Martin's Press, 1958), 464.

102   **"as French as the"**: Nicholas Atkin, *The Forgotten French: Exiles in the British Isles, 1940–44* (Manchester, UK: Manchester University Press, 2003), 190.

"**Everybody's goal**": Erik Hazelhoff Roelfzema, *Soldier of Orange* (London: Sphere, 1982), 39–40.

103   **"might find"**: Charles Drazin, *The Finest Years: British Cinema of the 1940s* (London: I. B. Tauris, 2007), 236.

"**Its reputation was such**": Ibid., 237.

104   **"Basically [the British]"**: Eric Sevareid, *Not So Wild a Dream* (New York: Atheneum, 1976), 176.

"**We were living**": Madeleine Albright, *Prague Winter: A Personal Story of Remembrance and War, 1937–1948* (New York: HarperCollins, 2012), 235.

"**degree of separation**": Ibid., 263.

"**a very pleasant**": Ibid.

"**because of the daily**": Lara Feigel, *The Love-charm of Bombs: Restless Lives in the Second World War* (New York: Bloomsbury, 2013), 69.

"**I hope you'll**": Roelfzema, *Soldier of Orange,* 101.

105   **"It was the kind"**: Ibid.

"Faced with the prospect": Charles de Gaulle, *The Complete War Memoirs* (New York: Carroll & Graf, 1998), 102.

106   **"I do not want"**: François Kersaudy, *Churchill and de Gaulle* (New York: Atheneum, 1982), 86.

"Without Anne": Jean Lacouture, *De Gaulle: The Rebel, 1890–1944* (New York: W.W. Norton, 1990), 108.

"We in this country": Atkin, *The Forgotten French*, 10.

"The generous kindness": de Gaulle, *The Complete War Memoirs*, 102.

107   **"I had been a spectator"**: Sevareid, *Not So Wild a Dream*, 168.

"You had the impulse": Ibid.

108   **"The Poles flying"**: Robert Post, "Poland's Avenging Angels," *New York Times Magazine,* June 29, 1941.

"The Polish aviators": Reynolds, *A London Diary*, 73.

109   **"one of the gayest"**: *The Tatler*, March 5, 1941.

"never to invite": Author's interview with Tadeusz Andersz.

"That is one": Adam Zamoyski, *The Forgotten Few: The Polish Air Force in the Second World War* (London: Hippocrene, 1995), 110.

110   **"My God, this"**: Author's interview with Ludwik Martel.

"No matter our varied": Erik Hazelhoff Roelfzema, *In Pursuit of Life* (Stroud, UK: Sutton, 2003), 153.

"There was a diffused": Elizabeth Bowen, *The Heat of the Day* (New York: Anchor, 2002), 102.

"Well," she replied: John Colville, *The Fringes of Power: 10 Downing Street Diaries, 1939–1955* (New York: W. W. Norton, 1985), 296.

111   **"And remember, keep away"**: Zamoyski, *The Forgotten Few,* 173.

"devoted her entire attention": Nancy Caldwell Sorel, *The Women Who Wrote the War* (New York: Arcade, 1999), 220.

"As for the women": Zamoyski, *The Forgotten Few*, 69.

"I think English women": Witold Urbanowicz, speech, National Air and Space Museum, Washington, DC, Nov. 17, 1981.

112   **BOMBING REICH THRILLS POLES**: Zamoyski, *The Forgotten Few,* 116.

"to get to know": Arkady Fiedler, *Squadron 303: The Polish Fighter Squadron with the RAF* (New York: Roy, 1943), 181.

"As Great Britain": Lara Feigel, *The Love-charm of Bombs: Restless Lives in the Second World War* (New York: Bloomsbury, 2013), 203.

"basic principles": Ibid., 79.

"I cried about": Ibid., 119–20.

CHAPTER 8: "THIS IS LONDON CALLING"

114   **"Nobody ever imagined"**: Tangye Lean, *Voices in the Darkness: The Story of the European Radio War* (London: Secker & Warburg, 1943), 153.

"People of France": Ibid., 122.

115   **"escape for a few minutes"**: Michael Stenton, "Introduction," *Conditions and Politics in Occupied Western Europe, 1940–1945*, http://www.gale.cengage.com/pdf/facts/POWE40-45.pdf.

"In a world": Asa Briggs, *The History of Broadcasting in the United Kingdom*, vol. 3: *The War of Words* (London: Oxford University Press, 1970), 10.

115   **"People who are almost"**: Tom Hickman, *What Did You Do in the War, Auntie?* (London: BBC Books, 1995), 126–27.

"The initials BBC": Briggs, *The War of Words*, 164.

"Assuming that the BBC": Piers Brendon, *The Dark Valley: A Panorama of the 1930s* (New York: Knopf, 2000), 58.

116   **"conspiracy of silence"**: A. M. Sperber, *Murrow: His Life and Times* (New York: Freundlich, 1986), 131.

"very angry": Richard Cockett, *Twilight of Truth: Chamberlain, Appeasement and the Manipulation of the Press* (New York: St. Martin's Press, 1989), 112.

"British expeditionary forces": Leland Stowe, *No Other Road to Freedom* (New York: Alfred A. Knopf, 1941), 281.

"an agreeable": Briggs, *The War of Words*, 20.

"an exquisitely bored": Charles J. Rolo, *Radio Goes to War* (New York: G. P. Putnam's Sons, 1942), 143.

"I want our programs": R. Franklin Smith, *Edward R. Murrow: The War Years* (Kalamazoo, MI: New Issues Press, 1978), 8.

117   **"Well, brothers"**: Sperber, *Murrow: His Life and Times*, 138.

"It seems to me": Hickman, *What Did You Do in the War, Auntie?*, 23.

118   **"one of the most industrious"**: Briggs, *The War of Words*, 163.

"I cannot but resent": Ibid., 178.

"Noel Newsome set": Hickman, *What Did You Do in the War, Auntie?*, 106.

"effective control": Briggs, *The War of Words*, 77.

119   **"was the enemy"**: Ibid., 303.

"one of the major neutrals": Ibid., 77.

"being a historian": Ibid., 21.

120   **"the rock"**: Hickman, *What Did You Do in the War, Auntie?*, 105.

"we were packed": Ibid., 103–4.

121   **"People don't work"**: Hickman, *What Did You Do in the War, Auntie?*, 104.

"halfway between a girls' school": Briggs, *The War of Words*, 20.

"Sorry, dear": John van der Kiste, *Northern Crowns: The Kings of Modern Scandinavia* (Stroud, Gloucestershire, UK: Sutton, 1996), 105.

"It's curious how": Hickman, *What Did You Do in the War, Auntie?*, 127.

122   **"suddenly shouted hurrah"**: Ibid., 106.

"We were very": Ibid., 107.

"ignorant, knowing nothing": Ibid.

123   **"The liberty and independence"**: Tim Greve, *Haakon VII of Norway: Founder of a New Monarchy* (London, Hurst, 1985), 152.

"Thou shalt obey": Richard Petrow, *The Bitter Years: The Invasion and Occupation of Denmark and Norway, April 1940–May 1945* (New York: William Morrow, 1974), 104.

"had been part": Erik Hazelhoff Roelfzema, *Soldier of Orange* (London: Sphere, 1982), 26.

"The Queen had been right": Ibid., 33.

123   **"the arch-enemy of mankind"**: Ibid.

"Her speeches were": Henri van der Zee, *The Hunger Winter: Occupied Holland, 1944–45* (London: Jill Norman and Hobhouse, 1982), 97.

"amazingly heated swear words": N. David J. Barnouw, "Dutch Exiles in London," in *Europe in Exile: European Exile Communities in Britain, 1940–1945*, ed. Martin Conway and José Gotovitch (New York: Berghahn Books, 2001), 231.

124 **"a country"**: R. H. Bruce Lockhart, *Jan Masaryk: A Personal Memoir* (London: Philosophical Library, 1951), 18.

125 **"Jan," said**: Claire Sterling, *The Masaryk Case* (New York: Harper & Row, 1969), 125.
"The hour of retribution": Ibid.

126 **"one of the old servants"**: Ibid., 31.
"Hear *The Tale of Honza*": Ibid., 32.
"If you have sacrificed": John W. Wheeler-Bennett, *Munich: Prologue to Tragedy* (New York: Duell, Sloan and Pearce, 1948), 171.
"the English know": Lockhart, *Jan Masaryk*, 33.

127 **"was handled"**: John Lukacs, *The Great Powers and Eastern Europe* (New York: American Book Co., 1953), 388–89.

CHAPTER 9: "AN AVALANCHE OF Vs"

128 **"There was no great"**: Tom Hickman, *What Did You Do in the War, Auntie?* (London: BBC Books, 1995), 108.
"it was undesirable": Jean Lacouture, *De Gaulle: The Rebel, 1890–1944* (New York: W. W. Norton, 1990), 223.

129 **"I, General de Gaulle"**: Ibid., 225.
"As the irrevocable words": Charles de Gaulle, *The Complete War Memoirs* (New York: Carroll & Graf, 1998), 84.

130 **"a feast of radio"**: Asa Briggs, *The History of Broadcasting in the United Kingdom*, vol. 3: *The War of Words* (London: Oxford University Press, 1970), 226.
"one of the wittiest": Hickman, *What Did You Do in the War, Auntie?*, 115.
"Generally at this time": Tangye Lean, *Voices in the Darkness: The Story of the European Radio War* (London: Secker & Warburg, 1943), 152.
"We were giving": A. M. Sperber, *Murrow: His Life and Times* (New York: Freundlich, 1986), 181.
"shattered and terribly fatigued": Ibid., 121–22.

132 **"the mike as an old"**: Briggs, *The War of Words*, 248.
"The French frequently": Hickman, *What Did You Do in the War, Auntie?*, 116.
"With a message": Lean, *Voices in the Darkness*, 157.

133 **"would rather see"**: Ibid., 160.
"eavesdropping in a French café": Ibid.
"the very soul of French wit": Ibid., 161.
"If only you": Briggs, *The War of Words*, 230.
"I want to be understood": Ibid., 228.

134 **"*C'est moi*"**: Martin Gilbert, *Winston S. Churchill*, vol. 6: *Finest Hour, 1939–1941* (Boston: Houghton Mifflin, 1983), 855.
"Every word you said": Ibid., 857.
"We are going to hear": Lean, *Voices in the Darkness*, 152.

135 **"the father of our defeat"**: Jane Baldwin, *Michel Saint-Denis and the Shaping of the Modern Actor* (Westport, CT: Praeger, 2003), 103.
"Tonight," he told his listeners: Herman Bodson, *Downed Allied Airmen and Evasion of Capture: The Role of Local Resistance Networks in World War II* (Jefferson, NC: McFarland, 2005), 35.

136 **"a multitude of little Vs"**: Briggs, *The War of Words*, 334.
"*an avalanche of Vs*": Ibid., 335.
"not a single empty space": Ibid.

136   *"Il ne faut pas"*: Lean, *Voices in the Darkness*, 187.
137   **"the first pan-European gesture"**: Hickman, *What Did You Do in the War, Auntie?*, 122.
      "surrounded by an immense crowd": Ibid.
138   **"a symbol of the unconquerable"**: Briggs, *The War of Words*, 340.
      "the intellectual invasion": Ibid., 334.
139   **"Soon, perhaps"**: Ibid., 341.
140   **"The [French] underground movement"**: Lean, *Voices in the Darkness*, 149.
      "virtually the entire working population": Louis de Jong, *The Netherlands and Nazi Germany* (Cambridge, MA: Harvard University Press, 1990), 45.
      "desire an anti-Nazi revolution": Briggs, *The War of Words*, 235.
      "great uprising": Ibid.
      "Every patriot a saboteur": Ibid., 236.
141   **"the unknown soldiers"**: Lean, *Voices in the Darkness*, 190.
      "When the British Government gives": Briggs, *The War of Words*, 337.
      "encourage, develop": Ibid.
      "Silly people": Ibid., 336.
142   **"Many speak of revolt"**: Ibid.
      "We were not": Hickman, *What Did You Do in the War, Auntie?*, 120.

CHAPTER 10: SPYING ON THE NAZIS

144   **"the finest in the world"**: Christopher Andrew, *Her Majesty's Secret Service: The Making of the British Intelligence Community* (New York: Viking, 1986), p. 448.
145   **"a young man"**: Robert Erskine Childers, *The Riddle of the Sands* (London: Smith, Elder and Co.), 33.
      "practically every [SOE] officer": David Stafford, *The Silent Game: The Real World of Imaginary Spies* (Athens: University of Georgia Press, 1991), 58.
146   **"the supreme country"**: Ibid., 108.
      "ready to aid": Callum MacDonald, *The Killing of Obergruppenführer Reinhard Heydrich, May 27, 1942* (London: Macmillan, 1989), 20–21.
      "the British Secret Service": Walter Schellenberg, *The Labyrinth: Memoirs of Walter Schellenberg* (New York: Harper & Brothers, 1956), 83.
      "minds untainted": Christopher Andrew, "Introduction," in *The Missing Dimension: Governments and Intelligence Communities in the Twentieth Century*, ed. Christopher Andrew and David Dilks (Urbana, IL: University of Illinois Press, 1984), 9.
147   **"these metropolitan young gentlemen"**: Andrew, *Her Majesty's Secret Service*, 461.
      "by and large pretty stupid": Ibid.
      "there was no need": Ibid., 9.
      "only people with foreign names": Anthony Cave Brown, *"C": The Secret Life of Sir Stewart Graham Menzies* (New York: Macmillan, 1987), 131.
148   **"He was not"**: Patrick Howarth, *Intelligence Chief Extraordinary: The Life of the Ninth Duke of Portland* (London: Bodley Head, 1986), 115.
      "a most unpleasant man": Ben Macintyre, *Double Cross: The True Story of the D-Day Spies* (New York: Crown, 2012), 44.
      "eyes of a hyperactive ferret": Ibid.
      "Everyone was scared": Anthony Read and David Fisher, *Colonel Z: The Secret Life of a Master of Spies* (New York: Viking, 1985), 12.
      "an utter shit": Ibid.

149   **"nothing should ever"**: Malcolm Muggeridge, *Chronicles of Wasted Time,* vol. 2: *The Infernal Grove* (London: Collins, 1973), 122.
"**Secrecy," Muggeridge recalled**: Ibid., 123.
"the conspiracies of self-protection": Stafford, *The Silent Game,* 205–6.

151   **"never been a spy"**: Andrew, *Her Majesty's Secret Service,* 378–79.

152   **"lamentably weak"**: Nelson D. Lankford, *OSS Against the Reich: The World War II Diaries of Col. David K. E. Bruce* (Kent, OH: Kent State University Press, 1991), 125.

155   **"Nefertiti-like beauty and charm"**: Marie-Madeleine Fourcade, *Noah's Ark: A Memoir of Struggle and Resistance* (New York: Dutton, 1972), 9.
"letting women run anything": M.R.D. Foot and J. L. Langley, *MI9: Escape and Evasion, 1939–1945* (London: Biteback Publishing, 2011), 80.

156   **"The Poles had"**: Douglas Dodds-Parker, *Setting Europe Ablaze: Some Account of Ungentlemanly Warfare* (Windlesham, UK: Springwood, 1983), 40.
"they have the best": Jan Stanisław Ciechanowski et al., eds., *Rejewski: Living with the Enigma Secret* (Bydgoszcz, Poland: Bydgoszcz City Council, 2005), 174.
"With generations of clandestine action": Dodds-Parker, *Setting Europe Ablaze,* 182.
"If you live": Read and Fisher, *Colonel Z,* 278.

157   **"the amount of information"**: Schellenberg, *The Labyrinth,* 99.
"One always has": Ibid., 137.

158   **"a man who lives"**: Macintyre, *Double Cross,* 34.
"the boss will be a Pole": Ibid., 37.

159   **"It was thanks to Ultra"**: Brown, *"C,"* 671.
"Intelligence did not decide": Andrew, *Her Majesty's Secret Service,* 487.
"would never have gotten": Gordon Welchman, *The Hut Six Story* (New York: McGraw-Hill, 1982), 305.

160   **"a bit of a character"**: Mavis Batey, *Dilly: The Man Who Broke Enigmas* (London: Biteback Publishing, 2011), Kindle edition, loc. 1435.
"stony silence": Robin Denniston, *Thirty Secret Years: A. G. Denniston's Work in Signals Intelligence, 1914–1944* (Worcestershire, UK: Polperro Heritage Press, 2012), Kindle edition, loc. 2062.
"It was only": Ibid.

161   **"very slow to admit"**: Ibid.
"grasped everything": Batey, *Dilly,* Kindle edition, loc. 1512.
"He can't stand it": Ciechanowski, *Rejewski,* 236.
"Marian and Dilly": Ibid., 69.

162   **"became his own bright self"**: Denniston, *Thirty Secret Years,* Kindle edition, loc. 2091.
*"**Nous marchons ensemble"**:* Batey, *Dilly,* Kindle edition, loc. 1512.
*"**Serdeznie dziękuję"**:* Władysław Kozaczuk, *Enigma: How the German Machine Cipher Was Broken, and How It Was Read by the Allies in World War Two* (Frederick, MD: University Publications of America, 1984), 60.

163   **"Polish treasure trove"**: Batey, *Dilly,* Kindle edition, loc. 1569.
"of almost unbelievably high quality": Tessa Stirling, Daria Nałęcz, and Tadeusz Dubicki, eds., *Intelligence Cooperation Between Poland and Great Britain During World War II* (London: Valentine Mitchell, 2005), 70.
"As 'C' quickly saw": Muggeridge, *Chronicles of Wasted Time,* 128.
"He would not have": Howarth, *Intelligence Chief Extraordinary,* 115.

164   **"golden eggs"**: Andrew, *Her Majesty's Secret Service,* 449.

164   **"The experience of these men"**: Christine Large, *Hijacking Enigma* (Hoboken, NJ: John Wiley & Sons, 2003), 248.

"profited gratuitously": Kozaczuk, *Enigma*, 59.

CHAPTER 11: "MAD HATTER'S TEA PARTY"

166   **"We shall aid"**: Martin Gilbert, *Winston S. Churchill*, vol. 6: *Finest Hour, 1939–1941* (Boston: Houghton Mifflin, 1983), 1109.

"the slave lands": Christopher Andrew, *Her Majesty's Secret Service: The Making of the British Intelligence Community* (New York: Viking, 1986), 476.

167   **"Only a country"**: Eve Curie, *Journey Among Warriors* (New York: Doubleday, 1943), 173.

"was not prepared": John Keegan, ed., *Churchill's Generals* (London: Cassell, 2007), Kindle edition, loc. 212.

"Fear never abated": Ronald C. Rosbottom, *When Paris Went Dark: The City of Light Under German Occupation, 1940–1944* (New York: Little, Brown, 2014), 299.

169   **"only the Poles"**: David Stafford, *Britain and European Resistance, 1940–1945* (Toronto: University of Toronto Press, 1980), Kindle edition, loc. 693.

"The French have no experience": M.R.D. Foot, *SOE in France: An Account of the Work of the British Special Operations Executive in France, 1940–1944* (London: HMSO, 1966), 120.

"For [people] who": Margaret Collins Weitz, *Sisters in the Resistance: How Women Fought to Free France, 1940–1945* (New York: John Wiley & Sons, 1996), 285.

170   **"people pass by"**: Julian Jackson, *France: The Dark Years, 1940–1944* (Oxford: Oxford University Press, 2001), 265.

171   **"an ungentlemanly body"**: Foot, *SOE in France*, 29.

"I think that": Ibid., 140.

"the bastards of Broadway": Leo Marks, *Between Silk and Cyanide: A Codemaker's War, 1941–45* (Stroud, UK: The History Press, 2013), 147.

"full-scale and dangerous brawls": Anthony Cave Brown, *"C": The Secret Life of Sir Stewart Graham Menzies* (New York: Macmillan, 1987), 333.

"Though SOE and MI6": Malcolm Muggeridge, *Chronicles of Wasted Time*, vol. 2: *The Infernal Grove* (London: Collins, 1973), 174.

172   **"to discourage the Poles"**: Terry Charman, "Hugh Dalton, Poland and SOE: 1940–42," in *Special Operations Executive: A New Instrument of War*, ed. Mark Seaman (London: Routledge, 2006), 70.

173   **"Of course, it was amateurish"**: David Stafford, *Secret Agent: The True Story of the Covert War Against Hitler* (Woodstock, NY: Overlook Press, 2003), 44.

174   **"I work for a"**: Marks, *Between Silk and Cyanide*, 313.

"Underground warfare was": Stafford, *Britain and European Resistance*, Kindle edition, loc. 619.

"Entry into SOE": Foot, *SOE in France*, 40.

"The idea that I": Roderick Bailey, *Forgotten Voices in the Secret War: An Inside History of Special Operations During the Second World War* (London: Ebury Press, 2008), Kindle edition, loc. 691.

"The name SOE": Ray Jenkins, *A Pacifist at War* (London: Arrow, 2010), 49.

175   **"Even hardened Norwegians"**: Stafford, *Secret Agent*, 107.

176   **"at the sorts of things"**: Jenkins, *A Pacifist at War*, 60.

"that was pretty useless": Ibid.

"very severe": Ibid.

176   **"Oh, I love your shoes"**: Bailey, *Forgotten Voices in the Secret War*, Kindle edition, loc. 1703.

177   **"absolutely appalling"**: Christopher J. Murphy, *Security and Special Operations: SOE and MI5 During the Second World War* (Basingstoke, UK: Palgrave Macmillan, 2006), 8.
　　"a sensitive, somewhat dreamy girl": Foot, *SOE in France*, 337.
　　"tends to give far": Sarah Helm, *A Life in Secrets: Vera Atkins and the Missing Agents of WWII* (New York: Anchor, 2007), 12–13.

178   **"if this girl's an agent"**: Marks, *Between Silk and Cyanide*, 311.
　　"the amateurish way": H. J. Giskes, *London Calling North Pole* (London: William Kimber, 1953), 10.

CHAPTER 12: FACTIONS, FEUDS, AND INFIGHTING

179   **"almost ready"**: Richard Petrow, *The Bitter Years: The Invasion and Occupation of Denmark and Norway, April 1940–May 1945* (New York: William Morrow, 1974), 122.

180   **"discouragement and disheartenment"**: Joseph Lash, *Roosevelt and Churchill, 1939–1941: The Partnership That Saved the West* (New York: W. W. Norton, 1976), 312.
　　"We are in that": Sir Alexander Cadogan, *The Diaries of Sir Alexander Cadogan, 1938–1945*, ed. David Dilks (New York: Putnam, 1971), 374.
　　"perfect example": Petrow, *The Bitter Years*, 123.

181   **"a military Sunday school"**: Ibid., 124.

182   **"the horrors of German reprisals"**: Ibid., 128.

183   **"Political émigrés"**: Josef Korbel, *Twentieth-Century Czechoslovakia: The Meanings of Its History* (New York: Columbia University Press, 1977), 184.
　　"Intrigues flourished": Erik Hazelhoff Roelfzema, *Soldier of Orange* (London: Sphere, 1982), 172.

184   **"They lived in a world"**: Ibid., 105.
　　"For the British": Mary Soames, *Clementine Churchill: The Biography of a Marriage* (Boston: Houghton Mifflin, 2002), 324–25.

185   **"In time, all realized"**: Arnfinn Moland, "Milorg and SOE," in *Britain and Norway in the Second World War*, ed. Patrick Salmon (London: HMSO, 1995), 146.

186   **"unshakable courage and resolution"**: Tim Greve, *Haakon VII of Norway: Founder of a New Monarchy* (London: Hurst, 1985), 183.
　　"We are all": Ibid.

188   **"Goodbye"**: Roelfzema, *Soldier of Orange*, 122.

189   **"to meet people"**: H. R. H. Wilhelmina, Princess of the Netherlands, *Lonely but Not Alone* (London: Hutchinson, 1960), 86.
　　"For the Queen": Henri van der Zee, *The Hunger Winter: Occupied Holland, 1944–45* (London: Jill Norman and Hobhouse, 1982), 94–95.
　　"The simplest sailor": Roelfzema, *Soldier of Orange*, 171.
　　"suffocating in the porridge": Ibid., 106.
　　"really too busy": Ibid.

190   **"Instead of unbalanced adventurers"**: Erik Hazelhoff Roelfzema, *In Pursuit of Life* (Stroud, UK: Sutton, 2003), 98.

191   **"the focal point"**: Ibid., 99.
　　"people would laugh": Wilhelmina, *Lonely but Not Alone*, 38.
　　"I got the impression": Roelfzema, *Soldier of Orange*, 128.

191 **"Everyone in Holland"**: Ibid., 129.

192 **"The men of Munich"**: R. H. Bruce Lockhart, *Comes the Reckoning* (London: Putnam, 1947), 60.

194 **"This must be put right"**: Ibid., 115.

195 **"Extreme weakness"**: Jean Lacouture, *De Gaulle: The Rebel, 1890–1944* (New York: W. W. Norton, 1990), 261.
"I am no man's subordinate": Ibid., 267.
"All the French émigrés": Harold Nicolson, *The War Years, Diaries & Letters, 1939–1945*, ed. Nigel Nicolson (New York: Atheneum, 1967), 112.
"One had to be": Dorothy Shipley White, *Seeds of Discord: De Gaulle, Free France, and the Allies* (Syracuse, NY: Syracuse University Press, 1964), 95.

196 **"We were constantly"**: Lacouture, *De Gaulle*, 253.

197 **"a loaded pistol"**: White, *Seeds of Discord*, 178.

198 **"the worst muddles"**: François Kersaudy, *Churchill and de Gaulle* (New York: Atheneum, 1982), 98.

199 **"the lowest depths"**: Ibid., 102.
"no intention whatever": Ibid.
"After Dakar": Lacouture, *De Gaulle*, 278.

200 **"I do not think"**: Kersaudy, *Churchill and de Gaulle*, 138.
"He felt it was": Ibid., 116.

201 **"He has clearly"**: Lacouture, *De Gaulle*, 305.
"De Gaulle's attitude": John Colville, *The Fringes of Power: 10 Downing Street Diaries, 1939–1955* (New York: W. W. Norton, 1985), 432.
"A genuine feeling": Wilhelmina, *Lonely but Not Alone*, 185.

CHAPTER 13: "RICH AND POOR RELATIONS"

205 **"financially and socially beyond reproach"**: Noel F. Busch, "Ambassador Biddle," *Life*, Oct. 4, 1943.

206 **"old sport" or "old boy"**: A. J. Liebling, "The Omnibus Diplomat," part 1, *New Yorker*, June 6, 1942.
"One rather expects": Ibid.
"previous career": Busch, "Ambassador Biddle."
"notable for their": Ibid.
"In five years": A. J. Liebling, "The Omnibus Diplomat," part 2, *New Yorker*, June 13, 1942.

208 **"the greatest disaster"**: Robert E. Sherwood, *Roosevelt and Hopkins: An Intimate History* (New York: Harper & Brothers, 1948), 501.

209 **"Papa is at a very"**: Mary Soames, *Clementine Churchill: The Biography of a Marriage* (Boston: Houghton Mifflin, 2002), 415.
"The United Nations": Franklin D. Roosevelt, national radio broadcast, Feb. 23, 1942, Franklin D. Roosevelt Presidential Library and Museum, Hyde Park, NY.
"if the concept": Jan Stanislaw Ciechanowski et al., eds., *Rejewski: Living with the Enigma Secret* (Bydgoszcz, Poland: Bydgoszcz City Council, 2005), 128.

210 **"have any territorial"**: Valentin Berezhkov, "Stalin and FDR," in Cornelis van Minnen and John F. Sears, eds., *FDR and His Contemporaries: Foreign Perceptions of an American President* (New York: St. Martin's, 1992), 50.
"can't live together": Oliver Lyttelton, *The Memoirs of Lord Chandos: An Unexpected View from the Summit* (New York: New American Library, 1963), 296–97.
"He allowed his thoughts": Ibid., 297.

210 **"He seemed to see himself"**: Anthony Eden, *The Reckoning* (Boston: Houghton Mifflin, 1965), 432.

211 **"There is no France"**: "The Presidency: There Is No France," *Time,* July 19, 1943.
"who had escaped": Jean Lacouture, *De Gaulle: The Rebel, 1890–1944* (New York: W. W. Norton, 1990), 335.
"convinced": Wallace Carroll, *Persuade or Perish* (Boston: Houghton Mifflin, 1948), 103.

212 **"must be given no role"**: Lynne Olson, *Citizens of London: The Americans Who Stood with Britain in Its Finest, Darkest Hour* (New York: Random House, 2010), 220.

213 **"France without an Army"**: Lord Moran, *Churchill at War, 1940–45* (New York: Carroll & Graf, 2002), 88.
"You will remember": Martin Gilbert, *Winston S. Churchill,* vol. 7: *Road to Victory, 1941–1945* (Boston: Houghton Mifflin, 1986), 248.
"I am still sorry": Ibid., 249.
"the gravity of the affront": Lacouture, *De Gaulle,* 396.
"French leaders": Kersaudy, *Churchill and Roosevelt,* 220.
"the ungracious attitude": Lacouture, *De Gaulle,* 329–30.
"I don't understand you!": François Kersaudy, *Churchill and de Gaulle* (New York: Atheneum, 1982), 23.

214 **"these words made"**: Ibid.
"There was no other": Margaret Collins Weitz, *Sisters in the Resistance: How Women Fought to Free France, 1940–1945* (New York: John Wiley & Sons, 1996), 266.
"exemplary, amiable and helpful": Julian Jackson, *France: The Dark Years, 1940–1944* (Oxford: Oxford University Press, 2001), 265.

215 **"I lower my head"**: Ibid.
"grew as naturally": Janet Teissier du Cros, *Divided Loyalties* (New York: Alfred A. Knopf, 1964), 241.

216 **"our passionate love"**: Weitz, *Sisters in the Resistance,* 93.
"In our war": Jonathan H. King, "Emmanuel d'Astier and the Nature of the French Resistance," *Journal of Contemporary History,* Oct. 1973.

217 **"The proportion of Jews"**: Ronald C. Rosbottom, *When Paris Went Dark: The City of Light Under German Occupation, 1940–1944* (New York: Little, Brown, 2014), 311.
"the lifeblood of the Resistance": Herman Bodson, *Downed Allied Airmen and Evasion of Capture: The Role of Local Resistance Networks in World War II* (Jefferson, NC: McFarland, 2005), 4.
"Women have such": Ibid., 154–55.

218 **"ninety-nine percent"**: Anthony Cave Brown, *"C": The Secret Life of Sir Stewart Graham Menzies* (New York: Macmillan, 1987), 500.

221 **"We knew that men"**: Douglas Porch, *The French Secret Services: From the Dreyfus Affair to the Gulf War* (New York: Farrar, Straus and Giroux, 1995), 185.
"The general appeared": M.R.D. Foot, *SOE in France: An Account of the Work of the British Special Operations Executive in France, 1940–1944* (London: HMSO, 1966), 181.
"the sort of natural authority": Jackson, *France: The Dark Years,* 409.

222 **"the main organizations"**: Foot, *SOE in France,* 180.
"There is a rising tide": Michael Stenton, *Radio London and Resistance in Occupied*

Europe: British Political Warfare, 1939–1943 (Oxford: Oxford University Press, 2000), 176.

224   **"For my part"**: Lacouture, *De Gaulle*, 349.
"The British in 1940": Eric Sevareid, *Not So Wild a Dream* (New York: Atheneum, 1976), 222.
"has produced violent reactions": Kersaudy, *Churchill and de Gaulle*, 225.
"their uncontested leader": Jackson, *France: The Dark Years*, 428.

225   **"Between Giraud and de Gaulle"**: Harold Nicolson, *The War Years, Diaries & Letters, 1939–1945,* ed. Nigel Nicolson (New York: Atheneum, 1967), 294.
"The people of France": Lacouture, *De Gaulle,* 479.

CHAPTER 14: "THE UGLY REALITY"

227   **"could afford to irritate"**: Edward Raczyński, *In Allied London* (London: Weidenfeld & Nicolson, 1963), 155.
"little short of genocide": George F. Kennan and John Lukacs, *George F. Kennan and the Origins of Containment, 1944–1946: The Kennan-Lukacs Correspondence* (Columbia: University of Missouri Press, 1997), 28.

228   **"with the precision"**: Sir Owen O'Malley, "Disappearance of Polish Officers in the Union of Soviet Socialist Republics," FO 371/34577, National Archives, London.

229   **"Whether you wish"**: Allen Paul, *Katyn: The Untold Story of Stalin's Polish Massacre* (New York: Scribner's, 1991), 159.
"one of the greatest": John Keegan, *Six Armies in Normandy: From D-Day to the Liberation of Paris* (New York: Viking, 1982), 262.
"neither sought nor cared": Raczyński, *In Allied London,* 100.

230   **"To survive is an obsession"**: "Czechoslovakia: The Art of Survival," *Time,* March 27, 1944.

231   **"The Poles are not troublesome"**: John Darnton, "The Polish Awakening," *New York Times,* June 14, 1981.
"The Czechs seem": A. J. Liebling, *The Road Back to Paris* (Garden City, NY: Doubleday, 1944), 149.

232   **"lost all realistic perspective"**: František Moravec, *Master of Spies: The Memoirs of General František Moravec* (Garden City, NY: Doubleday, 1975), 184.

233   **"those who kept"**: Madeleine Albright, *Prague Winter: A Personal Story of Remembrance and War, 1937–1948* (New York: HarperCollins, 2012), 294.

234   **"It was futile"**: Moravec, *Master of Spies*, 196.
"our whole situation": Vojtech Mastny, *The Czechs Under Nazi Rule: The Failure of National Resistance, 1939–1942* (New York: Columbia University Press, 1971), 177.
"a blond god": Callum MacDonald, *The Killing of Obergruppenführer Reinhard Heydrich, May 27, 1942* (London: Macmillan, 1989), 4.
"a predatory animal": Walter Schellenberg, *The Labyrinth: Memoirs of Walter Schellenberg* (New York: Harper & Brothers, 1956), 13.
"This man is": Laurent Binet, *HHhH* (New York: Farrar, Straus and Giroux, 2012), 39.
"had an ice-cold intellect": Schellenberg, *The Labyrinth,* 13.

235   **"orgy of massacre"**: Martin Gilbert, *The Second World War: A Complete History* (New York: Henry Holt, 1987), 5.

236   **"any infringement whatsoever"**: H. J. Giskes, *London Calling North Pole* (London: William Kimber, 1953), 25.

236   **"The epidemic of assassination"**: MacDonald, *The Killing of Obergruppenführer Reinhard Heydrich*, 163.
"**obviously large-scale resistance movement**": Mastny, *The Czechs Under Nazi Rule*, 186.

237   **"the Czechs at the moment"**: MacDonald, *The Killing of Obergruppenführer Reinhard Heydrich*, 162.
"**our trained paratroop commandos**": Moravec, *Master of Spies*, 196.
"**necessary for the good**": Ibid., 197.
"**spectacular assassination**": MacDonald, *The Killing of Obergruppenführer Reinhard Heydrich*, 118.

238   **"For everyone politically active"**: Ibid., 140.
"**had no intention**": Ibid., 141.
"**This assassination**": Ibid., 156.

239   **"Why should my Czechs"**: Binet, *HHhH*, 216.

240   **"[The Führer] foresees"**: MacDonald, *The Killing of Obergruppenführer Reinhard Heydrich*, 176.
"**It is our holy duty**": Ibid., 3.
"**They're completely mad**": Ibid., 177.
"**seemed almost insane**": M.R.D. Foot and J. L. Langley, *MI9: Escape and Evasion, 1939–1945* (London: Biteback Publishing, 2011), 166.

241   **"If future generations"**: MacDonald, *The Killing of Obergruppenführer Reinhard Heydrich*, 200.

242   **"I was in the U.S."**: Ibid.
"**In the delicate matter**": Ibid., 209.
"**the Czechs and all**": Ibid., 200.
"**In view of the trials**": Ibid.
"**Somebody else**": Roderick Bailey, *Forgotten Voices in the Secret War: An Inside History of Special Operations During the Second World War* (London: Ebury Press, 2008), Kindle edition, loc. 1938.

243   **"By his death"**: Mastny, *The Czechs Under Nazi Rule*, 221.
"**a complete fabrication**": MacDonald, *The Killing of Obergruppenführer Reinhard Heydrich*, 206.

CHAPTER 15: "THE ENGLAND GAME"

244   **"In London"**: Winston S. Churchill, *The Second World War*, vol. 4: *The Hinge of Fate* (Boston: Houghton Mifflin, 1950), 780.

245   **"No matter which"**: Leo Marks, *Between Silk and Cyanide: A Codemaker's War, 1941–45* (Stroud, UK: The History Press, 2013), 28.
"**the most valuable link**": M.R.D. Foot, *SOE in France: An Account of the Work of the British Special Operations Executive in France, 1940–1944* (London: HMSO, 1966), 102.

246   **"so familiar"**: "Leo Marks" (obituary), *Guardian*, Feb. 2, 2001.

247   **"Every code"**: "The Masterspy of Acton Town," *Evening Standard*, Jan. 8, 1999.
"**If some shit-scared**": Marks, *Between Silk and Cyanide*, 8.

248   **"performing with the precision"**: Ibid., 23.

251   **"The whole thing"**: Ibid., 16.
"**no one will blame you**": "Rivalry in London Led to Deaths of Agents," *SOE and the Resistance: As Told in* The Times *Obituaries*, ed. Michael Tillotson (London: Bloomsbury, 2011), Kindle edition, loc. 1583.

253 **"I cursed my stupidity"**: H. J. Giskes, *London Calling North Pole* (London: William Kimber, 1953), 178.

254 **"extremely skilled and dangerous opponents"**: M.R.D. Foot, *Holland at War Against Hitler: Anglo-Dutch Relations, 1940–1945* (London: Frank Cass, 1990), 147.
   "a deep love": Giskes, *London Calling North Pole*, 176.
   "no agent": Foot, *Holland at War*, 133.

255 **"this continuous negligence"**: Giskes, *London Calling North Pole*, 200.
   "famous for its long experience": Ibid., 10.
   "amateurs, despite their training": Ibid., 92.

257 **"conveyor-belt"**: Ibid., 122.
   "open attacks": Ibid., 113.

258 **"unknown criminal elements"**: Ibid., 104.
   "I was faced": Ibid., 107.
   "fairy tales": Ibid., 99.
   "all was not well": Nicholas Kelso, *Errors of Judgement: SOE's Disaster in the Netherlands, 1941–44* (London: Robert Hale, 1988), 196.
   "too bloody perfect": Roderick Bailey, *Forgotten Voices in the Secret War: An Inside History of Special Operations During the Second World War* (London: Ebury Press, 2008), Kindle edition, loc. 3334.

259 **"People seemed to get"**: Ibid.
   "They had a stock": Marks, *Between Silk and Cyanide*, 101.
   "Why were the Dutch": Ibid., 113.
   "to be in a prison cell": Ibid., 132.

260 **"God help these agents"**: Ibid., 123.
   "despite deaths by drowning": Ibid., 124.
   "certainly could not be ignored": Ibid., 133.
   "SOE will be ready": Ibid., 148.
   "I did my best": Ibid.

261 **"Only one, sir"**: Ibid., 336.

262 **"The attempt of"**: Giskes, *London Calling North Pole*, 136.

263 **"I'd worked too long"**: Marks, *Between Silk and Cyanide*, 593.
   "treachery on either": House of Commons debate, March 2, 1953, Hansard, vol. 512.
   "The truth is more mundane": M.R.D. Foot, "SOE in the Low Countries," in Mark Seaman, ed., *Special Operations Executive: A New Instrument of War* (London: Routledge, 2006), 83.

CHAPTER 16: "BE MORE CAREFUL NEXT TIME"

264 **"Rather lacking in dash"**: Ray Jenkins, *A Pacifist at War* (London: Arrow, 2010), 62.
   "a plodder who": Leo Marks, *Between Silk and Cyanide: A Codemaker's War, 1941–45* (Stroud, UK: The History Press, 2013), 199.
   "wasn't plodding at all": Ibid., 200.

265 **"tested the logic of it all"**: Ibid.
   "feeling very sorry": Ibid., 201.

266 **"I never believed"**: Jenkins, *A Pacifist at War*, 52.

267 **"I can tell you"**: Marks, *Between Silk and Cyanide*, 387.

268 **"during the year 1942"**: M.R.D. Foot, *SOE in France: An Account of the Work of the British Special Operations Executive in France, 1940–1944* (London: HMSO, 1966), 224.

268   "brought the optimism": Jenkins, *A Pacifist at War*, 51.

"there was nobody else": Sarah Helm, *A Life in Secrets: Vera Atkins and the Missing Agents of WWII* (New York: Anchor, 2007), 485.

"fears and excitements": Philippe de Vomécourt, *An Army of Amateurs* (New York: Doubleday, 1961), 185.

"He was not firm": Ibid.

"one of the most exceptional": Ibid., 10.

"those whose French accents": Ibid., 185.

269   "the all-important details": Ibid., 175.

"If there had been": Ibid., 126.

"almost to a man": Ibid., 108.

"There appeared to be": Jenkins, *A Pacifist at War*, 4.

270   "five or six young men": Ibid., 57.

"That kind of stupidity": Ibid., 74.

271   "From the moment": Ibid., 75.

"Without any": Ibid., 98.

272   "What has continuously": Ibid., 124.

"The resistance in France": Vomécourt, *An Army of Amateurs*, 18.

"The French people": Ibid., 36.

"I was very much awake": Roderick Bailey, *Forgotten Voices in the Secret War: An Inside History of Special Operations During the Second World War* (London: Ebury Press, 2008), Kindle edition, loc. 3069.

273   "You couldn't go ten meters": Douglas Porch, *The French Secret Services: From the Dreyfus Affair to the Gulf War* (New York: Farrar, Straus and Giroux, 1995), 236.

"At the start": de Vomécourt, *An Army of Amateurs*, 85.

274   "the Prosper folks": Jenkins, *A Pacifist at War*, 218.

"The entire Prosper organization": Helm, *A Life in Secrets*, 44.

"You have forgotten": Ibid., 37.

275   "never wanted to believe": Ibid., 38.

"Strategically France is": Ibid., 50.

276   "No one has": Ibid., 54.

"not overburdened with brains": Ibid., 13–14.

277   "The Germans seem": Robert Marshall, *All the King's Men: The Truth Behind SOE's Greatest Wartime Disaster* (London: Collins, 1988), 181.

278   "Déricourt's operation": Jenkins, *A Pacifist at War*, 102.

279   "Aren't you the organizer": Ibid., 221–22.

"You're not supposed to be": Bailey, *Forgotten*, Kindle edition, loc. 3221.

"since the armistice": Marshall, *All the King's Men*, 153.

"typical French backbiting": Ibid., 154.

280   "[Déricourt's] efficiency": Ibid., 272.

281   "formed the impression": Anthony Cave Brown, *"C": The Secret Life of Sir Stewart Graham Menzies* (New York: Macmillan, 1987), 508–9.

"Make no mistake": Marshall, *All the King's Men*, 121.

"could be suppressed": Brown, *"C,"* 508.

"A lot of nonsense": W. Somerset Maugham, *Ashenden: Or the British Agent* (London: Heinemann, 1928), 52.

282   "With delight": Brown, *"C,"* 512.

"resistance groups": Ibid., 552.

"the blood of the martyrs": Ibid., 553.

283   **"the most ridiculous"**: Marie-Madeleine Fourcade, *Noah's Ark: A Memoir of Struggle and Resistance* (New York: Dutton, 1972), 70.
      "operating admirably": Ibid., 86.
      "It's incredible, incredible!": Ibid., 121.

284   **"one of Dansey's"**: Anthony Read and David Fisher, *Colonel Z: The Secret Life of a Master of Spies* (New York: Viking, 1985), 297.
      "did not dim": Fourcade, *Noah's Ark,* 10.
      "Since September 16": Ibid., 272.
      "while the Gestapo": Ibid., 278.

285   **"Everyone I speak to"**: Jonathan H. King, "Emmanuel d'Astier and the Nature of the French Resistance," *Journal of Contemporary History,* Oct. 1973.
      "Never in my life": Fourcade, *Noah's Ark,* 278.

CHAPTER 17: "HEROISM BEYOND ANYTHING I CAN TELL YOU"

287   **"a girl of radiant integrity"**: M.R.D. Foot and J. L. Langley, *MI9: Escape and Evasion, 1939–1945* (London: Biteback Publishing, 2011), 80.

289   **"it takes less time"**: Herman Bodson, *Downed Allied Airmen and Evasion of Capture: The Role of Local Resistance Networks in World War II* (Jefferson, NC: McFarland, 2005), 12.
      "Nothing could have expressed": Airey Neave, *Saturday at M.I.9: The Classic Account of the WWII Escape Organisation* (London: Hodder and Stoughton, 1969), 25.

290   **"admiration of the girls"**: Paul Routledge, *Public Servant, Secret Agent: The Elusive Life and Violent Death of Airey Neave* (London: HarperCollins, 2002), Kindle edition, loc. 1749.
      "I'd never seen": "Airmen Remember Comet Line to Freedom," BBC News, Oct. 24, 2000.
      "the little cyclone": Airey Neave, *Little Cyclone: The Girl Who Started the Comet Line* (London: Biteback Publishing, 2013), Kindle edition, loc. 45.

291   **"a cross-section"**: Foot and Langley, *MI9,* 155.
      "The last decision": J. M. Langley, *Fight Another Day* (Barnsley, UK: Pen & Sword, 2013), 168.

292   **"always a very sore point"**: Ibid., 16.
      "parcels": Bodson, *Downed Allied Airmen,* 52.
      "It is not an easy": Langley, *Fight Another Day,* 50.

293   **"I loved them"**: "Airmen Remember Comet Line to Freedom," BBC News, Oct. 24, 2000.
      "I fell in love": *The Bulletin: The Newsweekly of the Capital of Europe,* Oct. 19, 2000.

294   **"I have nothing"**: Ibid.
      "They were afraid": Neave, *Saturday at M.I.9,* 132.
      "Belgian, Dutch, or French": Langley, *Fight Another Day,* 251.
      "kept to her own rules": Ibid., 136.

296   **"It seemed incredible"**: Ibid., 187.
      "one of the most colorful": Neave, *Saturday at M.I.9,* 183.
      "Pour une femme": Peter Morley, director, "Women of Courage," television documentary, 1978, Imperial War Museum, London.

297   **"looked much younger"**: Neave, *Saturday at M.I.9,* 188.
      "she was used to": Ibid.

297 **"fearlessness, independence"**: Ibid., 189.
"**My godfather died**": Peter Morley, director, "Women of Courage," television documentary, 1978, Imperial War Museum, London.

298 **"simply loved titled people"**: Ibid.
"**I found this completely ridiculous**": Neave, *Saturday at M.I.9*, 190.
"**endanger her own life**": Ibid., 192.

299 **"to use a battle-axe"**: Langley, *Fight Another Day*, 189.
"**I've nothing to say**": Neave, *Saturday at M.I.9*, 191.
"**Spare no effort**": Langley, *Fight Another Day*, 189.
"**I just wanted**": Neave, *Saturday at M.I.9*, 195.

300 **"We've got only one"**: Ibid., 201.
"***die arrogante Engländerin***": Peter Morley, director, "Women of Courage," television documentary, 1978, Imperial War Museum, London.

302 **"In four minutes"**: Foot and Langley, *MI9*, 230.
"**The airmen who come**": "Airmen Remember Comet Line to Freedom," BBC News, Oct. 24, 2000.

CHAPTER 18: A GIANT JIGSAW PUZZLE

303 **"were quite unable"**: H. J. Giskes, *London Calling North Pole* (London: William Kimber, 1953), 138.
"**As fast as**": Tessa Stirling, Daria Nałęcz, and Tadeusz Dubicki, eds., *Intelligence Cooperation Between Poland and Great Britain During World War II* (London: Valentine Mitchell, 2005), 558.

304 **"We had no illusions"**: Giskes, *London Calling North Pole*, 139.
"**practically all the Allied**": Noel F. Busch, "Ambassador Biddle," *Life*, Oct. 4, 1943.

305 **"We can state"**: Stirling, Nałęcz, and Dubicki, eds., *Intelligence Cooperation Between Poland and Great Britain*, 489.

306 **"do for Germany"**: Ben Macintyre, *Double Cross: The True Story of the D-Day Spies* (New York: Crown, 2012), 98.
"**his loyalty is entirely**": Ibid., 32.
"**From their knowledge**": Ibid., 418.

307 **"studied not only by"**: Ibid., 620.

308 **"the new weapons"**: R. V. Jones, *Most Secret War* (Ware, UK: Wordsworth Editions, 1998), 352.
"**no one could say**": Winston S. Churchill, *The Second World War*, vol. 6: *Triumph and Tragedy* (Boston: Houghton Mifflin, 1953), 43.
"**a bit of a tall order**": Stirling, Nałęcz, and Dubicki, eds., *Intelligence Cooperation between Poland and Great Britain*, 475.

309 **"as deafening as"**: Jones, *Most Secret War*, 351.
"**so sited**": Ibid.
"**this extraordinary report**": Ibid., 354.
"**one of the most effective**": David Ignatius, "After Five Decades, a Spy Tells Her Tale," *Washington Post*, Dec. 28, 1998.

310 **"teased them"**: Ibid.
"**had a far-reaching influence**": Winston S. Churchill, *The Second World War*, vol. 5: *Closing the Ring* (Boston: Houghton Mifflin, 1951), 207.

312 **"Were the Germans able"**: Stirling, Nałęcz, and Dubicki, eds., *Intelligence Cooperation Between Poland and Great Britain*, 476.

312   **"The man going home"**: Churchill, *Triumph and Tragedy,* 39.

   "as impersonal": Angus Calder, *The People's War: Britain, 1939–1945* (New York: Pantheon, 1969), 560.

   "aerial shooting gallery": Lynne Olson and Stanley Cloud, *A Question of Honor: The Kosciuszko Squadron: Forgotten Heroes of World War II* (New York: Knopf, 2003), 330.

313   **"Although we could do"**: Churchill, *Triumph and Tragedy,* 53.

   "excellence" and "gallantry": Ibid., 49.

   "lonesomeness": Jones, *Most Secret War,* xiv.

   "We have been working": R. V. Jones, *Reflections on Intelligence* (London: Heinemann, 1989), 218.

314   **"A substantial proportion"**: Jones, *Most Secret War,* 346.

   "a self-propelled projectile": Jones, *Reflections on Intelligence,* 43.

315   **"We will go"**: Sarah Helm, *Ravensbrück: Life and Death in Hitler's Concentration Camp for Women* (New York: Doubleday, 2015), 426.

   "to stir up old memories": Jones, *Most Secret War,* xiv.

   "a great personal experience": Ibid.

CHAPTER 19: "A FORMIDABLE SECRET ARMY"

316   **"YOU ARE TRYING"**: H. J. Giskes, *London Calling North Pole* (London: William Kimber, 1953), 135.

   "would gladly have murdered me": Patrick Howarth, *Intelligence Chief Extraordinary: The Life of the Ninth Duke of Portland* (London: Bodley Head, 1986), 175.

317   **"supreme objective"**: Giskes, *London Calling North Pole,* 146.

   "Despite severe setbacks": Ibid., 139–40.

318   **"Drops of agents"**: Ibid., 147.

   "a highly efficient espionage organization": Ibid., 150.

319   **"To a greater extent"**: Werner Warmbrunn, *The Dutch Under German Occupation 1940–1945* (Stanford, CA: Stanford University Press, 1983), 120.

320   **"In parts of Belgium"**: Giskes, *London Calling North Pole,* 154.

321   **"Well, of course"**: Roderick Bailey, *Forgotten Voices in the Secret War: An Inside History of Special Operations During the Second World War* (London: Ebury Press, 2008), Kindle edition, loc. 3228.

   "There was no shadow": Ray Jenkins, *A Pacifist at War* (London: Arrow, 2010), 102.

   "We thank you": Leo Marks, *Between Silk and Cyanide: A Codemaker's War, 1941–45* (Stroud, UK: The History Press, 2013), 521.

322   **"I knew that"**: Bailey, *Forgotten Voices,* Kindle edition, loc. 3415.

323   **"Without Churchill"**: M.R.D. Foot, *SOE in France: An Account of the Work of the British Special Operations Executive in France, 1940–1944* (London: HMSO, 1966), 352.

   "a man of the Scarlet": David Stafford, *Churchill and Secret Service* (London: John Murray, 1997), 277.

   "Like de Gaulle": Ibid.

   "Brave and desperate men": Ibid.

   "After Churchill": Marks, *Between Silk and Cyanide,* 29.

324   **"Tommy was always prepared"**: Ibid.

   "Our present puny efforts": Stafford, *Churchill and Secret Service,* 278.

   "carry messages": Ibid.

325 **"From January 1944"**: Foot, *SOE in France*, 356.
 "By now, France": Philippe de Vomécourt, *An Army of Amateurs* (New York: Doubleday, 1961), 176.
 "These are very difficult": Jenkins, *A Pacifist at War*, 119.
326 **"No one man"**: de Vomécourt, *An Army of Amateurs*, 184–85.

CHAPTER 20: "THE POOR LITTLE ENGLISH DONKEY"

327 **"lived by faith"**: Jan Nowak, *Courier from Warsaw* (Detroit: Wayne State University Press, 1982), 104.
 "the ideals": Ibid., 105.
328 **"a rage I could"**: Ibid., 268.
 "the forcible transfer": Martin Gilbert, *Winston S. Churchill*, vol. 7: *Road to Victory, 1941–1945* (Boston: Houghton Mifflin, 1986), 17.
 "found it convenient": Max Hastings, *Armageddon: The Battle for Germany, 1944–1945* (New York: Knopf, 2004), 508–9.
329 **"go[ing] to the peace conference"**: U.S. Department of State, *Foreign Relations of the United States*, vol. 3, *1943*, 15.
 "Here I sat": Paul D. Mayle, *Eureka Summit: Agreement in Principle and the Big Three at Tehran* (Newark, DE: University of Delaware Press, 1987), 24.
 "I am the leader": Gilbert, *Road to Victory*, 646.
330 **"As long as I live"**: Edward Raczyński, *In Allied London* (London: Weidenfeld & Nicolson, 1963), 141.
 "the only line of safety": Gilbert, *Road to Victory*, 389.
 "Force is on Russia's side": Allen Paul, *Katyn: The Untold Story of Stalin's Polish Massacre* (New York: Scribner's, 1991), 222.
331 **"This is the end of Poland"**: "Władysław Sikorski," Wikipedia, http://en.wikipedia.org/wiki/Władysław_Sikorski.
 "He was unmistakably": William Mackenzie, *The Secret History of S.O.E.: Special Operations Executive, 1940–1945* (London: St. Ermin's Press, 2000), 312.
332 **"regarded Sikorski"**: Ibid., 164.
 "a tremendous impact": Raczyński, *In Allied London*, 150.
 "was the only man": Harold Nicolson, *The War Years, Diaries & Letters, 1939–1945*, ed. Nigel Nicolson (New York: Atheneum, 1967), 303.
 "the Soviets didn't want": Harvey Sarner, *General Anders and the Soldiers of the Second Polish Corps* (Cathedral City, CA: Brunswick Press, 1997), 155.
 "with the Russians": Gilbert, *Road to Victory*, 671.
333 **"Neither SOE nor"**: E.D.R. Harrison, "The British Special Operations Executive and Poland," *Historical Journal*, Dec. 2000.
 "make-believe joint planning": Peter Wilkinson, *Foreign Fields: The Story of an SOE Operative* (London: I. B. Tauris, 1997), 124.
334 **"the desirability of preparing"**: Michael Alfred Peszke, *The Polish Underground Army, the Western Allies, and the Failure of Strategic Unity in World War II* (Jefferson, NC: McFarland, 2005), 106.
 "the time is fast approaching": Ibid., 110.
 "the Czechs would have found": Marcia Davenport, *Too Strong for Fantasy* (New York: Pocket Books, 1969), 273.
335 **"has been and is now"**: François Kersaudy, *Churchill and de Gaulle* (New York: Atheneum, 1982), 288.

335 **"this vain and even malignant man"**: Ibid., 275.
 "we would not only": Ibid., 279.

336 **"I am reaching the point"**: Ibid., 291.

337 **"An open clash"**: Dwight David Eisenhower, *Crusade in Europe* (Garden City, NY: Doubleday, 1974), 248.
 "The prime minister": Kersaudy, *Churchill and de Gaulle*, 338–39.

338 **"We are going to liberate"**: Charles de Gaulle, *The Complete War Memoirs* (New York: Carroll & Graf, 1998), 557.
 "I did not like": Gilbert, *Road to Victory*, 1083.
 "treason at the height": Lacouture, *De Gaulle*, 524.
 "FDR's pique": Jean Edward Smith, *FDR* (New York: Random House, 2007), 616.
 "in deathly silence": R. H. Bruce Lockhart, *Comes the Reckoning* (London: Putnam, 1947), 304.
 "Without a trace": Ibid.

339 **"I'll have trouble"**: Ibid.

CHAPTER 21: SETTLING THE SCORE

341 **"but the opening phase"**: Asa Briggs, *The History of Broadcasting in the United Kingdom*, vol. 3: *The War of Words* (London: Oxford University Press, 1970), 597.

342 **"In the undergrowth"**: Ibid., 405–6.
 "You make up a short message": Leo Marks, *Between Silk and Cyanide: A Codemaker's War, 1941–45* (Stroud, UK: The History Press, 2013), 487.
 "That was the first manifestation": Roderick Bailey, *Forgotten Voices in the Secret War: An Inside History of Special Operations During the Second World War* (London: Ebury Press, 2008), Kindle edition, loc. 1781.

343 **"active resisters were"**: Ibid., loc. 3044.

344 **"the entire French railway system"**: M.R.D. Foot, *SOE in France: An Account of the Work of the British Special Operations Executive in France, 1940–1944* (London: HMSO, 1966), 408.
 "among the most formidable": Max Hastings, *Das Reich: The March of the 2nd Panzer Division Through France, June 1944* (London: Zenith Press, 2013), 222.

345 **"They surrounded the Germans"**: Foot, *SOE in France*, 408.
 "in a state": Ibid.
 "must immediately pass": Hastings, *Das Reich*, 88.
 "obsession with retaining": Ibid., 39.

346 **"politically impossible"**: Ibid., 74.
 "Not even the most": Ibid., 75.
 "They burned, pillaged and killed": Philippe de Vomécourt, *An Army of Amateurs* (New York: Doubleday, 1961), 14.
 "a foolish": Ray Jenkins, *A Pacifist at War* (London: Arrow, 2010), 153.

347 **"never had a chance"**: Rick Atkinson, *The Guns at Last Light: The War in Western Europe, 1944–1945* (New York: Picador, 2013), 197.
 "the underwater obstacles": Douglas Porch, *The French Secret Services: From the Dreyfus Affair to the Gulf War* (New York: Farrar, Straus and Giroux, 1995), 558.

348 **"the Allies had never"**: Eric Sevareid, *Not So Wild a Dream* (New York: Atheneum, 1976), 439.
 "the shopkeepers": Ibid., 436.

348   **"the resistance reduced"**: Foot, *SOE in France*, 441.
      "yapping at their heels": Julian Jackson, *France: The Dark Years, 1940–1944* (Oxford: Oxford University Press, 2001), 535.
      "What the resistance achieved": David Stafford, *Secret Agent: The True Story of the Covert War Against Hitler* (Woodstock, NY: Overlook Press, 2003), 239.

350   **"of inestimable value"**: Ronald C. Rosbottom, *When Paris Went Dark: The City of Light Under German Occupation, 1940–1944* (New York: Little, Brown, 2014), 273.
      "In no previous war": Foot, *SOE in France*, 441.
      "showed as never before": Porch, *The French Secret Services*, 258.

351   **"One may ask"**: Ibid., 263.
      "in the end": Ibid., 262.
      "necessary myth": Ibid., 263.
      "de Gaulle had to convince": Ibid.
      "If there had been": Jackson, *France: The Dark Years*, 537.

352   **"a Resistance myth"**: Ibid., 535.
      "In recent years": Airey Neave, *Saturday at M.I.9: The Classic Account of the WWII Escape Organisation* (London: Hodder and Stoughton, 1969), 315.
      "While its military successes": de Vomécourt, *An Army of Amateurs*, 10.
      "A man whose army": Sevareid, *Not So Wild a Dream*, 434.

CHAPTER 22: "A TALE OF TWO CITIES"

354   **"standing at Warsaw's gates"**: Tadeusz Bór-Komorowski, *The Secret Army* (London: Victor Gollancz, 1950), 292.

355   **"National dignity and pride"**: Stefan Korbonski, *Fighting Warsaw* (New York: Funk & Wagnalls, 1968), 347–48.
      "in a world of illusion": Wacław Jędrzejewicz, ed., *Poland in the British Parliament, 1939–1945* (New York: Pilsudski Institute, 1946), 498.
      "we are ready to fight": Bór-Komorowski, *The Secret Army*, 206.

356   **"completely impossible"**: Ibid.
      "proclaim the insurrection": Michael Alfred Pezke, *The Polish Underground Army, the Western Allies, and the Failure of Strategic Unity in World War II* (Jefferson, NC: McFarland, 2005), 156.
      "We have no choice": Jan Nowak, *Courier from Warsaw* (Detroit: Wayne State University Press, 1982), 339.

357   **"Every inhabitant of Warsaw"**: Tadeusz Bielecki and Leszek Szymański, *Warsaw Aflame: The 1939–1945 Years* (Los Angeles: Polamerica Press, 1973), 137.

358   **"From the historical point"**: Ibid., 175.
      "maximum effort": Elisabeth Barker, *Churchill and Eden at War* (New York: St. Martin's Press, 1978), 253.
      "could only be provided": Martin Gilbert, *Winston S. Churchill*, vol. 7: *Road to Victory, 1941–1945* (Boston: Houghton Mifflin, 1986), 895.

359   **"Confident of the part"**: Bór-Komorowski, *The Secret Army*, 263.
      "WHEN IN 1940": Edward Raczyński, *In Allied London* (London: Weidenfeld & Nicolson, 1963), 334.

360   **"The Russian armies"**: Gilbert, *Road to Victory*, 871.
      "the Soviet Government could not": U.S. Department of State, *Foreign Relations of the United States*, vol. 3, *1944*, 1374.
      "We intend to have Poland": George F. Kennan, *Memoirs (1925–1950)* (New York: Bantam Books, 1969), 221.

360 **"My grandfather"**: Winston Churchill, speech, American Institute for Polish Culture, Miami, Jan. 28, 2001.

361 **"Thank you"**: Winston S. Churchill and Franklin D. Roosevelt, *Churchill and Roosevelt: The Complete Correspondence,* vol. 3, ed. Warren F. Kimball (Princeton, NJ: Princeton University Press, 1984), 294.

362 **"I suggest that he"**: Jean Lacouture, *De Gaulle: The Rebel, 1890–1944* (New York: W. W. Norton, 1990), 527.
   "On the barricades": Larry Collins and Dominique Lapierre, *Is Paris Burning?* (New York: Simon & Schuster, 1965), 172.

366 **"Tonight," he said**: Stanley Cloud and Lynne Olson, *The Murrow Boys: Pioneers on the Front Lines of Broadcast Journalism* (Boston: Mariner Books, 1996), 217.
   "For five weeks": Bielecki and Szymanski, *Warsaw Aflame,* 175.
   "heartbreaking": "A Tale of Two Cities," *Economist,* Aug. 26, 1944.

367 **"vastly bloodier"**: Ibid.

368 **"did not mean"**: Lynne Olson and Stanley Cloud, *A Question of Honor: The Kosciuszko Squadron: Forgotten Heroes of World War II* (New York: Knopf, 2003), 343.
   "terrible and even humbling": Winston S. Churchill, *The Second World War,* vol. 6: *Triumph and Tragedy* (Boston: Houghton Mifflin, 1953), 141.
   "The problem of relief": Ibid., 143–44.
   "Had the containers": Bór-Komorowski, *The Secret Army,* 350.

369 **"We have been free"**: Ibid., 376.

371 **"jammed elbow to elbow"**: Eric Sevareid, *Not So Wild a Dream* (New York: Atheneum, 1976), 472.
   "had learned": Ibid.

372 **"We all tremble"**: Kersaudy, *Churchill and de Gaulle,* 373.
   "magnanimous in victory": Ibid., 357.
   "it had to be seen": Ibid., 374.
   "were cheering": Lord Ismay, *The Memoirs of General Lord Ismay* (New York: Viking, 1960), 387.
   "For you": Charles de Gaulle, *The Complete War Memoirs* (New York: Carroll & Graf, 1998), 723.
   "incontestable leader": Kersaudy, *Churchill and de Gaulle,* 384.

373 **"We would not have seen"**: Ibid., 375–76.

CHAPTER 23: "I WAS A STRANGER AND YOU TOOK ME IN"

374 **"The joy of Paris"**: Rick Atkinson, *The Guns at Last Light: The War in Western Europe, 1944–1945* (New York: Picador, 2013), 231.

375 **"as much use"**: Ibid., 233.
   "At that moment": Max Hastings, *Armageddon: The Battle for Germany, 1944–1945* (New York: Knopf, 2004), 19.
   "refit, refuel, and rest": Cornelius Ryan, *A Bridge Too Far* (New York: Simon & Schuster, 1995), Kindle edition, loc. 719.

376 **"My excuse"**: Ibid., loc. 1291.
   "militarily, the war is won": Atkinson, *The Guns at Last Light,* 223.

377 **"bad mistake"**: Hastings, *Armageddon,* 21.
   "It was a flight": Henri van der Zee, *The Hunger Winter: Occupied Holland, 1944–45* (London: Jill Norman and Hobhouse, 1982), 21.
   "I wish to give": Ibid., 18.

377 **"The liberation"**: Ryan, *A Bridge Too Far,* Kindle edition, loc. 131.

378 **"This is not the marriage"**: "HRH Prince Bernhard of the Netherlands" (obituary), *Telegraph,* April 12, 2004.
"**exuded an aura**": Erik Hazelhoff Roelfzema, *In Pursuit of Life* (Stroud, UK: Sutton, 2003), 143.

379 **"I would have been"**: van der Zee, *The Hunger Winter,* 108.
"**played a vital**": "HRH Prince Bernhard of the Netherlands" (obituary), *Telegraph,* April 12, 2004.
"**adored him**": van der Zee, *The Hunger Winter,* 111.
"**a major thrust**": Ryan, *A Bridge Too Far,* Kindle edition, loc. 620.

380 **"I don't think"**: Ibid., loc. 1001.
"**Montgomery didn't believe**": Ibid.
"**I would rather**": Michael Alfred Peszke, *The Polish Underground Army, the Western Allies, and the Failure of Strategic Unity in World War II* (Jefferson, NC: McFarland, 2005), 224–25.
"**considered us a bunch**": Ryan, *A Bridge Too Far,* Kindle edition, loc. 1013.
"**I am just**": Ibid.

381 **"obsessed with the idea"**: Ibid., loc. 878.
"**Monty, you're nuts**": Ibid., loc. 1077.

382 **"It was absolutely impossible"**: Ibid., loc. 1563.
"**hysterical and nervous**": "A Life in Peace and War: Conversation with Sir Brian Urquhart," Institute of International Studies, University of California, Berkeley, March 19, 1996.
"**was the one awkward fact**": Atkinson, *The Guns at Last Light,* 264.

383 **"light-heartedness and inexperience"**: M.R.D. Foot, *Holland at War Against Hitler: Anglo-Dutch Relations, 1940–1945* (London: Frank Cass, 1990), 164.
"**inability to suffer fools**": Max Arthur, "Obituary: General Sir John Hackett," *Independent,* Sept. 10, 1996.
"**rather argumentative**": Ryan, *A Bridge Too Far,* Kindle edition, loc. 5032.
"**After harrying**": Foot, *Holland at War,* 164.

384 **"But the Germans"**: Ibid.
"**giggling like schoolboys**": Hastings, *Armageddon,* 36.
"**doesn't like being told**": Ryan, *A Bridge Too Far,* Kindle edition, loc. 6807.
"**German capability**": Foot, *Holland at War,* 165.

385 **"epic cock-up"**: Atkinson, *The Guns at Last Light,* 288.
"**We were prepared**": Ryan, *A Bridge Too Far,* Kindle edition, loc. 4157.

386 **"had an outstanding force"**: Ibid., loc. 5832.
"**absolutely invaluable**": Foot, *Holland at War,* 116.
"**figure of truly heroic proportions**": Ibid., 168.
"**Arnhem, one of the most**": Ryan, *A Bridge Too Far,* Kindle edition, loc. 4125.
"**kind, chivalrous, even comforting**": Atkinson, *The Guns at Last Light,* 275.
"**It was pretty dismaying**": Hastings, *Armageddon,* 60.

387 **"I still feel sick"**: Barry Paris, *Audrey Hepburn* (New York: Berkley, 1996), Kindle edition, loc. 769.

388 **"the heart and center"**: General Sir John Hackett, *I Was a Stranger* (London: Chatto & Windus, 1978), 61.
"**Thank God**": Ibid., 48.
"**passing to and fro**": Ibid., 65.

389 **"was now becoming"**: Ibid.

389 **"peace and industry"**: Ibid., 80.
"It would be": Ibid., 77.
390 **"it would not"**: Ibid., 78.
"Such loving kindness": Ibid.
"Someone in my house": Ibid., 68.
"the tidy gardens": Ibid., 115.
391 **"this mild and unassuming woman"**: Ibid., 140.
"When these ladies": Ibid., 82.
392 **"One or another"**: Ibid., 116.
"almost slinking away": Ibid., 91.
"the searches": Ibid., 141.
394 **"Everything is well"**: van der Zee, *The Hunger Winter*, 136.
395 **"heavy stone of sadness"**: Hackett, *I Was a Stranger*, 160.
"rare and beautiful": Ibid.
"expressed their loyalty": Foot, *Holland at War*, 113.
"It was like": Hackett, *I Was a Stranger*, 187–88.
"I was still both": Ibid., 185.
396 **"Good luck"**: Ibid., 192.
"Hullo, Shan": Ibid., 196.
"The gray goose has gone": Ibid., 201.
"lost three-quarters": Hastings, *Armageddon*, 60.
397 **"The worst thing"**: Peszke, *The Polish Underground Army*, 174.
"Between our front": Hastings, *Armageddon*, 196.
398 **"in order to hinder"**: van der Zee, *The Hunger Winter*, 29.
"it was the most important": Ibid.
"Don't worry": Ibid., 30.

CHAPTER 24: THE HUNGER WINTER

400 **"You saw them"**: Henri van der Zee, *The Hunger Winter: Occupied Holland, 1944–45* (London: Jill Norman and Hobhouse, 1982), 181.
"You look and feel": Ibid., 184.
"smoking ruins and deadly silence": Ibid., 46.
"People with their feet torn": Ibid., 75–76.
401 **"a quiet, oppressive apathy"**: Ibid., 147.
"the year of liberation": Ibid., 87.
"For the first time": Ibid.
402 **"By March"**: Janet Flanner, "Letter from Amsterdam," *New Yorker*, Feb. 15, 1947.
"Everyone tried to cook grass": Barry Paris, *Audrey Hepburn* (New York: Berkley, 1996), Kindle edition, loc. 671.
"shrunken bodies": van der Zee, *The Hunger Winter*, 158.
"My old, beautiful, and noble": Ibid., 209.
403 **"For the first time"**: Erik Hazelhoff Roelfzema, *Soldier of Orange* (London: Sphere, 1982), 233.
"Have you heard?": Ibid.
"living my self-satisfied life": Ibid.
"thanked God": Ibid.
"one of the most impressive": van der Zee, *The Hunger Winter*, 34.

403   "'Famine, floods'": "Wreckers at Work," *Newsweek,* Oct. 16, 1944.

404   "all the farm hands": Rick Atkinson, *The Guns at Last Light: The War in Western Europe, 1944–1945* (New York: Picador, 2013), 399.

"calamity as has not": van der Zee, *The Hunger Winter,* 171.

"military considerations": Hastings, *Armageddon,* 412.

405   "The Allies are admired": van der Zee, *The Hunger Winter,* 191.

"I felt as if I": H. R. H. Wilhelmina, Princess of the Netherlands, *Lonely but Not Alone* (London: Hutchinson, 1960), 181.

"I shall not forget": van der Zee, *The Hunger Winter,* 187.

"I must leave this": Ibid., 173.

"to have Holland cleared up": Martin Gilbert, *Winston S. Churchill,* vol. 7: *Road to Victory, 1941–1945* (Boston: Houghton Mifflin, 1986), 1081.

406   "this slaughter of the Dutch": van der Zee, *The Hunger Winter,* 188.

"thorough explanation": Ibid.

"deep regrets": Ibid.

407   "the Anglo-Americans": Julian Jackson, *France: The Dark Years, 1940–1944* (Oxford: Oxford University Press, 2001), 513.

"wanted to shriek out loud": Ray Jenkins, *A Pacifist at War* (London: Arrow, 2010), 157–58.

"Terrible things": Gilbert, *Road to Victory,* 784.

"had not fully realized": Ibid., 739.

"not prepared to impose": Lord Ismay, *The Memoirs of General Lord Ismay* (New York: Viking, 1960), 349.

408   "The inner door": Atkinson, *The Guns at Last Light,* 555.

"fight to the last man": van der Zee, *The Hunger Winter,* 210.

409   "One has nothing": Ibid., 229.

"decided to leave": Ibid., 244.

410   "The Dutch must": Ibid., 252.

"To the Dutch people": Ian Buruma, *Year Zero: A History of 1945* (New York: Penguin, 2013), 53.

"An old man": van der Zee, *The Hunger Winter,* 253.

"ran outside": Ibid., 253–54.

"The emotion and enthusiasm": Ibid., 254.

"If any emotions": Ibid., 254–55.

411   "Fear was finished": Ibid., 257.

"We are no longer isolated": Ibid.

"I saw people": Ibid., 277–78.

"It gives me a shock": Ibid., 278.

"Wilhelmus of Orange": Ibid., 276–77.

"expressed the longing": Ibid., 277.

412   "felt about the Allies": Ibid.

"We have been kissed": Buruma, *Year Zero,* 14.

413   "On first appearance": van der Zee, *The Hunger Winter,* 298.

"The babies were tragic": Ibid., 302.

414   "We had almost": Paris, Kindle edition, loc. 835.

"packets of tea": General Sir John Hackett, *I Was a Stranger* (London: Chatto & Windus, 1978), 209.

"A leaden pall": Ibid., 210.

415   "looking about them": Ibid.

"With the certainty": Ibid., 211.

415   **"whose gate"**: Ibid.
       "There was no surprise": Ibid.
       "Did you get": Ibid.
416   **"the sharp cold"**: Ibid., 212.
       "we sat down": Ibid., 214.
       "What is it?": Ibid.

CHAPTER 25: "THERE WAS NEVER A HAPPIER DAY"

417   **"in keeping with"**: H. R. H. Wilhelmina, Princess of the Netherlands, *Lonely but Not Alone* (London: Hutchinson, 1960), 227.
       "Rie, Peter, and I": Erik Hazelhoff Roelfzema, *Soldier of Orange* (London: Sphere, 1982), 247.
418   **"she ignored it"**: Ibid., 236.
       "At first, Peter": Ibid., 241.
419   **"In my garden"**: Ibid., 242.
       "I do not intend": Ibid., 241.
       "Captain, this is steak": Ibid.
420   **"synonymous as it was"**: Ibid., 253.
       "War brought Queen Wilhelmina": "The Woman Who Wanted a Smile," *Time*, Sept. 6, 1946.
       "dangers of leaving": Sir Peter Thorne, "Andrew Thorne and the Liberation of Norway," in *Britain and Norway in the Second World War*, ed. Patrick Salmon (London: HMSO, 1995), 209.
422   **"It is safe to say"**: Tim Greve, *Haakon VII of Norway: Founder of a New Monarchy* (London, Hurst, 1985), 170.
       "Everything was out of place": Ibid., 173.
423   **"the most beloved personage"**: "King Haakon Dead in Norway," *New York Times*, Sept. 22, 1957.
       "the situation was full": Sawyer to Secretary of State Edward R. Stettinius, March 29, 1945, U.S. State Department Records, National Archives, Washington, DC.
424   **"Were he to openly back"**: James H. Huizinga, *Mr. Europe: A Political Biography of Paul Henri Spaak* (New York: Praeger, 1961), 177.
425   **"It is not difficult"**: Ibid., 181.
       "gratuitously covering": Ibid., 184.
426   **"The question is not"**: Jan Velaers and Herman Van Goethem, *Leopold III* (Brussels: Lannoo, 2001), 955.
       "I have not had": Huizinga, *Mr. Europe*, 209.

CHAPTER 26: "WHY ARE YOU CRYING, YOUNG MAN?"

427   **"Iron Curtain of the next"**: Caleb Crain, "Almost History: Plzeň, May 1991," *New York Review of Books*, Aug. 21, 2013.
       "In our view": Martin Gilbert, *Winston S. Churchill*, vol. 7: *Road to Victory, 1941–1945* (Boston: Houghton Mifflin, 1986), 1322.
428   **"For God's sake, Brad"**: Carlo D'Este, *Patton: A Genius for War* (New York: HarperPerennial, 1995), 728.
429   **"Personally, and aside"**: Max Hastings, *Armageddon: The Battle for Germany, 1944–1945* (New York: Knopf, 2004), 486.

429   **"We Communists"**: Claire Sterling, *The Masaryk Case* (New York: Harper & Row, 1969), 17–18.

"No [Czech] citizen": Josef Korbel, *Twentieth-Century Czechoslovakia: The Meanings of Its History* (New York: Columbia University Press, 1977), 215.

430   **"Thank God"**: Ibid., 214.

"without enthusiasm": František Moravec, *Master of Spies: The Memoirs of General František Moravec* (Garden City, NY: Doubleday, 1975), 232.

"was moving": Ibid.

431   **"Beneš dealt"**: Korbel, *Twentieth-Century Czechoslovakia*, 197.

"I asked him": Moravec, *Master of Spies*, 229.

"we had no fascists": Ibid., 219.

"I was treated": Ibid., 233.

432   **"coming from America"**: Winston S. Churchill, *The Second World War*, vol. 6: *Triumph and Tragedy* (Boston: Houghton Mifflin, 1953), 367.

"fought like a tiger": Robert Payne, *The Rise and Fall of Stalin* (New York: Avon Books, 1966), 665.

433   **"Poland must be mistress"**: Gilbert, *Road to Victory*, 1184.

"Just think": M. Lisiewicz et al., eds., *Destiny Can Wait: The Polish Air Force in the Second World War* (Nashville, TN: Battery Press, 1949), 168–69.

"keep peace, dignity": Wacław Jędrzejewicz, ed., *Poland in the British Parliament, 1939–1945* (New York: Pilsudski Institute, 1946), 369.

434   **"On the one hand"**: Romuald Lipinski, Memoirs, Polish Combatants Association, http://www.execulink.com/jferenc.

"are starved, beaten": Gilbert, *Road to Victory*, 1243.

435   **"The Poles"**: Max Hastings, *Inferno: The World at War, 1939–1945* (New York: Vintage, 2012), 631.

"source of increasing political embarrassment": Cabinet minutes, Jan. 22, 1946, AIR 8/1157, National Archives, London.

436   **"cold and dispassionate attitude"**: Air Ministry memo, Jan. 17, 1946, FO 371/115, National Archives, London.

"strongest, the most loyal": Ibid.

"Throughout our history": Edward Raczyński, *In Allied London* (London: Weidenfeld & Nicolson, 1963), xii.

437   **"Why are you crying"**: Zamoyski, *The Forgotten Few*, 4.

438   **"Any one of these"**: Jones, *Reflections*, 217.

"What a windfall": Władysław Kozaczuk, *Enigma: How the German Machine Cipher Was Broken, and How It Was Read by the Allies in World War Two* (Frederick, MD: University Publications of America, 1984), 207.

439   **"Setting them to work"**: Władysław Kozaczuk and Jerzy Straszak, *Enigma: How the Poles Broke the Nazi Code* (New York: Hippocrene Books, 2004), 240.

"We cannot exclude": Ibid.

"to cooperate": Jan Stanislaw Ciechanowski et al., eds., *Rejewski: Living with the Enigma Secret* (Bydgoszcz, Poland: Bydgoszcz City Council, 2005), 199.

440   **"It is clear"**: Kozaczuk and Straszak, *Enigma*, 248.

"Such a theft": R. V. Jones, *Reflections on Intelligence* (London: Heinemann, 1989), 213–14.

"The credit I gave them": Ibid., 213.

"Until just before": Ibid., 214.

"would never have": Gordon Welchman, *The Hut Six Story* (New York: McGraw-Hill, 1982), 305.

441   **"nothing but depressing"**: Ciechanowski et al., eds., *Rejewski,* 88.
"**a barren existence**": Ibid., 42.
"**greatly assisted**": 240http://virtualglobetrotting.com/map/polish-memorial
-at-bletchley-park/view/google/.
"**as a whole**": "Poland's Overlooked Enigma Codebreakers," BBC News Magazine, July 5, 2014.

CHAPTER 27: "A COLLECTIVE FAULT"

442   **"There will be food"**: Henri van der Zee, *The Hunger Winter: Occupied Holland,
1944–45* (London: Jill Norman and Hobhouse, 1982), 286.
"**as close to destitution**": Rudy Abrahamson, *Spanning the Century: The Life of
W. Averell Harriman* (New York: William Morrow, 1992), 413.
443   **"Under the German occupation"**: Malcolm Muggeridge, *Chronicles of Wasted
Time,* vol. 2: *The Infernal Grove* (London: Collins, 1973), 224.
"**we [regarded ourselves]**": Paul Watkins, *Fellowship of Ghosts: A Journey Through
the Mountains of Norway* (New York: Picador, 2011), 202.
"**If you wanted**": Ibid.
444   **"one of the more"**: Muggeridge, *The Infernal Grove,* 224.
445   **"People who did not"**: Jan Karski, *Story of a Secret State* (Washington, DC:
Georgetown University Press, 2013), 243–44.
"**In the circumstances**": Tony Judt, *Postwar: A History of Europe Since 1945* (New
York: Penguin, 2005), 45.
447   **"needed and received"**: Louis de Jong, *The Netherlands and Nazi Germany* (Cambridge, MA: Harvard University Press, 1990), 7.
448   **"During the first two years"**: Julian Jackson, *France: The Dark Years, 1940–1944*
(Oxford: Oxford University Press, 2001), 350.
"**fear and their own worries**": Elsa van der Laaken, *Point of Reference* (Bloomington, IN: Xlibris, 2002), 138.
"**The criminal madness**": Marlise Simons, "Chirac Affirms France's Guilt in Fate
of Jews," *New York Times,* July 17, 1995.
449   **"give a 'message'"**: Kenneth Turan, " 'Sorrow and the Pity' Still Potent, Powerful," *Los Angeles Times,* July 7, 2000.
"**required the solidarity**": Jackson, *France: The Dark Years,* 360.
"**every [one of them]**": Tessa Stirling, Daria Nałęcz, and Tadeusz Dubicki, eds.,
Intelligence Cooperation Between Poland and Great Britain During World War II (London: Valentine Mitchell, 2005), 64.
450   **"It is not surprising"**: Michael R. Marrus, *The Holocaust in History* (Hanover,
NH: University Press of New England, 1987), 107.

CHAPTER 28: "THE WORLD COULD NOT POSSIBLY BE THE SAME"

451   **"Silence over Europe's recent past"**: Tony Judt, *Postwar: A History of Europe Since
1945* (New York: Penguin, 2005), 10.
"**through the entire war**": Erik Hazelhoff Roelfzema, *In Pursuit of Life* (Stroud,
UK: Sutton, 2003), 200.
"**a creature from another**": Ibid., 197.
452   **"The fearsome dangers"**: Ibid., 75.
"**skeptical, pragmatic practitioners**": Judt, *Postwar,* 82.
453   **"was the embodiment"**: Erik Hazelhoff Roelfzema, *Soldier of Orange* (London:
Sphere, 1982), 258.

453   **"To every Hollander"**: Ibid.

454   **"a victory"**: M.R.D. Foot, *Holland at War Against Hitler: Anglo-Dutch Relations, 1940–1945* (London: Frank Cass, 1990), 213.

      "the spokesman for all": "The Netherlands: Woman in the House," *Time,* May 13, 1946.

      "The idea that": Ian Buruma, *Year Zero: A History of 1945* (New York: Penguin, 2013), 8.

455   **"It is now obvious"**: David Dimbleby and David Reynolds, *An Ocean Apart: The Relationship Between Britain and America in the Twentieth Century* (New York: Random House, 1988), 188.

456   **"I left for Moscow"**: Marcia Davenport, *Too Strong for Fantasy* (New York: Pocket Books, 1969), 339.

457   **"What happened in Washington"**: Ibid., 342.

      "Facing an implacable foe": Josef Korbel, *Twentieth-Century Czechoslovakia: The Meanings of Its History* (New York: Columbia University Press, 1977), 247.

458   **"They have found out"**: Ibid., 246.

      "What died with him": Claire Sterling, *The Masaryk Case* (New York: Harper & Row, 1969), 1.

459   **"determination of the free countries"**: Foot, *Holland at War,* 190.

      "really made us": Jack Adams, *The Doomed Expedition: The Norwegian Campaign of 1940* (London: Leo Cooper, 1989), 35.

CHAPTER 29: "MY COUNSEL TO EUROPE . . . : UNITE!"

463   **"rather like one"**: James H. Huizinga, *Mr. Europe: A Political Biography of Paul Henri Spaak* (New York: Praeger, 1961), 75.

      "I'm often told": Ibid.

      "If the European Community": Robert W. Allen, *Churchill's Guests: Britain and the Belgian Exiles During World War II* (Westport, CT: Praeger, 2003), 181.

464   **"I have opened"**: Huizinga, *Mr. Europe,* 239.

      "When we have won": Douglas Dodds-Parker, *Setting Europe Ablaze: Some Account of Ungentlemanly Warfare* (Windlesham, UK: Springwood, 1983), 94.

      "Britain, our closest": Foot, *Holland at War,* 189.

      "consent to think": Huizinga, *Mr. Europe,* 233.

      "in which the barriers": Martin Gilbert, *Winston S. Churchill,* vol. 7: *Road to Victory, 1941–1945* (Boston: Houghton Mifflin, 1986), 239.

      "When the Nazi power": M.R.D. Foot, *Holland at War Against Hitler: Anglo-Dutch Relations, 1940–1945* (London: Frank Cass, 1990), 246.

465   **"solemn charades"**: Huizinga, *Mr. Europe,* 235.

      "I admire those": Ibid., 236.

      "We are with Europe": John Colville, *Winston Churchill and His Inner Circle* (New York: Wyndham Books, 1981), 261.

      "Every time we must": Charles de Gaulle, *The Complete War Memoirs* (New York: Carroll & Graf, 1998), 557.

466   **"renounce the insularity"**: Willy Brandt, *In Exile: Essays, Reflections and Letters, 1933–1947* (Philadelphia: University of Pennsylvania Press, 1971), 10.

      "moment of national reconciliation": Tony Judt, *Postwar: A History of Europe Since 1945* (New York: Penguin, 2005), 160.

      "as the last hurrah": Max Hastings, *Inferno: The World at War, 1939–1945* (New York: Vintage, 2012), 639.

467   **"an embrace so close"**: Antony Beevor and Artemis Cooper, *Paris: After the Liberation: 1944–1949* (New York: Penguin, 2004), 375.

"any British pretension": Beevor and Cooper, *Paris: After the Liberation*, 375.

468   **"If we try to remain"**: David Reynolds, "France, Britain and the Narrative of Two World Wars," in *Britain and France in Two World Wars: Truth, Myth and Memory*, ed. Robert Tombs and Emile Chabal (London: Bloomsbury Academic, 2013), 207.

"The price of British overdependence": Jeremy Paxman, *The English: A Portrait of a People* (Woodstock, NY: Overlook Press, 2000), 41.

469   **"on the fringe"**: Reynolds, "France, Britain and the Narrative of Two World Wars," 207.

"prosperity, security, health": Neal Ascherson, "Painful Lessons Must Be Learned by Europe," *Guardian*, July 18, 2015.

"What began as a drive": Roger Cohen, "For a Europe Remade, a Celebration in Uncertainty," *International Herald Tribune*, March 24, 2007.

470   **"Never again"**: *New York Times*, July 10, 1997.

"an unforgettable one": Bronisław Geremek, "Britain and Poland: The Neglected Friendship?," Polish International Institute, May 17, 2006.

"During the Second World War": Foot, *Holland at War*, 186.

"The essential": François Kersaudy, *Churchill and de Gaulle* (New York: Atheneum, 1982), 414.

"bordered on idolatry": Tom Hickman, *What Did You Do in the War, Auntie?* (London: BBC Books, 1995), 204.

"it was almost embarrassing": Ibid., 127.

471   **"During the long"**: Asa Briggs, *The History of Broadcasting in the United Kingdom*, vol. 3: *The War of Words* (London: Oxford University Press, 1970), 611.

"The British made me feel": Joachim Rønneberg, "The Linge Company and the British," in *Britain and Norway in the Second World War,* ed. Patrick Salmon (London: HMSO, 1995), 157.

"love which was neither": Ray Jenkins, *A Pacifist at War* (London: Arrow, 2010), 125.

472   **"the courage and compassion"**: General Sir John Hackett, *I Was a Stranger* (London: Chatto & Windus, 1978), 210.

473   **"the spiritual experience"**: John Waddy, "Shan Hackett at Arnhem," Airborne Assault, http://www.paradata.org.uk/articles/shan-hackett-arnhem-article-john-waddy.

"This was a battle": Ibid.

"50 years ago": "WW2 People's War," www.bbc.co.uk/history/ww2peoples war/stories/61/a5887461.shtml.

475   **"buried here"**: Robert Hardman, "Seventy Years On, Arnhem Has Never Forgotten Thousands of British and Polish Soldiers Who Gave Their Lives in Ill-Fated Allied Plan to Deliver Blow to Hitler," *Daily Mail*, Sept. 27, 2014.

Bibliography

ARCHIVAL MATERIAL

CHURCHILL COLLEGE, CAMBRIDGE
  Winston Churchill Papers
FRANKLIN D. ROOSEVELT PRESIDENTIAL LIBRARY,
  HYDE PARK
  Franklin D. Roosevelt Papers
NATIONAL ARCHIVES UK, KEW
  Air Ministry Papers
  Foreign Office Papers
POLISH INSTITUTE AND SIKORSKI MUSEUM,
  LONDON
  Squadron 303 Papers and Diaries
TUFTS UNIVERSITY
  Edward R. Murrow Papers

PUBLISHED MATERIAL

Abramson, Rudy. *Spanning the Century: The Life of W. Averell Harriman.* New York: William Morrow, 1992.
Adams, Jack. *The Doomed Expedition: The Norwegian Campaign of 1940.* London: Leo Cooper, 1989.
Adamson, Hans Christian, and Per Klem. *Blood on the Midnight Sun.* New York: W. W. Norton, 1964.
Albright, Madeleine. *Prague Winter: A Personal Story of Remembrance and War, 1937–1948.* New York: HarperCollins, 2012.

Allen, Robert W. *Churchill's Guests: Britain and the Belgian Exiles During World War II.* Westport, CT: Praeger, 2003.

Andrew, Christopher. *Her Majesty's Secret Service: The Making of the British Intelligence Community.* New York: Viking, 1986.

Andrew, Christopher, and David Dilks, eds. *The Missing Dimension: Governments and Intelligence Communities in the Twentieth Century.* Urbana, IL: University of Illinois Press, 1984.

Atkin, Nicholas. *The Forgotten French: Exiles in the British Isles, 1940–44.* Manchester, UK: Manchester University Press, 2003.

Atkinson, Rick. *The Guns at Last Light: The War in Western Europe, 1944–1945.* New York: Picador, 2013.

Baden-Powell, Dorothy. *Pimpernel Gold.* New York: St. Martin's Press, 1978.

Bailey, Roderick. *Forgotten Voices in the Secret War: An Inside History of Special Operations During the Second World War.* London: Ebury Press, 2008.

Baldwin, Jane. *Michel Saint-Denis and the Shaping of the Modern Actor.* Westport, CT: Praeger, 2003.

Barker, Elisabeth. *Churchill and Eden at War.* New York: St. Martin's Press, 1978.

Batey, Mavis. *Dilly: The Man Who Broke Enigma.* Biteback Publishing, Kindle edition.

Beevor, Antony, and Artemis Cooper. *Paris: After the Liberation: 1944–1949.* New York: Penguin, 2004.

Bell, P.M.H. *A Certain Eventuality: Britain and the Fall of France.* London: Saxon House, 1974.

Bielecki, Tadeusz, and Leszek Szymański. *Warsaw Aflame: The 1939–1945 Years.* Los Angeles: Polamerica Press, 1973.

Binet, Laurent. *HHhH.* New York: Farrar, Straus and Giroux, 2012.

Bloch, Marc. *Strange Defeat.* Important Books, 2013.

Bodson, Herman. *Downed Allied Airmen and Evasion of Capture: The Role of Local Resistance Networks in World War II.* Jefferson, NC: McFarland, 2005.

Bond, Brian. *Britain, France and Belgium, 1939–1940.* London: Brassey's, 1990.

Bor-Komorowski, Tadeusz. *The Secret Army.* London: Victor Gollancz, 1950.

Bowen, Elizabeth. *The Heat of the Day.* New York: Anchor, 2002.

Bradford, Sarah. *The Reluctant King: The Life and Reign of George VI, 1895–1942.* New York: St. Martin's Press, 1989.

Brandt, Willy. *In Exile: Essays, Reflections and Letters, 1933–1947.* Philadelphia: University of Pennsylvania Press, 1971.

Brendon, Piers. *The Dark Valley: A Panorama of the 1930s.* New York: Knopf, 2000.

Brian, Denis. *The Curies: A Biography of the Most Controversial Family in Science.* Hoboken, NJ: John Wiley & Sons, 2005.

Briggs, Asa. *The History of Broadcasting in the United Kingdom.* Vol. 3, *The War of Words.* London: Oxford University Press, 1970.

Brown, Alan. *Airmen in Exile: The Allied Air Forces in the Second World War.* Stroud, UK: Sutton, 2000.

Brown, Anthony Cave. *"C": The Secret Life of Sir Stewart Graham Menzies.* New York: Macmillan, 1987.

Bungay, Stephen. *The Most Dangerous Enemy: A History of the Battle of Britain.* London: Aurum Press, 2000.

Buruma, Ian. *Year Zero: A History of 1945.* New York: Penguin, 2013.

Cadogan, Sir Alexander. *The Diaries of Sir Alexander Cadogan, 1938–1945.* Ed. David Dilks. New York: Putnam, 1971.

Calder, Angus. *The People's War: Britain, 1939–1945.* New York: Pantheon, 1969.

Carroll, Wallace. *Persuade or Perish*. Boston: Houghton Mifflin, 1948.

Churchill, Winston S. *The Second World War*. Vol. 1: *The Gathering Storm*. Boston: Houghton Mifflin, 1948.

———. *The Second World War*. Vol. 2: *Their Finest Hour*. Boston: Houghton Mifflin, 1949.

———. *The Second World War*. Vol. 4: *The Hinge of Fate*. Boston: Houghton Mifflin, 1950.

———. *The Second World War*. Vol. 5: *Closing the Ring*. Boston: Houghton Mifflin, 1951.

———. *The Second World War*. Vol. 6: *Triumph and Tragedy*. Boston: Houghton Mifflin, 1953.

Churchill, Winston S., and Franklin D. Roosevelt. *Churchill and Roosevelt: The Complete Correspondence*, vol. 3. Ed. Warren F. Kimball. Princeton, NJ: Princeton University Press, 1984.

Ciechanowski, Jan. *Defeat in Victory*. Garden City, NY: Doubleday, 1947.

Ciechanowski, Jan Stanisław, et al., eds. *Rejewski: Living with the Enigma Secret*. Bydgoszcz, Poland: Bydgoszcz City Council, 2005.

Clark, Ronald. *Battle for Britain: Sixteen Weeks That Changed the Course of History*. New York: Franklin Watts, 1966.

Cloud, Stanley, and Lynne Olson. *The Murrow Boys: Pioneers on the Front Lines of Broadcast Journalism*. Boston: Mariner Books, 1996.

Cockett, Richard. *Twilight of Truth: Chamberlain, Appeasement and the Manipulation of the Press*. New York: St. Martin's Press, 1989.

Collier, Richard. *1940: The World in Flames*. London: Hamish Hamilton, 1979.

———. *Eagle Day: The Battle of Britain*. New York: Dutton, 1966.

Collins, Larry, and Dominique Lapierre. *Is Paris Burning?* New York: Simon & Schuster, 1965.

Colville, John. *The Fringes of Power: 10 Downing Street Diaries, 1939–1955*. New York: W. W. Norton, 1985.

———. *Winston Churchill and His Inner Circle*. New York: Wyndham Books, 1981.

Conway, Martin, and José Gotovitch, eds. *Europe in Exile: European Exile Communities in Great Britain, 1940–1945*. New York: Berghahn Books, 2001.

Cowles, Virginia. *Looking for Trouble*. New York: Harper, 1941.

Curie, Eve. *Journey Among Warriors*. New York: Doubleday, 1943.

Curtis, Rosme. *Winged Tenacity: The Polish Air Force, 1918–1944*. London: Kingston Hill, 1944.

Dahl, Per F. *Heavy Water and the Wartime Race for Nuclear Energy*. Bristol, UK: Institute of Physics Publishing, 1999.

Davenport, Marcia. *Too Strong for Fantasy*. New York: Pocket Books, 1969.

de Gaulle, Charles. *The Complete War Memoirs*. New York: Carroll & Graf, 1998.

de Jong, Louis. *The Netherlands and Nazi Germany*. Cambridge, MA.: Harvard University Press, 1990.

Denniston, Robin. *Thirty Secret Years: A.G. Denniston's Work in Signals Intelligence, 1914–1944*. Clifton-upon-Teme, UK: Polperro Heritage Press, 2012.

D'Este, Carlo. *Patton: A Genius for War*. New York: HarperPerennial, 1995.

Dimbleby, David, and David Reynolds. *An Ocean Apart: The Relationship Between Britain and America in the Twentieth Century*. New York: Random House, 1988.

Dodds-Parker, Douglas. *Setting Europe Ablaze: Some Account of Ungentlemanly Warfare*. Windlesham, UK: Springwood, 1983.

Dowling, Maria. *Czechoslovakia*. London: Bloomsbury, 2002.

Drazin, Charles. *The Finest Years: British Cinema of the 1940s*. London: I. B. Tauris, 2007.

Eden, Anthony. *The Reckoning*. Boston: Houghton Mifflin, 1965.

Eisenhower, Dwight D. *Crusade in Europe*. Garden City, NY: Doubleday, 1974.

Feigel, Lara. *The Love-charm of Bombs: Restless Lives in the Second World War*. New York: Bloomsbury, 2013.

Fiedler, Arkady. *Squadron 303: The Polish Fighter Squadron with the RAF*. New York: Roy, 1943.

Foot, M.R.D., ed. *Holland at War Against Hitler: Anglo-Dutch Relations, 1940–1945*. London: Frank Cass, 1990.

———. *SOE: An Outline History of the Special Operations Executive, 1940–1946*. Frederick, MD: University Publications of America, 1986.

———. *SOE in France: An Account of the Work of the British Special Operations Executive in France, 1940–1944*. London: HMSO, 1966.

Foot, M.R.D., and J. M. Langley. *MI9: Escape and Evasion, 1939–1945*. London: Biteback Publishing, 2011.

Fourcade, Marie-Madeleine. *Noah's Ark: A Memoir of Struggle and Resistance*. New York: Dutton, 1972.

Garby-Czerniawski, Roman. *The Big Network*. London: G. Ronald, 1961.

Gelardi, Julia. *Born to Rule: Five Reigning Consorts, Granddaughters of Queen Victoria*. New York: St. Martin's Press, 2005.

Gelb, Norman. *Scramble: A Narrative History of the Battle of Britain*. San Diego: Harcourt Brace Jovanovich, 1985.

Gilbert, Martin. *The Second World War: A Complete History*. New York: Henry Holt, 1987.

———. *Winston S. Churchill*. Vol. 6: *Finest Hour, 1939–1941*. Boston: Houghton Mifflin, 1983.

———. *Winston S. Churchill*. Vol. 7: *Road to Victory, 1941–1945*. Boston: Houghton Mifflin, 1987.

Giskes, H. J. *London Calling North Pole*. London: William Kimber, 1953.

Glass, Charles. *Americans in Paris: Life and Death Under Nazi Occupation*. New York: Penguin, 2009.

Greve, Tim. *Haakon VII of Norway: Founder of a New Monarchy*. London: Hurst, 1985.

Hackett, General Sir John. *I Was a Stranger*. London: Chatto & Windus, 1978.

Hambro, C. J. *I Saw It Happen in Norway*. New York: D. Appleton-Century, 1940.

Harpprecht, Klaus. *Willy Brandt: Portait and Self-Portrait*. London: Abelard-Schuman, 1975.

Harriman, Florence Jaffray. *Mission to the North*. Philadelphia: J. B. Lippincott, 1941.

Hastings, Max. *Armageddon: The Battle for Germany, 1944–1945*. New York: Knopf, 2004.

———. *Das Reich: The March of the 2nd Panzer Division Through France, June 1944*. Zenith Press, 2013.

———. *Inferno: The World at War, 1939–1945*. New York: Vintage, 2012.

———. *The Secret War: Spies, Codes and Guerrillas, 1939–1945*. London: William Collins, 2015.

Helm, Sarah. *A Life in Secrets: Vera Atkins and the Missing Agents of WWII*. New York: Anchor, 2007.

———. *Ravensbrück: Life and Death in Hitler's Concentration Camp for Women*. New York: Doubleday, 2015.

Henrey, Mrs. Robert. *The Incredible City*. London: J. M. Dent & Sons, 1944.

Hickman, Tom. *What Did You Do in the War, Auntie?* London: BBC Books, 1995.

Hough, Richard, and Denis Richards. *The Battle of Britain*. London: Hodder & Stoughton, 1989.

Howarth, Patrick. *Intelligence Chief Extraordinary: The Life of the Ninth Duke of Portland*. London: Bodley Head, 1986.

Huizinga, James H. *Mr. Europe: A Political Biography of Paul Henri Spaak*. New York: Praeger, 1961.

Irving, David. *The War Between the Generals: Inside the Allied High Command*. New York: Congdon & Lattes, 1981.

Ismay, Lord. *The Memoirs of General Lord Ismay*. New York: Viking, 1960.

Jackson, Julian. *France: The Dark Years, 1940–1944*. Oxford: Oxford University Press, 2001.

Jędrzejewicz, Wacław, ed. *Poland in the British Parliament, 1939–1945*. New York: Pilsudski Institute, 1946.

Jeffery, Keith. *The Secret History of MI6*. New York: Penguin Press, 2010.

Jenkins, Ray. *A Pacifist at War*. London: Arrow, 2010.

Jones, R. V. *Most Secret War*. Ware, UK: Wordsworth Editions, 1998.

———. *Reflections on Intelligence*. London: Heinemann, 1989.

Judt, Tony. *Postwar: A History of Europe Since 1945*. New York: Penguin, 2005.

Karski, Jan. *Story of a Secret State*. Washington, DC: Georgetown University Press, 2013.

Keegan, John, ed. *Churchill's Generals*. London: Cassell, 2007.

Kelso, Nicholas. *Errors of Judgement: SOE's Disaster in the Netherlands, 1941–44*. London: Robert Hale, 1988.

Kennan, George F. *Memoirs (1925–1950)*. New York: Bantam Books, 1969.

Kennan, George F., and John Lukacs. *George F. Kennan and the Origins of Containment, 1944–1946: The Kennan-Lukacs Correspondence*. Columbia: University of Missouri Press, 1997.

Kent, John A. *One of the Few*. London: Kimber, 1971.

Kersaudy, François. *Churchill and de Gaulle*. New York: Atheneum, 1982.

———. *Norway 1940*. Lincoln: University of Nebraska Press, 1998.

Ketchum, Richard M. *The Borrowed Years, 1938–1941: America on the Way to War*. New York: Random House, 1989.

Keyes, Roger. *Outrageous Fortune: The Tragedy of Leopold III of the Belgians, 1901–1941*. London: Secker & Warburg, 1984.

Knightley, Philip. *The First Casualty*. New York: Harcourt Brace Jovanovich, 1975.

Koht, Halvdan. *Norway: Neutral and Invaded*. New York: Macmillan, 1941.

Korbel, Josef. *Twentieth-Century Czechoslovakia: The Meanings of Its History*. New York: Columbia University Press, 1977.

Korbonski, Stefan. *Fighting Warsaw*. New York: Funk & Wagnalls, 1968.

Kozaczuk, Władysław. *Enigma: How the German Machine Cipher Was Broken, and How It Was Read by the Allies in World War Two*. Frederick, MD: University Publications of America, 1984.

Kozaczuk, Władysław, and Jerzy Straszak. *Enigma: How the Poles Broke the Nazi Code*. New York: Hippocrene Books, 2004.

Lacouture, Jean. *De Gaulle: The Rebel, 1890–1944*. New York: W. W. Norton, 1990.

Langley, J. M. *Fight Another Day*. Barnsley, UK: Pen & Sword, 2013.

Lankford, Nelson D. *OSS Against the Reich: The World War II Diaries of Col. David K. E. Bruce*. Kent, OH: Kent State University Press, 1991.

Large, Christine. *Hijacking Enigma*. Hoboken, NJ: John Wiley & Sons, 2003.

Lash, Joseph. *Roosevelt and Churchill, 1939–1941: The Partnership That Saved the West.* New York: W. W. Norton, 1976.

Lean, Tangye. *Voices in the Darkness: The Story of the European Radio War.* London: Secker & Warburg, 1943.

Liebling, A. J. *The Road Back to Paris.* Garden City, NY: Doubleday, 1944.

Lindbaek, Lise. *Norway's New Saga of the Sea: The Story of Her Merchant Marine in World War II.* New York: Exposition Press, 1969.

Lisiewicz, M., et al., eds. *Destiny Can Wait: The Polish Air Force in the Second World War.* Nashville, TN: Battery Press, 1949.

Lockhart, R. H. Bruce. *Comes the Reckoning.* London: Putman, 1947.

———. *Jan Masaryk: A Personal Memoir.* London: Philosophical Library, 1951.

Lukacs, John. *The Great Powers and Eastern Europe.* New York: American Book Co., 1953.

Lyttelton, Oliver. *The Memoirs of Lord Chandos: An Unexpected View from the Summit.* New York: New American Library, 1963.

MacDonald, Callum. *The Killing of Obergruppenführer Reinhard Heydrich, May 27, 1942.* London: Macmillan, 1989.

Macintyre, Ben. *Double Cross: The True Story of the D-Day Spies.* New York: Crown, 2012.

———. *A Spy Among Friends: Kim Philby and the Great Betrayal.* New York: Crown, 2014.

Mackenzie, William. *The Secret History of S.O.E.: Special Operations Executive, 1940–1945.* London: St. Ermin's Press, 2000.

Macmillan, Harold. *The Blast of War: 1939–1945.* New York: Harper & Row, 1967.

MacMillan, Margaret. *Peacemakers: The Paris Conference of 1919 and Its Attempt to End War.* London: John Murray, 2001.

Manchester, William. *The Last Lion: Winston Spencer Churchill: Alone, 1932–1940.* New York: Dell, 1988.

Marks, Leo. *Between Silk and Cyanide: A Codemaker's War, 1941–45.* Stroud, UK: The History Press, 2013.

Marrus, Michael R. *The Holocaust in History.* Hanover, NH: University Press of New England, 1987.

Marshall, Robert. *All the King's Men: The Truth Behind SOE's Greatest Wartime Disaster.* London: Collins, 1988.

Mastny, Vojtech. *The Czechs Under Nazi Rule: The Failure of National Resistance, 1939–1942.* New York: Columbia University Press, 1971.

Mayle, Paul D. *Eureka Summit: Agreement in Principle and the Big Three at Tehran.* Newark, DE: University of Delaware Press, 1987.

Moran, Christopher. *Classified: Secrecy and the State in Modern Britain.* New York: Cambridge University Press, 2012.

Moran, Lord. *Churchill at War, 1940–45.* New York: Carroll & Graf, 2002.

Moravec, František. *Master of Spies; The Memoirs of General František Moravec.* Garden City, NY: Doubleday, 1975.

Muggeridge, Malcolm. *Chronicles of Wasted Time, vol. 2: The Infernal Grove.* London: Collins, 1973.

Murphy, Christopher J. *Security and Special Operations: SOE and MI5 During the Second World War.* Basingstoke, UK: Palgrave Macmillan, 2006.

Neave, Airey. *Little Cyclone: The Girl Who Started the Comet Line.* London: Biteback Publishing, 2013.

———. *Saturday at M.I.9: The Classic Account of the WWII Escape Organisation*. London: Hodder and Stoughton, 1969.

Némirovsky, Irène. *Suite Française*. New York: Knopf, 2006.

Nicolson, Harold. *The War Years, Diaries & Letters, 1939–1945*. Ed. Nigel Nicolson. New York: Atheneum, 1967.

Nowak, Jan. *Courier from Warsaw*. Detroit: Wayne State University Press, 1982.

Olson, Lynne. *Citizens of London: The Americans Who Stood with Britain in Its Finest, Darkest Hour*. New York: Random House, 2010.

Olson, Lynne, and Stanley Cloud. *A Question of Honor: The Kosciuszko Squadron: Forgotten Heroes of World War II*. New York: Knopf, 2003.

Owen, James. *Danger UXB: The Heroic Story of the WWII Bomb Disposal Teams*. London: Abacus, 2010.

Panter-Downes, Mollie. *London War Notes: 1939–1945*. New York: Farrar, Straus and Giroux, 1971.

Paris, Barry. *Audrey Hepburn*. New York: Berkley, 1996.

Paul, Allen. *Katyn: The Untold Story of Stalin's Polish Massacre*. New York: Scribner's, 1991.

Payne, Robert. *The Rise and Fall of Stalin*. New York: Avon Books, 1966.

Paxman, Jeremy. *The English: A Portrait of a People*. Woodstock, NY: Overlook Press, 2000.

Peszke, Michael Alfred. *The Polish Underground Army, the Western Allies, and the Failure of Strategic Unity in World War II*. Jefferson, NC: McFarland, 2005.

Petrow, Richard. *The Bitter Years: The Invasion and Occupation of Denmark and Norway: April 1940–May 1945*. New York: William Morrow, 1974.

Polonsky, Antony. *Politics in Independent Poland, 1921–1939: The Crisis of Constitutional Government*. Oxford: Clarendon Press, 1972.

Porch, Douglas. *The French Secret Services: From the Dreyfus Affair to the Gulf War*. New York: Farrar, Straus and Giroux, 1995.

Raczyński, Edward. *In Allied London*. London: Weidenfeld & Nicolson, 1963.

Read, Anthony, and David Fisher. *Colonel Z: The Secret Life of a Master of Spies*. New York: Viking, 1985.

Reynolds, Quentin. *A London Diary*. New York: Random House, 1941.

Ritchie, Charles. *The Siren Years: A Canadian Diplomat Abroad, 1937–1945*. Toronto: Macmillan, 1974.

Roelfzema, Erik Hazelhoff. *In Pursuit of Life*. Stroud, UK: Sutton, 2003.

———. *Soldier of Orange*. Hastings-on-Hudson, NY: Holland Heritage Society, 1983.

Rolo, Charles J. *Radio Goes to War*. New York: G. P. Putnam's Sons, 1942.

Rosbottom, Ronald C. *When Paris Went Dark: The City of Light Under German Occupation, 1940–1944*. New York: Little, Brown, 2014.

Routledge, Paul. *Public Servant, Secret Agent: The Elusive Life and Violent Death of Airey Neave*. London: HarperCollins, 2002.

Ryan, Cornelius. *A Bridge Too Far*. New York: Simon & Schuster, 1995.

Salmon, Patrick, ed. *Britain and Norway in the Second World War*. London: HMSO, 1995.

Sarner, Harvey. *General Anders and the Soldiers of the Second Polish Corps*. Cathedral City, CA: Brunswick Press, 1997.

Schellenberg, Walter. *The Labyrinth: Memoirs of Walter Schellenberg*. New York: Harper & Brothers, 1956.

Seaman, Mark, ed. *Special Operations Executive: A New Instrument of War*. London: Routledge, 2006.

Sebag-Montefiore, Hugh. *Enigma: The Battle for the Code*. New York: John Wiley and Sons, 2000.

Sevareid, Eric. *Not So Wild a Dream*. New York: Atheneum, 1976.

Seymour-Jones, Carole. *She Landed by Moonlight: The Story of Secret Agent Pearl Witherington*. London: Hodder & Stoughton, 2013.

Sherwood, Robert E. *Roosevelt and Hopkins: An Intimate History*. New York: Harper and Brothers, 1948.

Smith, Jean Edward. *FDR*. New York: Random House, 2007.

Smith, R. Franklin. *Edward R. Murrow: The War Years*. Kalamazoo, MI: New Issues Press, 1978.

Soames, Mary. *Clementine Churchill: The Biography of a Marriage*. Boston: Houghton Mifflin, 2002.

Sorel, Nancy Caldwell. *The Women Who Wrote the War*. New York: Arcade, 1999.

Spears, Sir Edward. *Assignment to Catastrophe*, vol. 2: *The Fall of France, June 1940*. London: Heinemann, 1954.

Sperber, A. M. *Murrow: His Life and Times*. New York: Freundlich, 1986.

Stafford, David. *Britain and European Resistance, 1940–1945*. Toronto: University of Toronto Press, 1980.

———. *Churchill and Secret Service*. London: John Murray, 1997.

———. *Secret Agent: The True Story of the Covert War Against Hitler*. Woodstock, NY: Overlook Press, 2003.

———. *The Silent Game: The Real World of Imaginary Spies*. Athens: University of Georgia Press, 1991.

Stenton, Michael. *Radio London and Resistance in Occupied Europe: British Political Warfare, 1939–1943*. Oxford: Oxford University Press, 2000.

Sterling, Claire. *The Masaryk Case*. New York: Harper & Row, 1969.

Sterling, Tessa, Daria Nałęcz, and Tadeusz Dubicki, eds. *Intelligence Cooperation Between Poland and Great Britain During World War II*. London: Valentine Mitchell, 2005.

Stowe, Leland. *No Other Road to Freedom*. New York: Alfred A. Knopf, 1941.

Teissier du Cros, Janet. *Divided Loyalties*. New York: Alfred A. Knopf, 1964.

Tillotson, Michael, ed. *SOE and the Resistance: As Told in* The Times *Obituaries*. London: Bloomsbury, 2011.

Tombs, Robert, and Emile Chabal, eds. *Britain and France in Two World Wars: Truth, Myth and Memory*. London: Bloomsbury, 2013.

Tombs, Robert, and Isabelle Tombs. *That Sweet Enemy: Britain and France: The History of a Love-Hate Relationship*. New York: Vintage, 2008.

Undset, Sigrid. *Return to the Future*. New York: Alfred A. Knopf, 1942.

Van der Kiste, John. *Northern Crowns: The Kings of Modern Scandinavia*. Stroud, UK: Sutton, 1996.

Van der Laaken, Elsa. *Point of Reference*. Bloomington, IN: Xlibris, 2002.

Van der Zee, Henri. *The Hunger Winter: Occupied Holland, 1944–5*. London: Jill Norman and Hobhouse, 1982.

Velaers, Jan, and Herman Van Goethem. *Leopold III*. Brussels: Lannoo, 2001.

Vomécourt, Philippe de. *An Army of Amateurs*. New York: Doubleday, 1961.

Warmbrunn, Werner. *The Dutch Under German Occupation, 1940–1945*. Stanford, CA: Stanford University Press, 1983.

Watkins, Paul. *Fellowship of Ghosts: A Journey Through the Mountains of Norway*. New York: Picador, 2011.

Watt, Richard M. *Bitter Glory: Poland and Its Fate, 1918 to 1939*. New York: Simon & Schuster, 1979.

Weart, Spencer R. *Scientists in Power*. Cambridge, MA: Harvard University Press, 1979.

Weitz, Margaret Collins. *Sisters in the Resistance: How Women Fought to Free France, 1940–1945*. New York: John Wiley and Sons, 1996.

Welchman, Gordon. *The Hut Six Story*. New York: McGraw-Hill, 1982.

West, Nigel. *MI6: British Secret Intelligence Service Operations, 1909–1945*. New York: Random House, 1983.

Wheeler-Bennett, John W. *Friends, Enemies, and Sovereigns*. London: Macmillan, 1976.

———. *King George VI: His Life and Reign*. New York: St. Martin's Press, 1958.

———. *Munich: Prologue to Tragedy*. New York: Duell, Sloan and Pearce, 1948.

White, Dorothy Shipley. *Seeds of Discord: De Gaulle, Free France, and the Allies*. Syracuse, NY: Syracuse University Press, 1964.

White, Lewis M., ed. *On All Fronts: Czechs and Slovaks in World War II*. Boulder, CO: East European Monographs, 1991.

H. R. H. Wilhelmina, Princess of the Netherlands. *Lonely but Not Alone*. London: Hutchinson, 1960.

Wilkinson, Peter. *Foreign Fields: The Story of an SOE Operative*. London: I. B. Tauris, 1997.

Wynne, Barry. *The Scarlet Countess: The Incredible Story of Mary Lindell*. Kindle edition.

Zamoyski, Adam. *The Forgotten Few: The Polish Air Force in the Second World War*. London: Hippocrene, 1995.

Zumbach, Jan. *On Wings of War*. London: André Deutsch, 1975.

Index

Page numbers in italics refer to illustrations.